Gmelin Handbook of Inorganic Chemistry

8th Edition

Gmelin Handbook Volumes on "Silicon" (Syst. No. 15)

Main Volume

Part A 1 History – 1984 (present volume)

Part B*) Element and Compounds – 1959

Part C*) Organic Silicon Compounds – 1958

Supplement Volume

Part B 1 Silicon and Noble Gases. Silicon and Hydrogen (including SiH_n–Oxygen Compounds) – 1982

Part B 2 Properties of Crystalline Silicon Carbide. Diodes. Molecular Species in the Gas Phase. Amorphous Silicon – Carbon Alloys – 1984

*) in German

Gmelin Handbook of Inorganic Chemistry

8th Edition

Gmelin Handbuch der Anorganischen Chemie

Achte, völlig neu bearbeitete Auflage

Prepared and issued by | Gmelin-Institut für Anorganische Chemie der Max-Planck-Gesellschaft zur Förderung der Wissenschaften
Director: Ekkehard Fluck

Founded by | Leopold Gmelin

8th Edition | 8th Edition begun under the auspices of the Deutsche Chemische Gesellschaft by R. J. Meyer

Continued by | E. H. E. Pietsch and A. Kotowski, and by Margot Becke-Goehring

Springer-Verlag Berlin · Heidelberg · New York · Tokyo 1984

Gmelin Handbook of Inorganic Chemistry

8th Edition

Si

Silicon

Part A 1

History

AUTHOR

Karl Rumpf, Gmelin-Institut, Frankfurt am Main, Federal Republic of Germany (German manuscript)

EDITORS

Harry J. Emeléus, F.R.S., University Chemical Laboratory, Cambridge, U.K. (responsible for the English version)

Walter Lippert, Gmelin-Institut, Frankfurt am Main, Federal Republic of Germany

System Number 15

Springer-Verlag Berlin · Heidelberg · New York · Tokyo 1984

LITERATURE CLOSING DATE: 1983
IN SOME CASES MORE RECENT DATA HAVE BEEN CONSIDERED

Library of Congress Catalog Card Number: Agr 25-1383

ISBN 3-540-93508-8 Springer-Verlag, Berlin · Heidelberg · New York · Tokyo
ISBN 0-387-93508-8 Springer-Verlag, New York · Heidelberg · Berlin · Tokyo

Typesetting, printing, and bookbinding: LN-Druck Lübeck

Foreword

During the preparation for the writing of this history of the element silicon, it became apparent to begin with that the course of the discovery of the element silicon had in no way occurred so simply as it had previously been presented in the literature: It stretched out in fact over several decades with many kinds of experiments and their erroneous interpretation, until the great Berzelius understood the results clearly. This state of affairs was thus the incentive for reproducing in this part of the book the necessary old, today more than ever before difficultly accessible literature, in the original text and in translation, extensively enough that the reader can develop his own picture regarding the experiments and ideas of the investigators without being obligated to accept an opinion of the current author. It further developed that the element – one can say without exaggeration – lay for more than a century in many laboratories rather as a curiosity than as an object of scientific investigation. It was initially used only in metallurgy, where its significance had been recognized even before its first preparation. It is amazing that it has been found in antique bronzes. Only in our era has the element been prepared in a pure form and studied with precision, and the knowledge thus won begun to revolutionize our technology.

Quite similarly ran the development of our knowledge of several of the here-selected compounds of silicon: It became obvious that the widely distributed accounts of the discovery of several of them required a certain correction. Thus for these materials also, for which likewise only recent decades have brought sufficient recognition of their nature, it has been attempted by the kind of citations selected to satisfy the desire of the reader for knowledge of the facts of these matters.

Frankfurt am Main
July 1984

Karl Rumpf

Table of Contents

Silicon

Atomic Number 14 Atomic Weight 28.086

History of the Element and Its Compounds

General References:

M. E. Weeks, M. H. Leicester, Discovery of the Elements, 7th Ed., ACS, Easton, Pa., 1968, pp. 555/7.

J. R. Partington, A History of Chemistry, Vol. 4, MacMillan, London 1964, pp. 96, 150, 324, 498, 798, 911.

E. Enk in: W. Foerst, Ullmanns Encyklopädie der Technischen Chemie, 3rd Ed., Vol. 15, Urban-Schwarzenberg, München 1964, pp. 678/9.

E. Pilgrim, Entdeckung der Elemente mit Biographien der Entdecker, Enke, Stuttgart 1950, pp. 205/7.

R. Meyer, Vorlesungen über die Geschichte der Chemie, Akad. Verlagsges., Leipzig 1922, pp. 58, 89, 124.

C. Friedhelm in: F. Stohmann, B. Kerl, Encyklopädisches Handbuch der Technischen Chemie, 4th Ed., Vol. 7, Vieweg, Braunschweig 1900, p. 1735.

R. Jagneaux, Histoire de la Chimie, Vol. 1, Pt. 2, Masson, Paris 1891, pp. 707/10.

H. Kopp, Geschichte der Chemie, Vol. 4, Vieweg, Braunschweig 1847, p. 75.

Introduction

Silicon is of special interest in the series of chemical elements. Firstly, next to oxygen, it is the most abundant element in the form of silicates in building up the earth's crust. These silicates, however, are not only of geochemical significance as structural components of the earth but are of outstanding practical and cultural value in the form of man-made synthetic materials: bricks, earthenware, glass, enamel, white ware, porcelain and cement. A discussion of the history of these materials will not be given here. The history of man's use of natural "stones", such as flint, the most important material for making man's first implements, which were mined more than 5000 years ago and the history of quartz and of semi-precious stones ["Halbedelsteine"] which were used as ornaments even in the earliest times, is of cultural and archeological rather than of chemical significance and will also not be dealt with in what follows.

The element itself, the history of whose discovery has been the subject of many incorrect reports in the literature, has great technical significance, which is mainly metallurgical, owing to its metal-like properties. This interest has been considerably increased and broadened in very recent time as a result of the semiconducting properties of the element. Compounds of silicon with metals, the silicides, play an important rôle as alloy components in the metallurgy of the heavy and light metals. In keeping with the arrangement of the Handbook these also will not be described at this point.

1 History of the Element

1.1 Names, Nomenclature, Symbol

Silicon is one of the elements which got its name before being prepared in the elementary form since the identity of its oxide had been recognized earlier. Thus, it happened that Humphry Davy (1778–1829) in reporting to the Royal Society in London on 30th June, 1808, on his success with the electrolytic preparation of metals from the alkalis and alkaline earths [1], also mentioned a series of attempts to decompose "silex" [i.e., precipitated silica, SiO_2] and other "primary earths" ["primäre Erden"], which were unsuccessful in spite of the size of the voltaic batteries of 200 and 500 pairs of plates which were applied. Notwithstanding this, he felt himself in a position to sum up by asserting:

"From the general tenor of these results, and the comparison between the different series of experiments, there seems very great reason to conclude that alumine, zircone, glucine, and silex are, like the alkaline earths, metallic oxides, for on no other supposition is it easy to explain the phenomena that have been detailed ...".

He concluded with the words:

"Had I been so fortunate as to have obtained more certain evidences on this subject, and to have procured the metallic substances I was in search of, I should have proposed for them the names of silicium, alumium, zirconium, and glucium."

Clearly, in forming the word *silicium,* H. Davy had gone back to *silex,* for which a definite etymology is unknown [2], though it stems from the Latin, which was used by him in naming silica. The derivation proposed by A. Fick [37] which takes the Latin *silex* = rock to be almost identical with the old-german *chisiline,* differing only in accent, and to come from a root **gisilek* from which the Latin *silex* is derived through *ĝisilek* and *silek* as *ks* is not a Latin prefix, is held both by A. Ernout, A. Meillet [2] and by A. Walde, J. B. Hofmann [3] to be incorrect. However, the latter agree with A. Fick [37] on the relationship between *kiesel* and karisch-phrygian γίσσα [gissa] = rock, a connection which J. Pokorny rejects [44]. A. Walde, J. B. Hofmann [38] would like to see for the Latin *silex* a derivation from **(s)q(h)eliq-,* which occurs again in the Irish *scelec* = rock. According to H. Blümner [4] *silex* did not signify any particular kind of rock to the Romans, but rocks in general which were harder than the much-used travertin. But even in the case of Georg Agricola (1494–1555) the word *silex* was used only for flint [5], a significance which it retains today in many places [6].

The translators of Davy's publications into other languages varied in their attitudes to his nomenclature proposals. While the translator into French [7], from which they were rendered into Italian by a Sig. N. [8] and into German by A. F. Gehlen [9], adopted the new names without comment, they were opposed by L. W. Gilbert, another German translator, in whose *Annalen der Physik* Davy's publications also appeared [10]. He proposed, referring to "the artificial chemical language which I have fostered in these *Annalen*" ["die chemische Kunstsprache, wie ich sie in diesen *Annalen* gepflegt habe"] that: "... should Davy's conjectures on the nature of the earths be confirmed by him, the following shortened names be adopted: *kieselmetall, thonmetall, zirkonmetall, beryllmetall*" ["... falls Davy's Vermutung über die Natur der Erden von ihm bestätigt werden, statt dieser Namen die folgenden, abkürzenden aufzunehmen: *Kieselmetall, Thonmetall, Zirkonmetall, Beryllmetall*"]. In 1814, at the high-point of anti-Napoleon sentiment in Europe, Hans Christian Oersted (1777–1851) went still further in rejecting the names than Gilbert in his fight against the nomenclature of the French antiphlogistonists and their adherents by proposing a "common chemical nomenclature for all Scandinavian-germanic languages" ["allen Skandinavisch-Germanischen Sprachen gemeinsame chemische Nomenclatur"] [11]. In his "*Einladungsschrift zu dem dießjährigen Kopenhagener Universitätsfeste zum Andenken der Reformation des Religionswesens und der Universität unter Christian III*" (1534–1559), produced originally in Latin, the Danish investigator attacked the nomenclature in question very strongly on the grounds that it was contrary to all reasonable requirements for naming a substance. He also set up a quite new basis for naming substances, proposing a range of new names among which, for designating "silicon or kiesel metal", was the new word "*Flintär*", derived from *Flint*, "partly because it is shorter and partly because it sounds better" than the word "kiesel" and has the terminal syllable *-är*, which should show the metallic character expected in the element. There was not a large response to Oersted's nomenclature though four years later R. Brandes (1795–1842), the founder of the "Apothekerverein im nördlichen Deutschland", accepted it in principle [12], though he wanted to introduce the terminal syllable *-el* to denote the elemental character of a substance and in this way arrived at *Flintel* as the name for elemental silicon.

Jöns Jacob Berzelius (1779–1848) took an approach similar to that of Davy. Although he had to confess in 1808 [13] that his experiments to decompose silica electrolytically had failed and he could only hope now "to discover some sort of way to decompose these earths (namely silica-, alumina-, and yttria earth) also into a metallic body and oxygen" ["irgend einen Weg zu entdecken auch diese Erden (nämlich: Kieselerde, Tonerde, und Yttererde) in einen metallischen Körper und in Sauerstoff zu zerlegen"], he still spoke without comment three years later in his "*Versuch einer lateinischen Nomenclatur für die Chemie, nach electro-chemischen Ansichten*" [14] of a metal *silicium*, the word being applicable in this form for the Latin and German languages. Even L.W. Gilbert, the translator of this publication by Berzelius into German, who was formerly quite critical of this name (see above) now adopted the word without comment. Berzelius also appears to have used the name earlier, as is apparent in a letter to H. Davy on 30th June, 1809, where *silicium* is used for the "basis of kieselerde" [15], so that it was probably a matter of course, in announcing the first preparation of the element fifteen years later [16], to name it *silicium* without wasting a single word in discussing it. Berzelius was not alone in adopting Davy's name. Thus in 1816, for example, it was used by the mineralogist E. D. Clarke (1769–1822) in Cambridge with a specific reference to its originator when he thought he had been successful in isolating the "basis of silica" with the aid of the oxy-hydrogen blowpipe [17], see p. 20, and, in 1821, by the French chemist J.-B. Boussingault (1802–1887) when he found that platinum can absorb the element from the ashes [43], see p. 19. J. J. Berzelius always adhered to the original form of the word while H. Davy in his textbook [18] *Elements of Chemical Philosophy*, which appeared twelve years before the first preparation of the element was announced and which was certainly known to the Swedish chemist, changed the name to *silicum* and also spoke of *aluminum, glucinum*, and *zirkonum*. No reason for this change, which he made for the "earth metals" and only for them, is given by H. Davy. It went back to a proposal by T. Bergman in 1784 [19] that for the latin designation of metals the ending *-um* be used only, though he was not consistent in this. But J. J. Berzelius was not quite true to his own criterion of latin names for metals and headed the corresponding chapter of his *Lehrbuch* in 1833 in the third [20] and in 1835 in the fourth edition [21] with "*Kiesel (Silicium)*", commenting on this title in a footnote: "Since the word kiesel is not used alone but as an adjunct as, for example, in kieselerde, kieselstein etc., I felt that it was not right to borrow a word from a foreign language to describe the radical of kieselerde. We therefore understand by "Kiesel" the combustible substance in "Kieselerde" and from this "Kieselsäure", "kieselsaure Salze" etc. follow quite naturally instead of the unacceptable "Siliciumsäure" and "siliciumsaure Salze" ["Da das Wort Kiesel nicht für sich allein, sondern nur als Zusatz gebraucht wird, wie z. B. Kieselerde, Kieselsteine usw., so habe ich es für durchaus unrichtig gehalten, aus einer fremden Sprache ein Wort zur Bezeichnung des Radikales der Kieselerde zu entleihen. Wir verstehen folglich unter Kiesel den brennbaren Körper in der Kieselerde, und hieraus folgt ungezwungen Kieselsäure, kieselsaure Salze etc., statt des widerwärtigen: Siliciumsäure und siliciumsaure Salze"]. J. Liebig in 1843 [39] in his *Handbuch der Chemie* used exactly the opposite formulation: "Silicium, Kiesel, Symbol Si".

The attitude of German chemists towards the naming of the element silicon and its compounds also remained divided subsequently. Thus, in 1876, O. Dammer in his *Kurzes chemisches Handwörterbuch* [33] preferred the name "*Kiesel*" for the element to its Latin form and, under "*Silicium*", he referred to the German word, under which the element was discussed; in the same way, under "*Kiesel*", the main compounds were described while, under "*Silico*", silicoorganic compounds, which were already known at the time, were enumerated. The reverse was the case in H. v. Fehling's *Neues Handwörterbuch der Chemie* [34] which appeared only two years later: under the heading "*Kiesel*" reference was made to "*Silicium*" and all the compounds were also dealt with there. In the previous editions of this *Handwörterbuch* it was attempted to turn this twofold approach to naming the element to advantage. In the first edition, published by J. Liebig, J. C. Poggendorff, and F. Wöhler, the editor, H. Kolbe [35] in the fourth volume, which appeared in 1849, included the element and its compounds under the title "*Kiesel*" in an article by F. Varrentrapp, while in the seventh volume printed twelve years later under the editorship of H. v. Fehling and H. Kolbe [36] a detailed expansion from the pen of H. v. Fehling of "what had earlier been said under Kiesel (Vol. IV, p. 316)" may be read under the heading "*Silicium*".

Whether the proposals for changes in naming the element and its compounds were made on aesthetic, systematic or even nationalistic grounds, they largely related to the whole word, and about 100 years later, because the knowledge of Latin on the part of beginning students was for the most part always decreasing, a desire to have a more readily understandable chemical technical language let the Nobel Prize winner Wilhelm Ostwald (1853–1932) to propose in an article "*Chemische Weltliteratur*" the introduction of a synthetic language for naming the elements and their compounds [22]. This was the *Ido* of the Frenchman L. de Beaufront (a special form of *Esperanto* introduced in 1887 by the Polish doctor Dr. L. L. Zamenhof). In this the latin word silicium has to be changed to *siliko* in the artificial language. The proposal met with little approval, although W. Ostwald had intended it only as a stimulus which was put forward at the Paris World Exhibition of 1900 [23].

Silicon, the usual name for the element in English speaking regions, goes back to a proposal by Thomas Thomson (1773–1853) [25] according to J. R. Partington [24]. The change in the end syllable is meant to draw attention to the great similarity of the nonmetal silicium and the nonmetals Kohlenstoff (English *carbon*) and Bor (English *boron*). Interestingly, seven years before the appearance of Thomson's *Lehrbuch,* J. J. Berzelius [26], in considering the nature of the element silicium, anticipated the name and rejected it decisively: "Since the ability to conduct electricity and the property of having a metal glance are lacking for silicium in the state in which it has so far been prepared, it is clear that it cannot be ranged alongside the metals but that its properties appear to place it nearer to carbon and boron. Some methodical natural philosophers will therefore probably call it *silicon* so as to indicate the type of combustible substance among which silicium can be included by this ending. I hold that this designation is unnecessary, for there is no sharp boundary between metalloids and metals. Carbon has a metallic glance and conducts electricity and yet it is not considered as a metal. If silicium could be melted it might perhaps have properties which are lacking in the powder form. Uranium in this state can be distinguished from silicium in outward appearance only with difficulty but, on the other hand, when crystalline it exhibits a metallic glance on reflexion and also shows transparency by refraction in thin films. Tantalum and titanium are also like silicium in their chemical properties. What will science gain by their removal from the metals to the metalloids, i. e., to the nonmetallic combustible substances? By these remarks I wish only to show that no natural boundary exists between these bodies and that in science it can be quite immaterial if a combustible substance is included in the metals or not if one only understands the electrochemical relationships of the body correctly" ["Da die Eigenschaft des Metallglanzes und das Vermögen die Electricität zu leiten beim Silicium in dem Zustand, in dem es bis jetzt erhalten wurde, fehlen, so ist es klar, daß man es nicht den Metallen anreihen kann, sondern daß seine Eigenschaften es mehr der Kohle und dem Boron zu nähern scheinen. Einige methodische Naturforscher werden es deshalb wohl *Silicon* nennen, um mit der Endigung die Art von brennbaren Körpern zu bezeichnen, zu welcher das Silicium gerechnet werden muß. Ich halte indeß diese Bezeichnungen für unnöthig, denn es giebt zwischen Metalloïden und Metallen keine scharfe Gränze. Die Kohle hat Metallglanz und leitet die Electricität, sie wird aber nicht als ein Metall betrachtet; wird man einst Silicium zum Schmelzen gebracht haben, so besitzt es vielleicht Eigenschaften, die ihm in Pulverform fehlen. Uran, in diesem Zustande, kann schwerlich im Aeußern von Silicium unterschieden werden, in krystallisirtem dagegen zeigt es bei der Reflexion Metallglanz, und bei Refraction an dünnen Kanten Durchsichtigkeit; Tantal und Titan gleichen dem Silicium auch ihren chemischen Eigenschaften nach; was sollte wohl die Wissenschaft durch ihre Versetzung von den Metallen zu den Metalloïden, d.h. zu den nicht metallischen brennbaren Körpern gewinnen? Ich will mit diesen Bemerkungen nur zeigen, daß es keine natürliche Gränze zwischen diesen Körpern gebe, und daß es, wenn man in der Wissenschaft nur die electrochemischen Beziehungen der Körper richtig auffaßt, ganz gleichgültig seyn kann, ob ein brennbarer Körper unter die Metalle gestellt werde oder nicht"].

The Russian name for the element silicon is кремний [kremnii], derived from the Russian word for flint кремень (kremen') and кремня (kremnya) [40]. According to the *Soviet Enciclopedia* [45] this word was introduced in 1834. The word occurs again in Ukrainian, Old Church Slavonic, Bulgarian, Serbo-Croatian, Slovene, Czech, Polish, Upper Sorbian, Lower Sorbian, and Polabian, so that a Proto-Slavic word **kremy* can be deduced, cognate with the latter is the Latvian *krams, krems* = flint from which this (and the corresponding words in other Baltic languages) cannot be considered in any case as loan-words. There is possibly a relationship [41] between the Slavic root and Saxon *scram-sahs* = knife-like sword, and the Middle High German *schram* = sword wound, rockcleft. N. Figurowski [40] also attempts to establish that naming of an element is connected with an earlier use of it or its compounds (see in this connection Gmelin Handbook "Blei" A1, 1973, p. 2). He adds to the word group of the etymology [41, 42] the verb кресати [kresati] = "to strike sparks from a stone with iron", which philologists, however, regard in an entirely different word relationship.

The symbol *Si* was first proposed by J. J. Berzelius in 1819 in his "*Versuch über die Theorie der chemischen Proportionen*" [27] where he sets out in detail the significance and principles used in developing his symbols, which he had sketched briefly some years before [28]. It may be noted that the selection of the symbol took place five years before the element was first prepared so that we must assume with E. v. Meyer [29] that Berzelius was convinced of the existence of a "metallic substance" ["metallischen Körpers"] as the "basis of silica" ["Basis der Kieselerde"] ever since Davy's publications and his own experiments with Pontin [13], that is, since 1808. It may also be mentioned that J. Dalton [30] in 1810 in his "*New System of Chemical Philosophy*" entered silica, *Silex*, under simple, that is elementary, bodies and proposed as a symbol for it a triangle surrounded by a circle with the apex pointing upwards. He had taken over the triangle from alchemical symbols where, in various positions and with various diagonal lines [31], though usually with the apex pointing downwards [32], it symbolizes the "earth element" [das "Element Erde"].

References:

[1] H. Davy (Phil. Trans. Roy. Soc. [London] **1808** 333/70, 353), J. Davy (The Collected Works of Sir Humphry Davy, Vol. 5, London 1840, pp. 102/39, 122). – [2] A. Ernout, A. Meillet (Dictionnaire Étymologique de la Langue Latine, 4th Ed., Paris 1959, p. 625). – [3] A. Walde, J. B. Hofmann (Lateinisch-Etymologisches Wörterbuch, 3rd Ed., Vol. 2, Heidelberg 1954, p. 530). – [4] H. Blümner (Technologie und Terminologie der Gewerbe und Künste bei Griechen und

Römern, Vol. 3, Leipzig 1884, p. 59). – [5] G. Agricola (De Natura Fossilium Libri X, Book 7 in: Opera, Basel 1558, p. 413 [1st Ed. 1546]).

[6] A. Hauptmann (Feuerstein Hornstein Flint Chert Silex: Eine Begriffsbestimmung, in: G. Weisgerber, R. Slotta, J. Weiner, 5000 Jahre Feuersteinbergbau: Die Suche nach dem Stahl der Steinzeit, 2nd Ed., Bochum 1981, pp. 7/11). – [7] C. A. Prieur (Ann. Chim. [Paris] **70** [1809] 189/254, 254). – [8] H. Davy (Giorn. Fis. Chim. Storia Nat. **2** [1809] 446/61, 453). – [9] H. Davy (J. Chem. Physik Gehlen **9** [1810] 484/527, 507). – [10] H. Davy (Ann. Physik **32** [1809] 365/96, 396 footnote).

[11] H. C. Oersted (Schweiggers J. Chem. Physik **12** [1814] 113/54, 136). – [12] R. Brandes (Neues J. Pharm. **2** [1818] 3/21, 11). – [13] J. J. Berzelius, M. M. Pontin (Ekon. Ann. Kongl. Wetenskaps Acad. **6** No. 2 [1808]; German translation by S. P. Leffler in: Ann. Physik **36** [1810] 247/80, 279). – [14] J. J. Berzelius (J. Physique **1811** Octbr., German translation in: Ann. Physik **42** [1812] 37/89, 47). – [15] H. G. Söderbaum (Jac. Berzelius Lettres, Vol. 1, Pt. 2, Correspondance entre Berzelius et Sir Humphry Davy 1808–1825, Uppsala 1912, pp. 12/6, 15).

[16] J. J. Berzelius (Ann. Physik Chem. [2] **1** [1824] 169/230, 210). – [17] E. D. Clarke (Ann. Physik **55** [1817] 1/39, 34). – [18] H. Davy (Elements of Chemical Philosophy [1st Ed. 1812] in: J. Davy, The Collected Works of Sir Humphry Davy, Vol. 4, London 1840, pp. 268/9). – [19] T. Bergman (Meditationes de Systemate Fossilium Naturali **1784** in: Opuscula Physica et Chemica, Vol. 4, edided by E. B. G. Hebenstreit, Leipzig 1887, pp. 180/278, 261). – [20] J. J. Berzelius (in: F. Wöhler, Lehrbuch der Chemie [German translation of the author's Swedish manuscript], 3rd Ed., Vol. 1, Dresden – Leipzig 1833, p. 325).

[21] J. J. Berzelius (in: F. Wöhler, Lehrbuch der Chemie [German translation of the author's Swedish manuscript], 4th Ed., Vol. 1, Dresden – Leipzig 1835, p. 325). – [22] Wi. Ostwald (Z. Physik. Chem. **76** [1911] 1/20, 8). – [23] N. I. Rodnyi, Yu. I. Solowjew (Wilhelm Ostwald, German translation by H. Sommer, Leipzig 1977, pp. 329/40). – [24] J. R. Partington (A History of Chemistry, Vol. 4, MacMillan, London 1964, p. 150). – [25] T. Thomson (A System of Chemistry of Inorganic Bodies, 7th Ed., London – Edinburgh 1831, p. 222 from [24]).

[26] J. J. Berzelius (Kongl. Svenska Vetenskaps Acad. Handl. **1824** I 46/98, 67/8; Ann. Physik [2] **1** [1824] 169/230, 230). – [27] J. J. Berzelius (Versuch über die Theorie der chemischen Proportionen und über die chemischen Wirkungen der Elektricität nebst Tabellen über die Atomgewichte, German translation by K. A. Blöde, Dresden 1820, pp. 117/8 [French 1st Ed. 1819]). – [28] J. J. Berzelius (Ann. Physik **46** [1814] 131/75, footnote pp. 154/5). – [29] E. v. Meyer (Geschichte der Chemie von den ältesten Zeiten bis zur Gegenwart, 3rd Ed., Leipzig 1905, pp. 361/2). – [30] J. Dalton (A New System of Chemical Philosophy, Pt. 2, Manchester 1810, p. 547).

[31] Dr. F. Lüdy Jr. (Alchemistische und Chemische Zeichen, Mittenwald 1929, Charts 72, 77). – [32] W. Schneider (Lexikon Alchemistisch-Pharmazeutischer Symbole, Verlag Chem., Weinheim 1962, p. 55). – [33] O. Dammer (Kurzes chemisches Handwörterbuch, Berlin 1876, pp. 373, 691). – [34] H. v. Fehling (Neues Handwörterbuch der Chemie, Vol. 3, Braunschweig 1878, p. 965). – [35] J. Liebig, J. C. Poggendorff, F. Wöhler (Handwörterbuch der reinen und angewandten Chemie, Vol. 4, Vieweg, Braunschweig 1849, pp. 316/52).

[36] J. v. Liebig, J. C. Poggendorff, F. Wöhler (Handwörterbuch der reinen und angewandten Chemie, Vol. 7, Vieweg, Braunschweig 1859, pp. 970/87). – [37] A. Fick (Beitr. Kunde Indoger. Sprache **24** [1899] 292/305, 301). – [38] A. Walde, J. B. Hofmann (Lateinisches etymologisches Wörterbuch, 3rd Ed., Vol. 1, Heidelberg 1938, p. 145). – [39] J. Liebig (Handbuch der Chemie mit Rücksicht auf Pharmacie: 1. Abtheilung – Anorganische Chemie, Heidelberg 1843, p. 335). – [40] N. Figurowski (Die Entdeckung der chemischen Elemente und der Ursprung ihrer Namen, German translation by L. Korniljew, E. Lemke, Köln 1981, p. 186).

[41] M. Vasmer (Russisches etymologisches Wörterbuch, Vol. 1, Heidelberg 1953, p. 659). – [42] A. G. Preobrazhensky (Etymological Dictionary of the Russian Language, New York 1951, pp. 380/1). – [43] J.-B. Boussingault (Ann. Chim. Phys. [2] **16** [1821] 5/16). – [44] J. Pokorny (Indogermanisches etymologisches Wörterbuch, Vol. 1, Bern – München 1959, p. 356). – [45] Sovetskaya Entsiklopedia: Kratkaya Khimicheskaya Entsiklopedia, Vol. 2, Moscow 1963, p. 202).

1.2 The Discovery

Attempts to elucidate the true nature of terra vitrescibilis [vitrifiable earth = silica, SiO_2], which Johann Joachim Becher (1635–1682) had already recognized as a separate substance [1] and which, sixty years later, had then been clearly distinguished from the other "earths" by Johann Heinrich Pott (1692–1777) [2], extended over an extraordinarily long time and first led to the isolation of the "basis of silica" by an indirect route involving a compound with hydrofluoric acid during a search for the element fluorine.

References:

[1] J. J. Becher (Alphabetum Minerale in: F. Roth-Scholtz, Opuscula Chymica Rariora, Nürnberg-Altdorf 1719, p. 105 [1st Ed. 1682]). – [2] J. H. Pott (Chymische Untersuchungen, welche fürnehmlich von der Lithogeognosia handeln, 2nd Ed., Berlin 1757, p. 3 [1st Ed. 1746]).

1.2.1 Reduction Experiments with "Kieselerde" SiO_2

1.2.1.1 Supposed Preparation of the Element in 1790

R. Koch claims [1] without, however, quoting his source in his biography of Georg Ernst Stahl (1660–1734), the founder of the Phlogiston Theory, that "he also came very near to the discovery of silicon" ["auch der Entdeckung des Siliciums kam er sehr nah"]. After perusal of those of his publications which are relevant this is still a surprising assertion. Even so careful and experienced a reader of the chemical literature as J. R. Partington [2] was quite unable to support such a contention and indeed came to the conclusion: "Stahl found it impossible to convert any earth proper [and kieselerde, SiO_2, was counted among these] into a metal by adding phlogiston". H. Kopp [3] was also unaware of this and does not once mention G. E. Stahl in connection with the history of silica and silicon and likewise has nothing to say about Koch's claim in describing Stahl's significance in the history of chemistry [4].

In 1790, however, what for Stahl had been impossible appeared to have become possible: "Among the chemical discoveries of the past year (1790) probably none has excited greater interest than the discovery by Bergrath A. v. Ruprecht and Doctor Tondy at Schemnitz [now Banscá Štoviatnica, Slovakia] that the simple earths are nothing but metallic calxes" ["Unter den chemischen Neuigkeiten des vorigen Jahres (1790) hat wohl keine eine größere Aufmerksamkeit erregt, als die Entdeckung des Herrn Bergrath A. v. Ruprecht und des Herrn Doctor Tondy zu Schemnitz [heute: Banscá Štoviatnica, Slowakei], daß die einfachen Erden nichts anderes sind, als metallische Kalke"]. It was in this way that the Royal Hannoverian Commissioner for Mines, J. F. Westrumb [5], opens his paper on the *Geschichte der neuentdeckten Metallisirung der einfachen Erden,* going on to describe the astonishment with which the first report of this important discovery was everywhere received. "One admired the extraordinary skill of those chemists who had been successful where other had been at pains for centuries to prove the opposite. One marvelled at the prophetic mind of a Lavoisier and of others" [twenty years later J. J. Berzelius [6] mentions the names of T. Bergman and B. Pelletier in this connection in his *Elektrisch-chemische Versuche über die Zerlegung der Alkalien und der Erden*] "who had anticipated this discovery long before, and probably sometimes envied it. They rightly feared the total revolution of our concepts and systems by it and thought that they had found in these discoveries unshakeable proof of the new teaching on oxygen and the true nature of metallic calxes. Their report also made the greatest impression on me.... I recall with pleasure the moment when we, my friend Lasius [probably Georg Siegmund Otto Lasius, 1752–1833, Hannoverian Ingenieur-Lieutenant] and I, estimated the consequences which the metallizability of simple earths would have. We calculated what massive collapse now lay ahead for the systems of mineralogy and chemistry which had been built up in such an artificial manner and what could accrue to the art of smelting from this discovery. We considered whether they would be of use and profit to the followers of the German or French schools and if perhaps making gold would no longer be a secret and concluded that everything on the planet where we live – of organic or nonorganic nature – was nothing but metal and more metal, though in another form" ["Mit großem Erstaunen nahm man die erste Nachricht von dieser wichtigen Entdeckung überall auf. Man bewunderte die außerordentliche Kunstfertigkeit jener Chemiker, denen etwas gelang, von dem man seit Jahrhunderten das Gegentheil zu beweisen bemüht war. Man bewunderte den prophetischen Geist eines Lavoisier, und einiger anderer" [J. J. Berzelius [6] nennt 20 Jahre später bei seinen *Elektrisch-chemische Versuche über die Zerlegung der Alkalien und der Erden* in diesem Zusammenhang noch die Namen T. Bergman und B. Pelletier], die diese Entdeckung lange vorher ahndeten; beneidete auch wohl mitunder die Entdeckung derselben. Befürchtete mit Recht, gänzliche Umwälzung unserer Begriffe und Systeme von ihr und glaubte in dieser Entdeckung den unerschütterlichsten Beweis, für die neue Lehre vom Oxygen, und der eigentlichen Beschaffenheit metallischer Kalke zu finden. Auch auf mich machte die Nachricht von ihr den größesten Eindruck ... Mit Vergnügen erinnere ich mich jenes Augenblicks, da wir, mein Freund Lasius [wohl: Georg Siegmund Otto Lasius, 1752 bis 1833, hannöverscher Ingenieur-Lieutenant] und ich, die Folgen berechneten, welche die Metallisirbarkeit der einfachen Erden haben würde. Berechneten, welch ein mächtiger Sturz, den so künstlich erbaueten Systemen der Mineralogie und Chemie, nun bevorstehe, und welche Vortheile der Schmelzkunst aus jener Entdeckung zuwachsen könnten. Untersuchten, ob sie den Anhängern der deutschen, oder französischen Schule, zu Nutz und Frommen gereichen würde; wähnten, nun sey Goldmachen vielleicht kein Geheimnis mehr; und fanden, daß nun alles auf dem Planeten, den wir bewohnen, organisirte und nichtorganisirte Wesen, nichts sind, als Metall und abermals Metall, nur in einer anderen Form"] [5].

F. A. C. Gren, the editor of the *Journal der Physik* und professor of natural sciences at the University of Halle an der Saale commented more forcibly on the report of the metallization of earths, which L. Crell [7] had communicated to him briefly: "If this discovery is confirmed it is as important to physical history of the 18th century as attempts in the political world to reduce kings" ["Bestätigt sich diese Entdeckung, so ist sie für die physische Geschichte des 18. Jahrhunderts so wichtig, als die Versuche in der politischen Welt es sind, die Könige zu reduziren"].

From the predictions of a metal content in the so-called primitive earths mentioned above the clearest, namely that of A. Lavoisier [8] may be selected; in his *Traité Élémentaire de Chimie,* which appeared in 1789, the following passage occurs: "It is most likely that we are only partially acquainted with the metallic substances which exist in nature; all of these, for example, which have more affinity to oxygen than to carbon, are not susceptible to reduction or conversion to a metallic state, they are not visible to our eyes but exist as oxides identical for us with the earths. It is very probable that baryta [barium oxide], which we class as an earth, is one of these cases. One sees it in many experimental details, and they approach characteristics of many metallic substances. It should be possible to rigorously state that all materials to which we give the name of earths are only the oxides of metals, irreducible by the techniques which we use" ["Il est probable que nous ne connoissons qu'une partie des substances métalliques qui existent dans la nature; toutes celles, par exemple, qui ont plus d'affinité avec l'oxygène qu'avec le carbone, ne sont pas susceptibles d'être réduites ou ramenées à l'état métallique, elles ne doivent se présenter à nos yeux que sous la forme d'oxides qui se confondent pour nous avec les terres. Il est très-probable que la baryte que nous venons de ranger dans la classe des terres, est dans ce cas; elle présente dans le détail des expériences des caractères qui la rapprochent beaucoup des substances métalliques. Il seroit possible à la rigueur que toutes les substances auxquelles nous donnons le nom de terres, ne fussent que des oxides métalliques, irréductibles par les moyens que nous employons"]. Later, in setting out his table of "simple bodies" ["substances simples"] in which the "earths" lime, magnesia, heavy earth, alumina, and silica are shown as "Substances simples salifiable terreuses" he again makes his ideas very clear: "It should be accepted that shortly the earths will cease to be classed with the simple substances, they are the only ones of this class which show no tendency to combine with oxygen, and I derive from this lack of reactivity toward oxygen, if I may be allowed to surmise, a degree of saturation. Earths behaving in this manner would be the simple substances, perhaps metal oxides oxygenated to a certain point. Besides, it is only a simple conjecture which I present here. I hope that the reader will certainly not be confused from what I give as truth, by fact and experience, which is still only hypothetical" ["Il est à présumer que les terres cesseront bientôt d'être composée au nombre des substances simples; elles sont les seules de toute cette classe qui n'aient point de tendance à s'unir à l'oxygène, et je suis bien porté à croire que cette indifférence pour l'oxygène, s'il m'est permis de me servir de cette expression, tient à ce qu'elles en sont déjà saturées. Les terres, dans cette manière de voir, seraient des substances simples, peut-être des oxydes métallique oxygénés jusqu'à un certain point. Ce n'est, au surplus, qu'une simple conjecture qui je présente ici. J'espère que le lecteur voudra bien de ne pas confondre ce que je donne pour des vérités de fait et d'expérience avec ce qui n'est encore qu'hypothétique"].

The earliest results on the "metallization of earths" ["Metallisierung der Erden"] which seem to have appeared in July and August of 1790, became known largely from published extracts from private letters. The first came from a letter by Doctor Savaresi, who was involved in the work, to Hofrat I. v. Born (1742–1791) in Wien [9] and is concerned chiefly with results of experiments on "heavy, bitter, and calcareous earths [BaO, MgO, CaO]" ["Schwer-, Bitter- und Kalk-Erde [BaO, MgO und CaO]"]. In the earliest of the letters by A. v. Ruprecht [10], who discovered the process, to I. v. Born only the question of tungsten-, molybdenum-, and uranium-reguli together with the question of "heavy earth reguli" ["Schwererdekönig"], is discussed but it contains a reference to an earlier and apparently unpublished exchange of letters which appeared to I. v. Born to justify [11] "taking away calcareous-, bitter-, and heavy-earths from the group of earths and calling them oxidized metals" ["die Kalk-, Bitter- und Schwererde aus der Classe der Erden wegzunehmen und ihnen den Namen der oxidirten Metalle beyzulegen"]. His procedure, which was also used for siliceous earth, is described in detail by A. v. Ruprecht in the case of the preparation of the heavy earth regulus [10]: "Calcined crystals [of nitrate] are finely ground in a glass mortar (in order to keep out everything which could result in accidental contamination with metal of any sort, which must be avoided in such experiments), an eighth part of fine carbon dust is added and a paste is made from this with linseed oil, a lump of it being then placed in a Hessian crucible, covered with fine carbon. The latter is put into a large crucible and surrounded with carbon. The second lump was stuck on the inner surface of the outer crucible (which was situated opposite the blast); this was near the wall of the small crucible with one of its surfaces on the inner wall of the outer crucible. The remaining space was filled with coarse carbon dust and the top of this was covered with sieved bone ash. The crucible prepared in this way was placed on a pottery support and then in a chimney which was exactly closed by fine pottery bricks placed on top of one another so that the blast had to act on the base of the outer crucible. As soon as the heaped carbon was glowing completely and the vessel had been sufficiently heated, I allowed the double blast, which carried 80 pounds, to come into operation and to operate for 1¾ hours while carbon was continually added and the surface was frequently damped down with water. In this way at the end not only the support but also the pottery brick, the upper part of the crucible and even the bone ash had melted to a spongy white slag which gave sparks when struck with steel" ["Man rieb sodann die verkalkten Krystallen [des Nitrats] in einem gläsernen Mörser (um alles, was eine zufällige Beymischung irgend eines Metalls, die auf das sorgfältigste bey dergleichen Versuchen vermieden werden muß, entfernt zu halten) ganz fein, setzte den achten Theil feinen Kohlenstaubes hinzu, und machte daraus mit Leinöhl einen Teig, von dem man ein Klümpchen in einen kleinen heßischen Tiegel, mit einer dünnen Kohle bedeckt, in einen größern Tiegel mit Kohlenstaub umgeben, setzte; das zweyte Klümpchen aber an die innere Fläche (die der Windseite des Gebläses entgegengesetzt war), des äußern Tiegels (und zunächst am Rande des kleinen, mit einer seiner Flächen an die innere Fläche des äußern Tiegels genau anliegenden Tiegels) anklebte; den übrigen Raum mit gröblichen Kohlenstaub, und diesen zu oberst, mit gesiebter Beinasche ½ Zoll hoch bedeckte. – Den auf diese Art vorgerichteten Tiegel, setzten wir auf eine Hafnerzeller Unterlage, so in die, rund um mit fünffach übereinander gestellten Hafnerzeller Ziegeln genau geschlossene, Esse, daß der Wind auf dem Boden des äußern Tiegels wirken mußte. Sobald die angehäuften Kohlen vollends glühten, und das Gefäß schon genugsam erhitzt war, ließ ich das Doppelgebläse, das mit 80 Pf. beschwert war, in Bewegung setzen, und 1¾ Stunden zublasen; während daß stets Kohlen nachgetragen, und von oben öfters mit Wasser gedämpft wurden; wodurch dann am Ende nicht allein die Unterlage, sondern auch die Hafnerzeller Ziegel, dann der obere Theil des Tiegels und selbst die Beinasche zu einer weißen schwammigen und am Stahl feuerschlagenden Schlacke schmolz"].

A silica regulus was first mentioned in a letter from A. v. Ruprecht to I. v. Born, which L. Crell printed in the September issue of *Chemische Annalen* [12]: "Snow white soft silica which has been purified with aqua regia behaves in the same way [as lime, mentioned earlier], giving a small regulus which was attracted [by magnets] and a metallic coating [on the crucible wall]. Consequently this must be repeatedly checked" ["Ebenso (wie die zuvor abgehandelte Kalkerde) verhielt sich auch eine mit Königswasser gereinigte, zarte, schneeweiße Kieselerde, die ebenfalls einen kleinen [vom Magneten] anziehbaren König und metallischen Anflug [an den Tiegelwänden] lieferte, und daher auch noch wiederhohlt geprüft werden muß"]. As A. v. Ruprecht remarked in his next letter to I. v. Born [13], this checking was delayed: "The purified silica has recently given us only a single regulus, which was, however, very strongly attracted [by magnets]. I will therefore attempt the preparation [reduction] again tomorrow or next monday with the softest and purest silica, which I obtained by decomposing glass in the course of preparing hydrofluoric acid and had washed completely, and then I will report upon the outcome" ["Die gereinigte Kieselerde hat uns zwar neulich einen einzigen, doch [vom Magneten] sehr anziehbaren König gegeben; ich will also morgen oder nächsten Montag mit der zartesten und reinesten Kieselerde, die ich aus der Zerlegung des Glases bei Erzeugung der Flußspathsäure erhalten und vollkommen ausgesüßt habe, die Wiederherstellung [Reduktion] versuchen und Ihnen den Ausschlag seiner Zeit einberichten"]. Meanwhile no further positive results on the reduction of silica were reported from Schemnitz. On the contrary, one of the Bergrat's Polish students, who had successfully repeated Ruprecht's experiments with other substances, was unable to obtain any regulus from silica, as Microszewski and Bienkowski reported to Paris in 1791 [14]. Bergrat J. F. W. Widenmann in Stuttgart, on 20th December, 1790, was even in a position to announce [15] that "from the most recent reports [from Wien and Schemnitz] siliceous earths cannot be reduced but only fuse to a white porcelain-like material" ["nach den neueren Nachrichten [aus Wien und Schemnitz] die Kieselerde sich nicht reduciren lassen solle, sondern immer nur in eine weiße porcellainähnliche Masse zusammenschmelze"].

Ruprecht's experiments were understandably repeated by many investigators and these mostly reported their lack of success with the alkaline earths without giving any information on silica. Only J. F. A. Göttling (1755–1809), professor in Jena, made known his lack of success with silica [16]. He must already have carried out his experiments in August-September 1790 immediately after the first announcement of "metallization" ["Metallisirung"], as J. F. Westrumb [17] concluded from the fact that Göttling's communication was already included in the *Taschenbuch für Scheidekünstler* for 1791 which was on sale at the Leipzig Michaelis [October] Fair of 1790. In letters to J. F. Westrumb [17], Göttling's lack of success was seen as due either to an insufficiently high temperature during the heating or to an insufficient exclusion of air during the experiments.

The discovery of "metallization of the earths" ["Metallisirung der Erden"] gave rise to objections at once, as J. F. Westrumb said in a letter on 1st October, 1790, to L. Crell, the Editor of *Chemische Annalen*, who first published it at the end of the year [18]: "Herrn von Ruprecht's discovery merits every conceivable attention, and he has earned the esteem and thanks of all patriotically minded Germans as soon as it has been confirmed. But allow me a small question of doubt: Has Herr v. R. already examined these newly obtained reguli from earths and sedative salt [i.e., boric acid] closely? Is it not, perhaps, the iron and the pyrolusite of the reducing agent? Or is it not the phosphoric acid of the latter in combination with one of the former? It is only too possible to be deceived in this" ["Herrn von Ruprechts Entdeckung verdient alle ersinnliche Aufmerksamkeit und Er, so bald sie sich bestätigen sollte, die Achtung und den Dank aller patriotisch gesinnten Deutschen. Aber erlauben Sie mir die kleine zweifelnde Frage: Hat Herr v. R. diese neuen aus Erden und dem Sedativsalze [das ist: Borsäure] erhaltenen Könige schon genau geprüft? ists nicht etwa das Eisen und der Braunstein des Reducirmittels? oder ists nicht die Phosphorsäure der letztern, in Verbindung mit einem der erstern? Nur zu leicht ist hier Täuschung ... möglich"].

At the end of October 1790 M. H. Klaproth [19] announced that in the course of his own experiments he had received a letter from Schemnitz by "Herr Savaresi, Chef der 6 Königl. Neapolitanischen Pensionäre" [that is, stipendiaries of whom the Dr. Tondi mentioned at the beginning was one] which showed that blank experiments indicated that the new reguli arose not from the "earths" ["Erden"] but from impurities in the carbon. M. H. Klaproth reported to the Royal Academy in Berlin [20] on 2nd February, 1791, on his own experiments made in common with Bergassessor D. L. G. Karsten, Hofapotheker S. Hermbstädt, Münzwardein Frik, and Bergsekretär Wähler, stating that "the new metal reguli ... are neither more nor less than iron which is reduced from the iron-containing material of the Hessian melting crucible used at a strong heat, this then melting to fine granules, the greater part of which react with phosphoric acid in the bone ash to form "hydrosiderum" [water iron, i.e., iron phosphide]" ["die neuen Metallkönige ... nichts mehr und nichts weniger sind als Eisen, welches sich aus der eisenhaltigen Masse der dazu angewendeten Hessischen Schmelztiegel bei heftigem Feuer reducirt und in kleinen Körnern ausschwitzt; davon der größte Theil mit der in der Beinasche enthaltenen Phosphorsäure sich zu Hydrosiderum [Wassereisen, das ist: Eisenphosphid] bildet"].

Even before this the same observation had been announced by J. F. Westrumb [21] when, in November, 1790, he had put forward as a conjecture what had now been demonstrated experimentally [22]. K.u.K. Artillerieoffizier Tihavski in Wien, while attempting to repeat Ruprecht's results in presence of J. F. v. Jacquin (1766–1839), the professor of chemistry and botanist, had come to the same results as the North German investigators [23]. The widespread controversy which arose in the course of this work may be passed over here as the details have to do for the most part only with the alkaline earths, alumina, and the "König des Sedativsalzes" [i.e., boron], as well as uranium, molybdenum, and tungsten.

References:

[1] R. Koch (in: G. Bugge, Das Buch der großen Chemiker, Vol. 1, Berlin 1929 [Reprint: Verlag Chem., Weinheim/Bergstr. 1955], pp. 192/203, 198). – [2] J. R. Partington (A History of Chemistry, Vol. 2, MacMillan, London 1961, pp. 653/86, 670). – [3] H. Kopp (Geschichte der Chemie, Vol. 4, Vieweg, Braunschweig 1847, pp. 69/75). – [4] H. Kopp (Geschichte der Chemie, Vol. 1, Vieweg, Braunschweig 1843, pp. 187/93). – [5] J. F. Westrumb (Geschichte der neuentdeckten Metallisirung der einfachen Erden nebst Versuchen und Beobachtungen, Hannover 1791, pp. 3/5).

[6] J. J. Berzelius, M. M. Pontin (Ann. Physik **36** [1810] 247/80, 259/60). – [7] L. Crell (J. Physik Gren **2** [1790] 160). – [8] A. Lavoisier (Traite Elementaire de Chimie, Bd. 1, Masson, Paris 1789, pp. 174, 194/5). – [9] I. v. Born (Chem. Ann. Crell **1791** I 3/10, 5th footnote). – [10] A. v. Ruprecht (Chem. Ann. Crell **1790** II 3/14).

[11] I. v. Born (Catalogue Méthodique et Raisonné de la Collection des Fossiles de Mlle. Éléonore de Raab, Pt. 2, Wien 1790, Preface, pp. 1, 4, 5 from F. Westrumb, Geschichte der neuentdeckten Metallisirung der einfachen Erde nebst Versuchen und Beobachtungen, Hannover 1791, pp. 32/3). – [12] A. v. Ruprecht (Chem. Ann. Crell **1790** II 195/202, 197). – [13] A. v. Ruprecht (Chem. Ann. Crell **1790** II 291/5, 293). – [14] Microszewski, Bienkowski (Ann. Chim. [Paris] **9** [1791] 51/3). – [15] J. F. W. Widenmann (Bergmänn. J. **3** II [1790] 501/2).

[16] J. F. A. Göttling (J. Physik Gren **3** [1791] 216/9; Taschenbuch für Scheidekünstler und Apotheker, 12th Ed., Weimar 1791, pp. 161/76, 163/4). – [17] J. F. Westrumb (Geschichte der neuentdeckten Metallisirung der einfachen Erden nebst Versuchen und Beobachtungen, Hannover 1791, pp. 48/9, footnote 32, pp. 51/2, footnote 38). – [18] J. F. Westrumb (Geschichte der neuentdeckten Metallisirung der einfachen Erden nebst Versuchen und Beobachtungen, Hannover 1791, pp. 80/1; Chem. Ann. Crell **1790** II 522). – [19] M. H. Klaproth (Intelligenzbl. Allgem. Litteraturztg. [Jena] Nr. 146 [1790] 6th Nov. from F. Westrumb, Geschichte der neuentdeckten Metallisirung der einfachen Erden nebst Versuchen und Beobachtungen, Hannover 1791, pp. 52/8). – [20] M. H. Klaproth (J. Physik Gren **3** [1791] 197/212).

[21] J. F. Westrumb (J. Physik Gren **3** [1791] 44/6). – [22] J. F. Westrumb (J. Physik Gren **2** [1790] 332/4). – [23] J. F. v. Jacquin (Ann. Chim. [Paris] **9** [1791] 54/5).

1.2.1.2 Early Attempts at Electrolytic Preparation

On 7th November, 1807, H. Davy [1] reported his great success in decomposing "fixed alkalies" ["fixe Alkalien"] electrolytically to the Royal Society in London. He also described experiments with barytes and strontites which promised success since colored, luminescent effects were observed at one of the poles. He concluded: "Barytes and strontites have the strongest relations to the fixed alkalies of any of the earthy bodies; but there is a chain of resemblances, through lime, magnesia, glucina, alumina, and silex. And by the agencies of batteries sufficiently strong, and by the application of proper circumstances, there is no small reason to hope, that even these refractory bodies will yield their elements to the methods of analysis by electrical attraction and repulsion." Publication followed in the first volume of the *Philosophical Transactions of the Royal Society* for 1808 and, as J. Freiherr v. Jacquin reported in Vienna on 5th February, 1808 [2], this was taken up by the widely circulated *Moniteur Scientific* in Paris and the *Jenaer Allgemeine Litteratur-Zeitung* at least in the form of an abstract. It is therefore not surprising that these sensational experiments were repeated in many places. In the *Intelligenz-Blatt,* the tenth issue of the last mentioned Zeitung for 27th February, 1808, a communication appeared [3] reporting that "Dr. [J. T.] Seebeck (1770–1831) in Jena had found that baryt-, kalk-, thon-, thalk-, and kiesel-earths became "combustible" under the action of the Volt[aic] column, in the same way as potash and soda. They showed, like them, luminous phenomena around the pile and appear to be "decomposed" in the same way as potash . . ." ["daß Dr. [J. T.] Seebeck (1770 bis 1831) zu Jena auch Baryt-, Kalk-, Thon-, Thalk- und Kieselerde ebenso "verbrennlich" durch die Action der Volt[aischen] Säule gefunden habe als Kali und Natron. Sie hatten ihm ähnliche feurige Phänomene im Kreise der Säule gewährt wie diese, schienen ihm ebenso "zersetzt" zu werden wie das Kali . . ."]. In a letter to J. W. Ritter (1776–1810) dated 30th March, 1808, J. T. Seebeck was more cautious in his statement: "I have investigated if it is possible to obtain with mercury [on electrolysis] amalgams such as we obtained with potash and soda . . . and got amalgams which were partly liquid and partly solid from kalk- and baryt- as well as from thalk- and clay-earths but not from silica. These also give gas in water . . ." ["Ich habe nämlich versucht, ob sich wohl aus den Erden [bei der Elektrolyse] vermittels des Quecksilbers ähnliche Amalgame gewinnen ließen, wie wir aus Kali und Natron erhalten . . . und ich habe wirklich aus Kalk und Baryt, auch aus Thalk- und Thonerde, nur nicht aus Kieselerde, ähnliche, theils flüssigere, theils festere Amalgame erhalten, die in Wasser ebenso Gas geben . . ."] [3].

J. J. Berzelius, in collaboration with the Royal Swedish physician M. M. Pontin [4] immediately after the announcement of Davy's research, repeated it, extending it in his own work; see Gmelin Handbuch, Calcium A 1, 1950, p. 68. Writing about his experiments with the "true earths" he says: "When we attempted to decompose the *true earths* in the same way as the alkaline earths our attempts were unsuccessful. We have examined *clay earth, yttria,* and *silica* without observing any trace of decomposition, even when we stirred the freshly precipitated and well washed earths with a little acid, which dissolved a part, thus bringing them into the sphere of action of the battery. Notwithstanding this failure we hope still to find some way to decompose these earths also to a metallic substance and oxygen, for this seems extremely probable by analogy. This failure appears to stem from the total insolubility of the earths in water and the fact that in the nonignited state they contain so much admixed water that the battery strength becomes quite exhausted in decomposing it" ["Als wir versuchten, die *eigentlichen Erden* eben so wie die alkalischen Erden zu zerlegen, blieben unsere Versuche ohne Erfolg. Wir haben *Thonerde, Yttererde* und *Kieselerde* der Prüfung unterworfen, ohne jedoch irgend eine Spur von Zerlegung wahrzunehmen, selbst nicht, wenn wir die eben gefällte und gut ausgelaugte Erde mit ein wenig Säure anrührten, welche einen Theil derselben auflösen und sie dadurch in den Wirkungskreis der Batterie hineinführen konnte. Ungeachtet dieses Mißgelingens hoffen wir aber doch noch, irgend einen Weg zu entdecken, auch diese Erden in einen metallischen Körper und in Sauerstoff zu zerlegen; denn die Analogie macht dieses äußerst wahrscheinlich. Es scheint dieses Fehlschlagen hauptsächlich von der gänzlichen Unauflöslichkeit der Erden im Wasser und davon herzurühren, daß sie im nicht gebrannten Zustande, des Wassers mechanisch so viel eingemengt enthalten, daß sich die Kraft der Batterie durch die Zerlegung desselben gänzlich erschöpft"].

Berzelius reported his and Pontin's successes and failures by letter [5] to H. Davy in June 1808, while the latter was engaged in carrying his electrolytic experiments further. On June 30th, 1808, H. Davy was already able to report his work fully to the Royal Society: "I tried the methods of electrization and combination with quicksilver, and the common metals, by which I had succeeded in decomposing the alkaline earths, on alumine and silex; but without gaining distinct evidences of their having undergone any change in the processes. – Obliged to seek for other means of acting upon them, it was necessary to consider minutely their relations to other bodies, and to search for analogies by which the principles of research might be guided. – Alumine very slowly finds its point of rest at the negative pole, in the electrical circuit; but silex, even when diffused in its gelatinous state through water, rests indifferently at the negative or positive poles. From this indifference to positive and negative electrical attractions, following the general order of facts, it might be inferred, that if these bodies be compounds, the electrical energies of their elements are nearly in equilibrium; and that their state is either analogous to that of insoluble neutral salts, or of oxides nearly saturated with oxygen .... On the idea that silex might be an insoluble neutrosaline compound, containing an unknown acid or earth, or both, and capable of being resolved into its secondary elements, in the same manner as sulphate of barytes, or fluate of lime, I made the following experiments. Two gold cones, connected by moistened amianthus, were filled with pure water, and placed in the electrical circuit, a small quantity of carefully prepared and well washed silex was introduced into the positive cone: the action was kept up from a battery of two hundred plates, for some hours till nearly half of the fluid in each cone was exhausted; the remainders were examined; the fluid in the cone containing the silex was strongly acid; that in the opposite cone was strongly alkaline; the two fluids were passed through bibulous paper, and mixed together, when a precipitate fell down, which proved to be silex. – On the first view of the subject, it appeared probable that this silex had been formed by the union of the acid and the alkaline matter in the two cones, and that the experiment demonstrated a decomposition and recomposition of silex; but before such a conclusion could be made, many points were to be determined. ..."

His consideration of the problem led him, however, to the conviction: "that there is no reason to suppose that silex had been changed in this experiment." These and similar results with alumina led him to change his experiments: "... I fused a mixture of one part of silex and six of potash in a platina crucible, and preserved the mixture fluid, and in ignition over a fire of charcoal; the crucible was rendered positive from the battery of five hundred, and a rod of platina, rendered negative, was brouhgt in contact with the alkaline menstruum. At the moment of contact there was a most intense light; when the rod was plunged into the liquid an effervescence took place, and globules which burnt with a brilliant flame rose to the surface, and swam upon it in a state of combustion. In a few minutes, when the mixture was cool, the platina bar was removed: after as much as possible of the alkali and silex had been detached from it by a knife, there remained brilliant metallic scales round it, which instantly became covered with a white crust in the air, and some of which inflamed spontaneously. The platina appeared much corroded, and of a darker tint than belongs to the pure metal. When it was plunged into water it strongly effervesced: the fluid that came from it was alkaline; when a few drops of muriatic acid were added to the solution, a white cloudiness occurred, which various trials demonstrated, depended upon the presence of silex. ... I endeavoured to procure an alloy of potassium, and the bases of the earths from mixtures of potash, silex, and alumine, fused by electricity, and acted on by the positive and negative surfaces in the same manner as pure potash, in experiments for the decomposition of that substance; but I obtained no good results. When the earths were in quantities equal to one-fourth or one-fifth of the alkali, they rendered it so highly nonconducting, that it was not easy to affect it by electricity; and when they were in very minute portions, the substance produced had the characters of pure potassium. ... I shall now mention the last trials that I made with respect to this object. – Potassium, amalgamated with about one-third of mercury, was electrified negatively under naphtha, in contact with silex very slightly moistened, by the power of five hundred; after an hour the result was examined. The potassium was made to decompose water, and the alkali formed neutralized by acetous acid; a white matter, having all the appearance of silex precipitated, but in quantity too small for accurate examination. – From the general tenor of these results, and the comparison between the different series of experiments, there seems very great reason to conclude that alumine, zircone, glucine, and silex are, like the alkaline earths, metallic oxides, for on no other supposition is it easy to explain the phenomena that have been detailed. The evidences of decomposition and composition, are not, however, of the same strict nature as those that belong to the fixed alkalies and alkaline earths; for it is possible, that in the experiments in which the silex, alumine, and zircone appeared to separate during the oxidation of potassium and sodium, their bases might not actually have been in combination with them, but the earths themselves, in union with the metals of the alkalies, or in mere mechanical mixture. And out of an immense number of experiments which I made of the kind last detailed, a very few only gave distinct indications of the production of any earthy matter; and in cases when earthy matter did appear, the quantity was such as rendered it impossible to decide on the species."

Here followed finally the paragraph quoted above, p. 2, which became so important in naming the metals hidden in the "true earths". In the following year on 16th November H. Davy [6] was able to report to the Royal Society on his partial success after he had obtained what was not the element but probably its alloy with iron:

"I have tried a number of experiments with the hope of gaining the same distinct evidences of the decomposition of the common earths, as those afforded by the electro-chemical processes on the alkalies, and on the alkaline earths. I find that when iron wire ignited to whiteness by the power of 1000 double plates, is negatively electrified, and fused in contact with either silex, alumine, or glucine, slightly moistened and placed in hydrogen gas, the iron becomes brittle and whiter, and affords by solution in acids, an earth of the same kind as that which has been employed in the experiment."

References:

[1] H. Davy (Phil. Trans. Roy. Soc. [London] **98** [1808] 1/44 in: J. Davy, The Collected Works of Sir Humphry Davy, Vol. 5, London 1840, pp. 57/101, 99; Ann. Physik **31** [1809] 113/75, 172). – [2] J. Freiherr von Jacquin (Ann. Physik **28** [1808] 132/4). – [3] J. T. Seebeck (Intelligenzbl. Allgem. Litteraturztg. [Jena] No. 10 Feb. 27 [1808] from J. W. Ritter, J. Chem. Physik Gehlen **5** [1808] 439/82, 448/9, 481/2; Ann. Physik **28** [1808] 367/8). – [4] J. J. Berzelius, M. M. Pontin (Ekon. Ann. Kongl. Wetenskaps Acad. **6** No. 2 [1808] [German translation by S. P. Leffler, Ann. Physik **36** [1810] 247/80, 279). – [5] H. Davy (Phil. Trans. Roy. Soc. [London] **98** [1808] 333/70 in: J. Davy, The Collected Works of Sir Humphry Davy, Vol. 5, London 1840, pp. 102/39, 109, 116/22).

[6] H. Davy (Phil. Trans. Roy. Soc. [London] **100** [1810] 16/74 in: J. Davy, The Collected Works of Sir Humphry Davy, Vol. 5, London 1840, pp. 225/83, 267; Ann. Physik **37** [1811] 155/207, 186).

1.2.1.3 Reduction Experiments with Metallic Potassium

In his experiments to apply the reducing power of metallic potassium in the case of the so-called earths also, H. Davy often came across brown amorphous silicon without recognizing what had happened. Right at the beginning of his communication which was read before the Royal Society in London on 30th June, 1808, and from the final sentence of which the name of the element stemmed, see p. 2, H. Davy [1] had to report that the metallic potassium acted not only on the alkaline earths present but also on the glass of the vessel, i.e., on its silica, coloring it brown. He was unable, however, to recognize the reaction which had occurred since he expected "metallic globules" in the event of reduction: "I heated potassium in contact with dry pure lime, barytes, strontites, and magnesia, in tubes of plate glass; but as I was obliged to use very small quantities, and as I could not raise the heat to ignition without fusing the glass, I obtained in this way no good results. The potassium appeared to act upon the earths and on the glass, and dark brown substances were obtained, which evolved gas from water; but no distinct metallic globules could be procured. . . ." Experiments with other earths led to just the same result: "I heated small globules of potassium, in contact with silex and alumine, in tubes of plate glass filled with the vapor of naphtha: the potassium seemed to act at the same time upon the glass and the earths, and a grayish opaque mass, not possessed of metallic splendor was obtained, which effervesced in water, depositing white clouds. Here it was possible that the potash had been converted wholly or partly into protoxide, by its action upon the earths; but as no globule was obtained, and as the plate glass alone might have produced the effect, no decided inference of the decomposition of the earths can be drawn from the process." H. Davy reported in the Bakerian Lecture, 1809, published in the following year [2], on further reduction experiments with silica and metallic potassium: "I have passed potassium in vapor through each of these earths, heated to whiteness in a platina tube: the results were remarkable, and perhaps not unworthy of being fully detailed. When silex was employed, being in the proportion of about ten grains to four of potassium, no gas was evolved, except the common air of the tube mingled with a little inflammable gas, not more than might by referred to the moisture in the crust of alkali formed upon the potassium. The potassium was entirely destroyed; and glass with excess of alkali was formed in the lower part of the tube. When this glass was powdered, it exhibited dark specks, having a dull metallic character, not unlike that of the protoxide of iron. When the mixture was thrown into water, there was only a very slight effervescence; but on the addition of muriatic acid to the water, globules of gas were slowly liberated, and the effect continued for nearly an hour; so that there is great reason to believe that the silex had been either entirely or partially deoxygenated, and was slowly reproduced by the action of the water, assisted by the slight attraction of the acid for the earth. When the potassium was in the quantity of six grains, and the silex of four grains, a part of the result inflamed spontaneously as it was taken out of the tube, though the tube was quite cool, and left, as the result of its combustion, alkali and silex. The part which did not inflame, was similar in character to the matter which has been just described; it did not act upon water, but effervesced with muriatic acid."

He did not at the time go more closely into the "dark specks having a dull metallic character". He first ventured to give an interpretation of his earlier observations at the meeting of the Royal Society on 8th July, 1813, in connection with his attempts to isolate the element fluorine, though he was mistaken in his statement of the times as seen from the dates given [3]: "In the Bakerian Lecture for 1810, I have given an account of the action of potassium upon pure silica. In this process the potassium acquires oxygen, and a combustible substance, which consists either of the basis of silica, or the basis of silica combined with potassium appears."

More than half a year later at the meeting of the Royal Society on 13th February, 1814, H. Davy reported on new experiments on the reduction of silica with metallic potassium to which he came after vain attempts to determine the oxygen content of the silica by other methods [4]: "I have endeavoured to separate the siliceous basis in a pure form, with the view of making synthetical experiments on its nature by combustion in oxygen, and my results, though not perfectly satisfactory, yet seem worthy of notice, and may lead to more successful attempts. – I decomposed silica by passing potassium in excess through it, in a heated tube of platinum: the result consisted chiefly of alkali containing a dark colored powder, the basis of silica diffused through it. I fused the whole mass with sulphur, which, in combining with the dry alkali, produced ignition. I attempted to detach the sulphuret of potassa by water: in this case the dark particles separated, but during their separation and after, they acted upon the water of the solution producing gas, and in attempting to collect them by the filter, I failed to procure sufficient for examination, for they were principally converted into silica. – I heated the substance procured in another experiment of this kind with hydrate of potassa; in this case there was a copious effervescence, and silica appeared to be reproduced and dissolved by the alkali. – I heated a portion of a similar result in strong lixivium of potassa; the solution gained a tint of olive, but there was scarcely any effervescence; from this it seems probable, that the inflammable basis of silica, like boron, is soluble in alkaline solutions without decomposing them. – Indeed this body, in its general characters, appears very analogous to boron. It appears to be neither volatile nor fusible; its oxide exerts, like boracic acid, a neutralizing power on the alkalies, though of a feebler kind, and forms, like boracic acid, vitreous bodies with the alkaline earths, and, like boron, the siliceous basis in combination with fluorine constitutes a powerful acid. – In my first views of the nature of the boracic and siliceous bases, I thought it probable that they would both appear as metals, if they could be entirely freed from oxygen: but it now seems more probable, that they form a class by themselves, offering a kind of link in the chain of natural bodies, when arranged according to their analogies, between charcoal, and sulphur and phosphorus. – It seems worthy of an experimental inquiry, whether the siliceous basis may not be obtained pure by heating the result procured from silica by potassium with pure sulphuric acid, which might possibly detach the potassa to form acid sulphate of potassa, without being decomposed by the inflammable basis."

In a footnote [5] John Davy, the editor of the *Collected Works of Sir Humphry Davy* drew attention to the fact that his brother's views on the siliceous basis ". . . has been confirmed by the later researches of Berzelius."

In fact, J. J. Berzelius had undertaken experiments on the reduction of silica by potassium and, in his publication on the preparation of the element in 1824, fully described and thus confirmed Davy's old results [6]. In a letter to H. Davy dated 21st April of that year he summarized his results briefly and recognized the priority of Davy's experiments [7]: "One can, as you have found long before me, reduce the silicon of silica by means of potassium, but one achieves thereby only one of the two results: either one obtains silicon saturated with potassium which dissolves in water in a few instants, or if the temperature is sufficient to expel the excess of potassium, the silicate of potassium which is formed fuses and envelopes the silicon in a vitrified dross, which can be dissolved by hydrofluoric acid. The quantity of silicon obtained in this way is very small" ["On peut, come vous l'avez trouvé longtemps avant moi, réduire le silicium de la silice moyennant la potasse, mais alors il arrive l'un des deux, ou que l'on obtient le silicium saturé de potassium qui se dissout dans l'eau en peu d'instants, ou, si la température suffit pour chasser l'excès de potassium, le siliciate de potasse formé se fond et enveloppe le silicium dans une scorie vitrifiée, laquelle on peut dissoudre moyennant l'acide fluorique. La quantité de silicium ainsi gagnée est très petite"].

References:

[1] H. Davy (Phil. Trans. Roy. Soc. [London] **98** [1808] 333/70 from J. Davy, The Collected Works of Sir Humphry Davy, Vol. 5, London 1840, pp. 102/39, 105; Ann. Physik **32** [1809] 365/96, 370). – [2] H. Davy (Phil. Trans. Roy. Soc. [London] **100** [1810] 16/74 from J. Davy, The Collected Works of Sir Humphry Davy, Vol. 5, London 1840, pp. 225/83, 267/8). – [3] H. Davy (Phil. Trans. Roy. Soc. [London] **103** [1813] 263/79 from J. Davy, The Collected Works of Sir Humphry Davy, Vol. 5, London 1840, pp. 408/24, 418). – [4] H. Davy (Phil. Trans. Roy. Soc. [London] **104** [1814] 62/73 from J. Davy, The Collected Works of Sir Humphry Davy, Vol. 5, London 1840, pp. 425/36, 430/1). – [5] J. Davy (The Collected Works of Sir Humphry Davy, Vol. 5, London 1840, p. 431 footnote).

[6] J. J. Berzelius (Ann. Physik Chem. [2] **1** [1824] 169/230, 224). – [7] H. G. Söderbaum (Jac. Berzelius Lettres, Vol. 1, Pt. 2, Correspondance entre Berzelius et Sir Humphry Davy 1808–1825, Uppsala 1912, pp. 67/73, 71).

1.2.1.4 Reduction Experiments with 'Ordinary Chemical Agents'

"Since silica is not sufficiently decomposed in the circuit of the electric pile in our experiments for one to conclude with certainty for or against the possibility of its reduction to a metallic body, which is the case with the other earths, I adopted another way. I exposed it, mixed with powdered carbon and iron filings, in a luted crucible [i.e., one strengthened with a layer of clay] to the temperature needed to melt the iron, when I imagined that iron would exert much the same effect on it as mercury with the other earths during their decomposition in the circuit of the electric pile" ["Da die Kieselerde bei unsern Versuchen in der Kette der elektrischen Säule nicht so deutlich zerlegt wurde, daß man daraus etwas für oder gegen die Möglichkeit ihrer Reduction zu einem metallischen Körper mit Sicherheit schließen konnte, wie das bei den übrigen Erden der Fall war, so schlug ich bei ihr einen anderen Weg ein. Ich setzte sie, mit Kohlepulver und Eisenspänen gemischt, in einen lutirten [d.h. durch eine Lehmschicht verstärkten] Tiegel einer zur Schmelzung des Eisens nöthigen Temperatur aus, wobei ich mir vorstellte, das Eisen werde auf sie ungefähr die nämliche Wirkung äußern, als das Quecksilber auf die anderen Erden bei der Zersetzung derselben in der Kette der elektrischen Säule"]. It was in this way that J. J. Berzelius [1] in 1809 described his experiments in the *decomposition of the silicious earths by ordinary chemical agents.* In these he obtained silvery reguli which dissolved only slowly in dilute sulfuric acid, in spite of warming, leaving a substance which "was in the shape of the iron globules, some of the grains being snow-white but others black. . . . I ignited them in the open furnace and only silica remained. . . . They amounted to 3½ percent of the weight of iron" ["welche die Gestalt der Eisenkugeln hatten, und von der einige Körner schneeweiß, andere aber schwarz waren. . . . Ich glühete sie in offenem Feuer aus, und nun blieb Kieselerde übrig. . . . Sie betrug 3½ Procent von dem Gewicht des Eisens"]. He reached higher contents (up to 19%) by a small change of the procedure. He then continues: "I believe myself justified by these experiments in coming to the following conclusion: the silica is reduced by carbon to a substance which combines with the iron and which must be *metallic* in character since it does not destroy the malleability of the iron. Several objections against this conclusion did indeed occur to me, the silica might only be mixed mechanically with the iron; but this is directly opposed to the picture, based on all our experience of the behavior of a molten metal towards a powdered substance: the metal shows no chemical relationship to it" ["Durch diese Versuche glaube ich mich nun zu folgendem Schlusse berechtigt: Die Kieselerde wird vermittelst der Kohle zu einem Körper reducirt, der sich mit Eisen vereinigt, und der, da er die Geschmeidigkeit des Eisens nicht zerstört, von *metallischer* Natur seyn muß. Zwar wendeten mir Mehrere gegen diesen Schluß ein, die Kieselerde möge sich mit dem geschmolzenen Eisen wohl nur mechanisch mengen; dieses widerspricht aber geradesweges den Begriffen, die wir uns nach allen unseren Erfahrungen von dem Verhalten eines geschmolzenen Metalls zu pulverförmigen Körpern machen; auf diese übt das Metall keine chemische Verwandtschaft aus"]. In spite of this he carried out a cementation experiment at a temperature at which the iron filings did not melt and found an SiO_2 content of 6% in them. The carbon which was always present interfered with the determination of the "silica basis" ["Kiesel-Basis"] content of the alloys, so that Berzelius also prepared alloys with copper in the same way, hoping to obtain these carbon-free. He was, however, disappointed and regretted that, because of lack of time, he was not able to prepare alloys with silver and gold for which he assumed the absence of absorbed carbon to be certain. F. Stromeyer (1776–1835) [5], who discovered cadmium, thought that Berzelius had been prompted in his experiments with silica by the report by L. J. Gay-Lussac and L. J. Thenard to the Institut de France on 7th March, 1808 [6], on the reduction of soda and potash by carbon in presence of iron, but it is more likely that they were the result of an observation made by him in the analysis of cast iron which he reported to H. Davy by letter on 30th June, 1809 [2]: "I have lately finished an analysis of cast iron. I succeeded in finding a means of determining with sufficient accuracy the amount of carbon which makes up the composition; I found that the cast iron does not contain oxygen, as has been presumed for a long time; but, on the contrary, it contains a small amount of the base of pyrolusite and a quite considerable amount of the base of silica. . . ." ["J'ai nouvellement fini une analyse du fer fondu. J'ai réussi à trouver un moyen de déterminer avec assez d'exactitude la quantité de carbon qui entre dans sa composition: j'ai trouvé que le fer fondu ne contient pas d'oxygène, comme on a longtemps supposé; mais qu'au contraire il contient un peu de la base de magnésie [manganese was meant] et une portion assez considérable de la base de la silice . . ."]. H. Davy mentioned this letter in his Bakerian Lecture in November 1809 [3] adding: "In the same communication this able chemist informed me, that he had succeeded in decomposing the earths by igniting them strongly with iron and charcoal." There must, however, be a slip in Davy's memory here for this observation does not appear in the letter in question though another may have been lost.

Berzelius' experiments were repeated by F. Stromeyer in Göttingen immediately after they became known because of their significance in steel making [4, 5]; he obtained alloys which gave between 5 and 22% SiO_2 on analysis. He, too, was convinced of the metallic character of the "basis of silica or silicium" ["Basis der Kieselerde oder Silicium"] especially since, like Berzelius previously, he had observed that the alloys evolved far more hydrogen on dissolution in acids than corresponded with their iron content. He also found it impossible to make "a more precise determination of the composition of the silica" ["eine schärfere Bestimmung der Zusammensetzung der Kieselerde"] when he established that "reduction experiments set up with silica, pine soot, and silver (instead of iron) . . . gave only ternary compounds consisting of silver, silicon, and carbon" ["mit Kieselerde, Kienruß und Silber (anstatt des Eisens) angestellte Reductionsversuche . . . [nur] dreifache, aus Silber, Silicium und Kohlenstoff bestehende Verbindungen ergaben"].

That platinum was also able to take up silicon was observed by J.-B. Boussingault (1801–1887) [6] in 1821, which is shortly before the first preparation of the pure element, see p. 29. He covered platinum completely with carbon dust and was able to melt it, whereas this was not successful with the uncovered metal. He attributed the phenomenon to the uptake of a foreign substance, as shown by the weight increase and was able to show that silicon and not carbon was involved. He was also able to determine silicon in steels, which, according to his report, L. Clouet (1751–1801), Professor of Chemistry at the Genie-school in Mezières had already observed: "the iron combines with glass" ["le fer se combine au verre"]. – Molten copper also is able to take up elemental silicon; see Section 1.3 on p. 37.

References:

[1] J. J. Berzelius (Ann. Physik **36** [1810] 89/102, 89/93). – [2] H. G. Söderbaum (Jac. Berzelius Lettres, Vol. 1, Pt. 2, Correspondance entre Berzelius et Sir Humphry Davy 1808–1825, Uppsala 1912, pp. 12/6, 15). – [3] H. Davy (Phil. Trans. Roy. Soc. [London] **100** [1810] 16/76 from J. Davy, The Collected Works of Sir Humphry Davy, Vol. 5, London 1840, pp. 225/82, 272 footnote). – [4] F. Stromeyer (Ann. Physik **37** [1811] 335/9). – [5] F. Stromeyer (Comment. Soc. Sci. Gottingensis **1** No. 6 [1808/11] 1/24; Ann. Physik **38** [1811] 321/30).

[6] J.-B. Boussingault (Ann. Chim. Phys. [2] **16** [1821] 1/16, 9, 11).

1.2.1.5 Supposed Reduction in the Oxy-Hydrogen Flame in 1816

Following the various attempts to reduce silica and set free its "basis" which are described above and which did not proceed satisfactorily, a new route for the direct thermal reduction of the "simple earths" ["einfache Erden"] with the aid of the newly-found high energy oxy-hydrogen blowpipe appeared to open up in 1816. The first step towards increasing the effectiveness of the blowpipe flame had already been taken in 1802 by Robert Hare (1781–1858), Professor of Physics at Philadelphia with his "hydrostatic blowpipe" ["hydrostatisches Lötrohr"]. In this, oxygen under pressure was blown from a gasometer into the flame of a torch.

In fusion tests on many substances with this device it was not observed that the hot flame had reducing properties. At all events, silica could only be fused to a white enamel [1]. In 1816, the "physical instrument maker J. Newman in London" improved this blowpipe, first of all by mechanisation of the pressure contrivance [2]. He then went over to a completely new construction, burning an oxy-hydrogen mixture at a fine jet [3]. E. D. Clarke (1769–1822), the Professor of Mineralogy at Cambridge, in the course of extensive melting experiments with this new device, demonstrated the reducing action of the flame on some oxides, initially for "heavy earth" ["Schwererde"; BaO], the metal of which he obtained, proposing to call it plutonium [3]. This result could not be confirmed when it was checked in London at the Royal Institution [4]; there were, however, observers of Clarke's experiments, who confirmed the results [5].

E. D. Clarke, in any case, wrote in connection with his preparation of plutonium: "When I later (on 13th September, 1816) also repeated the same experiment (introducing the "earth" made into paste with carbon and oil on charcoal into the flame) with silica, I obtained in one instance a lustrous grain of a pure white metal, which was more shiny and whiter than the purest silver which I have called silicium for the same reason [the naming of the "earth metal" by H. Davy in 1808, see p. 2], but I am not in a position to produce this metal again now" ["Als ich darauf (am 13. September 1816) dasselbe Verfahren (die mit Kohle und Öl angeteigte "Erde" auf eine Holzkohle vor die Flamme zu bringen) auch mit Kieselerde wiederholte, erhielt ich in einem Falle ein glänzendes Korn eines rein weißen Metalls, das glänzender und weißer als das reinste Silber war, welches ich aus derselben Ursach [nämlich die Benennung der "Erdmetalle" durch H. Davy im Jahre 1808, s. S. 2] Silicium genannt habe; aber ich bin nicht im Stande, dieses Metall jetzt wieder zu erzeugen"] [6]. In a letter to T. Thomson of October 5th, 1816, E. D. Clarke [7] once again went into his experiments with silica and described some successful melting experiments; he then also examined the reduction in presence of metallic iron: "*Silica* and *iron* fused together in the blowpipe flame to a bead which was a black, schisty brittle substance which could be polished and then showed a metallic glance somewhat whiter than pure iron" ["Zu einem Kügelchen geschmelzte *Kieselerde* und *Eisen* verbanden sich miteinander vor der Flamme des Gebläses zu einem schwarzen, schieferähnlichen, spröden Körper, der sich aber feilen ließ und dann Metallglanz von etwas mehr Weiße, als reines Eisen, zeigte"]. E. D. Clarke reported a repetition of this experiment more precisely on 23rd January, 1817: "When *silica* melted to a bead in a carbon crucible in the blast is added to a small piece of *iron* of the same size in the blowpipe flame, the iron removes all the oxygen from the silica at this high temperature and a white metal makes its appearance consisting almost entirely of *silicon* which contains only a little iron combined with it. This alloy is very similar in color to pure silicium obtained (in September 1816) by reduction of silica" ["Bringt man in einem Kohlentiegel vor dem Gebläse zu einem Kügelchen geschmelzte *Kieselerde* mit einem gleich großen Stückchen *Eisen* in die Flamme des Gebläses, so entreißt das Eisen in dieser hohen Temperatur der Kieselerde allen Sauerstoff, und es kömmt ein weißes Metall zum Vorschein, welches fast ganz aus *Silicium* besteht, und nur wenig damit verbundenes Eisen enthält. An Farbe ist diese Legirung dem durch Reduction von Kieselerde (im September 1816) erhaltenen reinen Silicium ganz gleich"] [8].

The experiments with the new blowpipe were checked not only in England but were repeated in many places. Thus, for example, J. B. van Mons, Professor of Chemistry at the University of Löwen reported [9] that not only his colleague L. G. Brugnatelli in Pavia, but particularly a Marquis Ridolphi in Florence had experimented with the "earths" but "alumina, silica, and lime and also calcium carbonate and barium carbonate were (only) fused and not once could they be transformed to metal" ["Thonerde, Kieselerde, Kalk sowie auch kohlensaurer Kalk und kohlensaurer Baryt (nur) geschmolzen wurden, allein niemals konnten sie in Metall verwandelt werden"]. The reason for such frequent failures of attempts to obtain silicon from silica in the oxy-hydrogen blowpipe appeared to have been found in December 1817 by

Josuah Mantell from Lewes (Sussex, England) [10]: "... he had soon convinced himself that too prolonged duration of the strong heating in the blowpipe flame was responsible and that during this the metal was at once transformed again into a gray substance. Now, however, he left the silica mixed with lamp oil in the gas flame *for only a few seconds* ... and subsequently always obtained silicon. The silica could be put on charcoal or introduced into a tobacco pipe. He enclosed some of the grains obtained on the previous evening in a letter [to Dr. T. Thomson, the editor of *Annals of Philosophy*]" ["... er habe sich aber bald überzeugt, daß daran die zu lange Dauer der großen Erhitzung vor der Flamme des Gebläses Schuld sey, bei der das Metall sogleich wieder in eine grauliche Substanz verwandelt werde. Jetzt aber lasse er die mit Lampenöl vermengte Kieselerde *nur wenige Sekunden* in der Gasflamme ... und erhalte seitdem immer Silicium, die Kieselerde möge auf Kohle oder auf eine Tabakspfeife gebracht werden. Einige am vergangenen Abend erhaltene Körnchen legte er dem Brief [an Dr. T. Thomson, den Herausgeber der *Annals of Philosophy*] bey"]. However: "Dr. Thomson stated that the sample obtained certainly had a perfect metallic glance ... but the metallic substance formed only a *very thin coating* on the silica, the whole being so light that it floats on water, and water, nitric acid, hydrochloric acid or aqua regia did not change the metallic glance, even when the specimen was left for 24 hours in them. Thus the substance with the metallic glance could *not be silicium,* should a metal of silica exist at all; for, from Davy's research, this decomposes water and all other liquids with which it comes into contact and cannot therefore be prepared as a single substance" ["Dr. Thomson bezeugte, die erhaltene Probe besitze allerdings den Metallglanz in großer Vollkommenheit ... der metallische Körper bilde aber nur einen *sehr dünnen Überzug* über die Kieselerde, das Ganze sey so leicht, daß es auf dem Wasser schwimme, und weder Wasser, noch Salpetersäure, noch Salzsäure, noch Königswasser, verändern den Metallglanz, selbst nicht, wenn man die Probestücke 24 Stunden lang in ihnen lasse. *Silicium* könne also der metallischglänzende Körper *nicht* seyn, gäbe es anders überhaupt ein Metall der Kieselerde; denn nach Davy's Versuchen zersetze dieses das Wasser und alle anderen Flüssigkeiten, mit denen es in Berührung kömmt, sogleich, daher es sich nicht einzeln herstellen lasse"] [10]. T. Thomson was able to detect iron in the coating: "When he melted some of the material received with potash in platinum tongs with the blowpipe he obtained a bead which had a greenish tinge and contained some iron and this was the only metal that could be detected. Dr. Thomson added that he believed the metallic glance to come more from the oil as he had frequently obtained carbon from animal oil which had a metallic glance which was not much less than that which the sample sent to him possessed" ["Als er etwas von der überschickten Materie vor dem Lötrohr mit Kali in einer kleinen Platinzange schmelzte, erhielt er ein Korn, das einen Stich ins Grünliche hatte, und etwas Eisen enthielt; und das war das einzige darin zu entdeckende Metall. Er halte dafür, setzt Dr. Thomson hinzu, der Metallglanz komme vielmehr aus dem Öhle, da er häufig von thierischem Öhle eine Kohle erhalten habe, die einen nicht viel geringeren Metallglanz als die ihm überschickten Probestücke gehabt habe"] [10]. The discussion following this suggestion need not be considered in detail here. It terminated with an E. D. Clarke's letter of June, 1818: "The metallic glance with which pure silica becomes coated when melted in the gas flame of the Newman torch often stems from the carbon, which serves as a carrier for it. The same phenomenon is seen under the same conditions on melting corundum and other refractory bodies, for example magnesia and lime, in the blowpipe. But he had also seen the formation of silvery metallic beads on melting silica with borax in the blast. It was, however, difficult to obtain them as they almost always volatilized at the instant of their formation" ["Es rühre manchmal der Metallglanz, mit welchem reine Kieselerde beim Schmelzen vor der Gasflamme des Neumann'schen Gebläses bekleidet werde, von der Kohle her, die als Träger derselben diene, und dieselbe Erscheinung zeige sich unter gleichen Umständen beim Schmelzen von Corund und anderen widerspenstigen Körpern, zum Beispiel Magnesia und Kalk vor dem Gebläse. Aber er habe auch beim Schmelzen von Kieselerde mit Borax vor dem Gasgebläse Kügelchen eines dem Silber ähnlichen Metalls entstehen sehen, sie zu erhalten, sey indeß schwierig, weil sie sich fast in demselben Augenblick verflüchtigen, in welchem sie sich bilden"] [10]. In what was at the start an almost schoolmasterish epilogue, L. W. Gilbert [11] finally gives a factual summary of the results of Clarke's research: "Mr. Clarke appears actually to have prepared *baryta metal,* if it was not an illusion that an alloy of it with silver decomposes in the air to dust; but what case there is for the metallic glance in the last of the experiments reported here is still to be adequately explained" ["*Barytmetall* scheint Hr. Clarke wirklich dargestellt zu haben, war es anders keine Täuschung, daß eine Legirung desselben mit Silber an der Luft zu Staub zerfiel; was es aber mit dem Metallglanze in den zuletzt hier erzählten Versuchen . . . für eine Bewandtniß gehabt habe, das ist erst noch genügend aufzuklären"].

References:

[1] P. A. Adet (Ann. Chim. [Paris] **45** [1801/02] 113/38). – [2] J. Newman (J. Roy. Inst. Gt. Brit. **1** [1812] 65 from J. Chem. Physik Schweigger **18** [1816] 225/7). – [3] E. D. Clarke (J. Roy. Inst. Gt. Brit. **3** [1814] 104 from J. Chem. Physik Schweigger **18** [1816] 228/55, 249). – [4] F. M. (J. Roy. Inst. Gt. Brit. **4** [1815] 461 from J. Chem. Physik Schweigger **18** [1816] 337/9). – [5] E. D. Clarke (Ann. Physik **62** [1819] 339/43, 342).

[6] E. D. Clarke (Ann. Physik **55** [1817] 1/39, 34). – [7] E. D. Clarke (Ann. Physik **62** [1819] 339/98, 391/2). – [8] E. D. Clarke (Ann. Physik **62** [1819] 339/93, 392). – [9] J. B. van Mons (Phil. Mag. [1] **49** [1817] 308/9; J. Chem. Physik Schweigger **20** [1818] 218/20). – [10] J. Mantell (Ann. Physik **62** [1819] 393/8).

[11] L. W. Gilbert (Ann. Phys. [2] **62** [1819] 398 footnote).

1.2.2 Reduction Experiments with Silicon-Fluorine Compounds

1.2.2.1 Unrecognized Observations of Amorphous Silicon

In 1808, both Louis Jacques Thenard (1777–1857) and Louis Joseph Gay-Lussac (1778–1850) in Paris as well as H. Davy in London were engaged in elucidating the composition of hydrofluoric acid (or what was thought to be it). In both places silicon tetrafluoride served as the real starting material. This Gay-Lussac and Thenard called "gaz fluorique silicé" and also frequently "gaz fluorique" while H. Davy spoke for the most part briefly of "fluoric acid", understanding by this "fluoric acid gas, which had been in contact with glass", that is to say likewise silicon tetrafluoride which perhaps still contained residual hydrofluoric acid.

The French researchers wrote about their first researches in their book *Recherches Physico-Chimiques* [1], which appeared in 1811 and collected together their own and also foreign research work which had become known: "On 2nd May 1808 we made the report to the Institut de France, and we published in the *Moniteur [universel],* for Friday, the 27th of May, and then in the *Nouveau Bulletin de la Société Philomatique de Paris* in June 1808 on page 156, that *potassium,* when heated, burns vigorously in the completely dry 'gaz fluorique' and that it probably decomposes this, since it completely absorbs it; and that the product thus formed, which was brown, contained potassium fluoride but no more *potassium*" ["Le 2 mai 1808, nous avons annoncé à l'Institut, et nous avons imprimé dans le *Moniteur [universel]* pour le vendredi 27 mai, et dans le *Nouveau Bulletin de la Société Philomat.[ique] de Paris* juin 1808, page 156, que le *potassium,* à l'aide de la chaleur, brûloit vivement dans le gaz fluorique le plus sec, et que probablement il le décomposoit, puisqu'il l'absorboit complètement; que le produit qui en résultoit et qui étoit brun, contenoit de fluate de potasse et ne contenoit plus de *potassium*"]. Eight weeks later, on 30th June, 1808, H. Davy in London reported to the Royal Society some quite similar experiments in connection with a more extensive study on the behavior of potassium metal with gaseous acids, among which was "fluoric acid gas, which had been in contact with glass." He obtained results with it which differed from those of the Paris chemists [2]: "On fluoric acid gas, which had been in contact with glass, the potassium produced a similar effect [to that of hydrogen chloride]; but the quantity of hydrogen generated was only one-sixth or one-seventh of the volume of gas, and a white mass was formed, which principally consisted of fluate of potash and silex, but which emitted fumes of fluoric acid when exposed to air." Clearly H. Davy had not excluded atmospheric oxygen in these experiments so that silicic acid occurred as a reaction product. The fact that he obtained so little hydrogen, shows the chief component of his gas had been tetrafluorosilane. In new experiments, when he purged his reaction vessel with hydrogen, he obtained quite different results, which were comparable with those of the French chemists. After a letter to M. A. Pictet in Geneva [3], he reported on them on 12th January, 1809, in a Bakerian Lecture, which was published a little later [4]. Here he described his work in full under the title: *Analytical Inquiries respecting Fluoric Acid,* by which he again meant the gas mixture noted above: "After this combustion, either the whole or a part of the fluoric acid, according as the quantity of potassium is great or small, is found to be destroyed or absorbed. A mass of a chocolate colour remains in the bottom of the retort; and a sublimate, in some parts chocolate, and in others yellow, is found round the sides, and at the top of the retort. . . . I have endeavoured to collect large quantities of the chocolate-coloured substance for minute examination; but some difficulties occurred. . . . When the product was examined by a magnifier, it evidently appeared consisting of different kinds of matter; a blackish substance, a white, apparently saline substance, and a substance having different shades of brown and fawn colour. . . . The mass did not conduct electricity, and none of its parts could be separated, so as to be examined as to this property. . . . When a portion of it was thrown into water, it effervesced violently, and the gas evolved had some resemblance in smell to phosphuretted hydrogen, and was inflammable. When a part of the mass was heated in contact with air, it burnt slowly, lost its brown colour, and became a white saline mass. When heated in oxygen gas, in a retort of plate glass, it absorbed a portion of oxygen, but burnt with difficulty, and required to be heated nearly to redness; and the light given out was similar to that produced by the combustion of liver of sulphur. The water which had acted upon a portion of it was examined; a number of chocolate-coloured particles floated in it. When the solid matter was separated by the filter, the fluid was found to contain fluate of potash, and potash. The solid residuum was heated in a small glass retort in oxygen gas; it burnt before it had attained a red heat, and became white. In this process, oxygen was absorbed, and acid matter produced. The remainder possessed the properties of the substance formed from fluoric acid gas holding siliceous earth in solution, by the action of water. . . . The decomposition of the fluoric acid by potassium, seems analogous to that of the acids of sulphur and phosphorus. In neither of these cases are the pure bases, or even the bases in their common form evolved; but new compounds result, and in one case sulphurets, and sulphites, and in the other phosphurets, and phosphites of potash, are generated. As silex was always obtained during the combustion of the chocolate-coloured substance obtained by lixiviation, it occurred to me that this matter might be a result of the operation, and that the chocolate substance might be a compound of the siliceous and fluoric bases in a low state of oxigenation, with potash; and this idea is favoured by some trials that I made to separate silex from the mass, by boiling it in concentrated fluoric acid; the substance did not seem to be much altered by this process, and still gave silex by combustion."

H. Davy was very near to the discovery of silicon in its amorphous form in these experiments. That he did not draw the same conclusions from his experiments as J. J. Berzelius 16 years later, see p. 30, was defended by his brother, John Davy, in a footnote when he published the *Collected Works of Sir Humphry Davy* [5], where he referred to his brother's later publications in 1813/14 [6, 7], adding: ". . . at this time, the chemical history of them [the fluorine compounds] was very imperfect, – as is indicated by the terms applied to them; – it was not then known that silica, or its basis, is an essential ingredient of the gas which in the text is named fluoric acid gas." J. R. Partington [8] renders this excuse invalid when he showed that volatile silicon-fluorine compounds had already been distinguished from simple hydrofluoric acid in 1781, see p. 103, and expressed the view: "Davy's mistake was due to his hasty work at this time."

Only a few days after H. Davy in London, on 23rd January, 1809, L. J. Gay-Lussac and L. J. Thenard announced more precise details to the Institut de France of their experience with the reaction of potassium metal with "siliceous hydrofluoric acid gas" ["kieseliges flußsaures Gas"] i.e., silicon tetrafluoride, which they had made in connection with their studies on (pure) hydrogen fluoride, but the true nature of the brown substance obtained was hidden from them too [9]: "If *potassium metal* is brought into contact with *siliceous hydrofluoric acid* [SiF_4] it does not change appreciably at ordinary temperatures and tarnishes only at the surface. On the other hand, if it is melted in this gas it soon thickens and burns briskly with evolution of much heat and light. During this burning a great deal of hydrofluoric acid is taken up and very little hydrogen gas released. The melt disappears and a solid with a reddish brown colour is produced. If this substance is treated with cold water, hydrogen is evolved although the material appears to contain no more metal. If it is then washed once again with hot water some further hydrogen is obtained, though much less than before. Overall it appears that in total barely ⅓ as much hydrogen appears as the potassium metal itself would have given with water. If the water is collected and evaporated one obtains from it only potassium fluoride with excess of potash. After the residue has been well washed it remains red-brown and is characterized by the following properties. When thrown into a silver crucible glowing at a red heat it burns vigorously, releases a little acid gas and is, instead of being insoluble in water, now partially soluble in it. The part of the residue which dissolves is potassium fluoride. The part which is insoluble is a ternary compound of hydrofluoric acid, potash and silica. If the experiment instead of in a red hot silver crucible is carried out in a small glass bell jar which is filled with oxygen and is bent (recourbée) and is gradually heated, combustion is more vigorous than in atmospheric air. Much oxygen gas is taken up, the residual gas being pure oxygen and a little hydrofluoric acid. The combustion product, as in the previous experiment, consists of a ternary compound of hydrofluoric acid, potash and silica" ["Wenn man das *Kali-Metall* mit *kieseligem flußsauren Gas* [SiF_4] in Berührung bringt, so verändert es sich in der gewöhnlichen Temperatur nicht merklich, und läuft bloß an der Oberfläche an. Wird es dagegen in diesem Gas geschmelzt, so verdickt es sich bald, und brennt lebhaft, unter Entwickelung von viel Wärme und Licht. Bei diesem Verbrennen wird sehr viel Flußsäure verschluckt, und sehr wenig Wasserstoffgas entbunden; das Metall verschwindet, und es entsteht ein fester Körper von röthlich brauner Farbe. Behandelt man diesen Körper mit kaltem Wasser, so entwickelt sich Wasserstoffgas, obgleich dieser Körper kein Metall mehr zu enthalten scheint, und wäscht man ihn darauf noch einmahl mit heißem Wasser, so erhält man noch etwas Wasserstoffgas, doch viel weniger als zuvor; überhaupt erscheint zusammengenommen kaum ⅓ so viel Wasserstoffgas, als das Kali-Metall selbst mit Wasser gegeben haben würde. Gießt man das Wasser zusammen, und dampft es ab, so erhält man daraus bloß flußsaures Kali mit Uebermaß an Kali. Der *Rückstand* bleibt, nachdem man ihn gut gewaschen hat, röthlich braun, und charakterisirt sich durch folgende Eigenschaften. Wirft man ihn in einen silbernen Tiegel, der kirschroth glüht, so verbrennt er lebhaft, entbindet ein wenig saures Gas, und ist, statt daß er zuvor unauflöslich im Wasser war, jetzt zum Theil darin auflöslich. Der Theil dieses Rückstandes, der sich auflöst, ist flußsaures Kali; der Theil, der sich nicht auflöst, ist eine dreifache Verbindung von Flußsäure, Kali und Kieselerde. Stellt man den Versuch, statt in einem kirschroth glühenden silbernen Tiegel, in einer kleinen Glasglocke voll Sauerstoffgas an, die gekrümmt ist (recourbée), und die man allmählig erhitzt, so ist das Verbrennen lebhafter als in der atmosphärischen Luft; es wird viel Sauerstoffgas verschluckt; der Gasrückstand ist reines Sauerstoffgas und ein wenig Flußsäure. Das Product des Verbrennens ist fest, wie in dem vorigen Versuche, und besteht aus einer dreifachen Verbindung von Flußsäure, Kali und Kieselerde"].

In their book mentioned at the outset the two researchers gave a summary of their studies on the brown compound which they designated briefly as 'Matière A' [10], about which they say from their preliminary investigations: "... All these properties tend to demonstrate the presence of a combustible substance in this material: However, how to obtain it pure? In this one has never succeeded, whether one treats it for this purpose with water, hydrochloric acid or fluoric acid ..." ["... Toutes ces propriétés tendent à démontrer l'existence d'un corps combustible dans cette matière: mais comment l'obtenir pur? On n'a jamais pu y parvenir, quoiqu'on l'ait traitée à cet effet par l'eau, par les acides muriatique et fluorique ..."]. In conclusion they summarize their opinion of the material as follows: "... if one considers with attentiveness all of the facts which have just been reported, it becomes very probable that this substance is the radical of hydrofluoric acid; for, since on the burning of *potassium* in the "gaz fluorique silicé" none at all or almost no hydrogen is evolved, one can in no case attribute this combustion to water, which one could suspect to be in this gas. It is also not probable that the destruction of the potassium is brought about by the silica" ["... si on considère avec attention tous les faits qu'on vient de rapporter, il deviendra très-probable que ce corps est le radical de l'acide fluorique; car, puisqu'en brûlant du *potassium* dans le gaz fluorique silicé, il ne se dégage point ou presque point d'hydrogène, on ne peut point attribuer cette combustion à de l'eau qu'on supposeroit dans ce gaz. Il n'est pas probable non plus que la destruction du potassium soit due à la silice"]. They repeat their view at the conclusion of the book where they compare the points of difference between their own ideas and those of H. Davy on all points in dispute in the current chemistry.

The investigations of the chemists on the two sides of the Channel aroused great interest on the part of the physicist and mathematician André Marie Ampère (1775–1836) in Paris who was engaged at the time in preliminary work on attempts to classify "simple substances", particularly because of the possibility of finding in the anticipated and sought for "simple new substance" from fluorspar a way of expanding the elements oxygen and chlorine which "support combustion" into a triad. He therefore turned to H. Davy [11] on 1st November, 1810, in a letter in which he wrote: "... Is the supposition that the boric acid and the oxide of silicium (silica) are dissolved in the gaseous state in this problematic acid not in contrast to all analogies, and would it not be probable that these phenomena belong to a third "combustible substance"? ... This [the hydrofluoric acid HF] namely must, when brought in contact with the oxide of silicon, bring about the formation of water, and the silicon would, combined with the "oxifluorique" [Ampère's proposal for naming the element fluorine] give this gas, which one names "acide fluorique silicé", and that one must call according to this hypothesis "acide silico-fluorique", and that would be an analog to the other gas-forming acids. – Likewise, if one heats calcium fluoride with boric acid, a part of this acid must be decomposed, in order to convert the calcium into lime, and the deoxygenated boron must combine with the fluorine; the result would then still be a gaseous acid, which one in fact obtains under these conditions, and which, according to this hypothesis, should be named "acide boro-fluorique". One would without doubt also soon discover the "acide sulfuro-fluorique" and the "acide phosphoro-fluorique", if one could obtain the elemental fluorine" ["... La supposition que l'acide boracique et l'oxyde de silicium (silice) sont dissous à l'état de gaz dans cet acide problématique n'est-elle pas contraire à toutes les analogies, et ne serait-il pas probable que ces phénomènes sont due à un troisième corps comburant? ... Celui-ci [hydrofluoric acid HF] mis en contact avec l'oxyde de silicium, il y aurait formation d'eau, et le silicium uni à l'oxy-fluorique [Ampère's proposal for naming the element fluorine] donnerait ce gaz qu'on nomme acide fluorique silicé , que dans cette hypothèse on devrait appeler acide silicio-fluorique, et serait analogue aux autres acides gazeux. – De même, lorsqu'on chauffe le fluate de chaux avec l'acide boracique, une partie de cet acide serait décomposée pour changer le calcium en chaux, et le bore désoxydé se combinant avec l'oxy-fluorique; il en résulterait encore un acide gazeux, celui qu'on obtient, en effet, dans cette circonstance, et qui devrait, dans cette hypothèse, être appelé acide boro-fluorique. On découvrirait sans doute bientôt les acides sulfuro-fluorique, phosphoro-fluorique, si l'on pouvait obtenir l'oxy-fluorique . . ."].

In his answering letter of 8th February, 1811, which came into the hands of the addressee only after 18 months because of Napoleon's tight continental embargo and the English counter blockade [11], H. Davy [12] did not go into this "analogy" between "silico-fluoric acid" and "boro-fluoric acid" proposed by Ampère to the discoverer of boron. The short letter contained only the following that was material [12]: ". . .You have pointed out the analogies between the fluoric and muriatic gases in a masterly manner. I know of but one objection to the views you propose. This is that potash seems to be formed by burning potassium in silicated fluoric gas; which seems to imply that there is some substance in it which contains oxygen." Ampère did not allow himself to be persuaded to deviate from his ideas, the essentials of which he finally published [15] in 1816. He answered Davy on 25th August, 1812, giving his ideas in detail, the most interesting section – and here it must not be overlooked that this letter became known publicly only in 1885 [11] – came to the surprising proposition that both experiments with the siliceous gas could only have been capable of producing elementary silicon:

". . . I have thereafter reflected on that which you told me concerning the nature of the hydrofluoric acid. It is evident that in the hypothesis according to which it is supposed to be formed, like hydrochloric acid, from *hydrogen* and a substance analogous to *oxygen* and *chlorine,* which I, in order to be able to express my thougths without paraphrasing, will name *fluorine* in analogy, that then only the gases which one calls in France "gaz *fluo-borique*" and "gaz *fluorique silicé*" are formed, the first only from boron and *fluorine,* the second from silicium and *fluorine,* and thus *potassium* oxide could never be formed in the combustion of this metal in the one or the other of the two gases, but only a compound of *potassium* and *fluorine* with the *boron* in the first case and the silicon in the second, if one does not add water or when the water included in the mixture is not decomposed. Now, however, one obtains in the one as in the other combustion a brown-black product. This product is formed during the combustion in the "gaz fluo-borique" and is composed, according to the experience of the MM. Thenard and Gay-Lussac and conforming with my hypothesis, of boron and a substance which dissolves in water and hydrochloric acid, either directly or because the *hydrogen* of the water or of the hydrochloric acid forms with the fluorine again hydrofluoric acid, while there is produced at the same time *potassium* oxide if it is with water or potassium chloride if it is with hydrochloric acid. For my hypothesis to be able to survive, it would be necessary that on the combustion in the "gaz fluorique silicé" the brown-black product be made of nothing other than *silicium* and the substance soluble in water and hydrochloric acid, so that after successive washings in one of the two liquids only *silicium* remains, the purer with the more carefully one has washed it. Now, I have read all of the studies of MM. Thenard and Gay-Lussac concerning the combustion of potassium in the "gaz fluorique silicé" and I have found nothing which contradicts this hypothesis nor which indicates that there is *potassium* oxide in the brown-black product before one puts it into the water. Since one in France does not possess your investigations concerning the same product, I do not know whether you have done anything with it which proves that the potassium oxide in it is completely formed. So far it appears to me that my hypothesis can be valid and that it corresponds better than all others to the properties of the "gaz fluo-borique" and the "gaz fluorique silicé", since it groups these with the other gases which are formed from combustible substances and oxygen, like the gases of carbonic and sulfuric acid, while one cannot see in the old hypotheses how hydrofluoric acid could form these gases, with bases as fixed as are silica and boric acid. One must only add:

1) That the *silicon* after the decomposition of its oxide is present in the form of a blackish powder like molybdenum, which supports an idea that appears to me rather probable, namely that silicon represents an intermediate class between the metals and the other combustible substances such as *boron, phosphorus,* and *carbon,* etc., in the same way as its oxide, silica, is a substance (so to speak) intermediate between the metallic oxides and the alkalies and the acids.

2) That the white substance obtained by chemists just cited on burning in gaseous oxygen the brown-black product – obtained from the "gaz fluorique silicé" by means of potassium, and washed with water or hydrochloric acid – was solely silica and that the "gaz fluorique silicé", which forms again in small quantities in this combustion would come from the fact that either the potassium did not completely separate the *silicium* from *fluorine* with which it is combined in the "gaz fluorique silicé", or from the fact that after the oxidation of the *potassium* by the wash water the fluorine had in part left this last-named metal in order to combine anew with *silicium,* but in much smaller quantity than in the "gaz fluorique silicé"; this would not occur if boron should replace the *silicium,* since the former has a lesser affinity for fluorine than the latter, in as much as boric acid does not decompose hydrofluoric acid as it does the silica.

If one interprets the phenomena in this manner, then it would follow that the combustion of potassium in "gaz fluorique silicé" would probably lead to the obtaining of pure silicium, which appears to have been obtained up to now only in combination with iron . . ."

["... J'ai réfléchi ensuite sur ce que vous me dites sur la nature de l'acide fluorique. Il est évident que dans l'hypothèse où il serait formé, comme l'acide muriatique, *d'hydrogène* et d'un corps analogue aux gaz, *oxygène* et *chlorine,* corps que je nommerai ici par analogie *fluorine,* pour pouvoir exposer ma pensée sans périphrase, les gaz qu'on nomme en France *fluo-borique* et *fluorique silicé,* étant formés uniquement le premier de bore et de *fluorine,* le second de silicium et de *fluorine,* il ne pourrait jamais se former d'oxyde de *potassium* par la combustion de ce métal dans l'un ou l'autre de ces deux gaz, mais seulement une combinaison de *potassium* et de *fluorine* avec du *bore* dans le premier cas, et du silicium dans le second, tant qu'on n'y joindrait pas d'eau, ou tant que l'eau unie au mélange ne serait pas décomposée. Or dans l'une et l'autre combustion on obtient un produit brun noirâtre. Ce produit, lors de la combustion dans le gaz fluo-borique, est formé, d'après les expériences de MM. Thenard et Gay-Lussac, et conformément à mon hypothèse, de bore et d'une substance qui se dissout dans l'eau et l'acide muriatique, soit immédiatement, soit parce que *l'hydrogène* de l'eau ou de l'acide muriatique formé de nouveau avec le fluorine de l'acide fluorique, tandis qu'il se produit en même temps de l'oxyde de *potassium* si c'est de l'eau, et du muriate de potasse si c'est de l'acide muriatique. Pour que mon hypothèse pût subsister, il faudrait que, lors de la combustion dans le gaz fluorique silicé, il n'y eût de même dans le produit brun noirâtre que du *silicium* et la même substance soluble dans l'eau et l'acide muriatique, en sorte que, par des lavages successifs dans un des deux derniers liquides, il ne restât que du *silicium* d'autant plus pur qu'on l'aurait lavé avec plus de soin. Or, j'ai relu toutes les expériences faites par MM. Thenard et Gay-Lussac sur la combustion du potassium dans le gaz fluorique silicé, et je n'ai rien trouvé qui contredît cette hypothèse, ni qui indiquât qu'il y a de l'oxyde de *potassium* dans le produit brun noirâtre avant qu'on y ait mis de l'eau. Comme on n'a point en France vos expériences sur ce même produit, je ne sais point si vous en avez fait qui prouvent que l'oxyde de potassium y est tout formé. Jusque-là, il me semble que mon hypothèse peut être admise et qu'elle rend raison mieux que tout autre des propriétés du gaz fluoborique et du gaz fluorique silicé, en les assimilant aux autres gaz formés de corps combustibles et d'oxygène, comme sont les gaz acides carbonique et sulfurique, au lieu qu'on ne voit point dans l'ancienne hypothèse comment l'acide fluorique pourrait former des gaz avec des bases aussi fixes que le sont la silice et l'acide borique. Il faut seulement admettre:

1° Que le *silicium* se présente après la décomposition de son oxyde, sous la forme d'une poudre noirâtre, comme le molybdène, ce qui vient à l'appui d'une idée qui me semble assez vraisemblable, savoir que le silicium fait une sorte de nuance entre les métaux et les autres corps combustibles, tels que le *bore*, le *phosphore*, le *carbone*, etc., de même que son oxyde, la silice, est un corps en quelque sorte intermédiaire entre les oxydes métalliques aux alcalins et les acides;

2° Que la substance blanche obtenue par les chimistes que je viens de citer, en brûlant dans le gaz oxygène le produit brun noirâtre obtenu du gaz *fluorique silicé* par le moyen du *potassium*, et lavé dans l'eau ou l'acide muriatique, était seulement de la silice et que le gaz *fluorique silicé* qui s'est reproduit en petite quantité dans cette dernière combustion venait ou de ce que le *potassium* ne sépare pas complètement le *silicium* du *fluorine* avec lequel il est combiné dans le *gaz fluorique silicé*, ou de ce que, l'eau des lavages ayant fourni de l'oxygène au *potassium*, le *fluorine* avait en partie quitté ce dernier métal pour s'unir de nouveau avec le *silicium*, mais en bien moindre quantité que dans le gaz *fluorique silicé;* ce qui n'aurait pas lieu dans le cas où le *bore* remplacerait le *silicium*, parce qu'il a moins d'affinité que lui pour le fluorine, l'acide borique ne décomposant pas l'acide fluorique comme le fait la silice.

Il suivrait de cette manière de concevoir les phénomènes que la combustion du *potassium* dans le gaz *fluorique silicé* conduirait probablement à obtenir pur le silicium, qu'on paraît n'avoir obtenu jusqu'à présent qu'en combinaison avec le fer..."].

In his answer of 6th March, 1813, H. Davy [12] went into all the other points which are omitted here, but not into the suggestion that elementary silicon is set free in the experiments described. He wrote [12, 14]: "Your ingenious views respecting fluorine may be confirmed. I have every reason to conclude from my experiments that there is no oxygen combined with the potassium in the experiment on the combustion of potassium in silicated fluoric gas, and that the first view which I formed on the subject is incorrect. I have many new experiments on the subject, but I have only a moment in which I can make this communication." In his paper on "Some Experiments and Observations on the Substances Produced in Different Chemical Processes on Fluorspar" [13], which was read before the Royal Society on 8th July, 1813, he made only a very brief mention of the Ampère letter and passed over its suggestion that the element silicon had already been liberated in his earlier experiments: "In the second hypothesis, that which I have alluded to in the beginning of this paper, and that adopted by M. Ampère, the silicated fluoric acid is conceived to consist of a peculiar undecompounded principle, analogous to chlorine and oxygen, united to the basis of silica, or *silicium;* the fluoboric acid of the same principle united to boron; and the pure liquid fluoric acid as this principle united to hydrogen." He was convinced that his old opinion was correct: ". . . and I concluded from the phenomena, that the acid gas was decomposed in the process, that oxygen was probably separated from it by the potassium, and that the combustible substance was a compound of the siliceous and fluoric bases" [13].

Davy's intransigent behavior did not hinder Ampère in any way from expressing his conviction: "Boron and silicon having been discovered only a little while ago . . ." ["Le bore et le silicium, n'étant découverts que depuis peu . . ."] in his book *"Classification Naturelle des Corps Simples"* though he did not repeat his line of reasoning [15].

References:

[1] L. J. Gay-Lussac, L. J. Thenard (Rech. Phys. Chim. [Paris] **1** [1811] 240, 313/4, **2** [1811] § 265, §323, 1/73, 65/73). – [2] H. Davy (Phil. Trans. Roy. Soc. [London] **98** [1808] 333/70 from J. Davy, The Collected Works of Sir Humphry Davy, London 1840, Vol. 5, pp. 102/39, 113 footnote). – [3] L. J. Gay-Lussac, L. J. Thenard (Rech. Phys. Chim. [Paris] **2** [1811] 72). – [4] H.

Davy (Phil. Trans. Roy. Soc. [London] **99** [1809] 39/104 from J. Davy, The Collected Works of Sir Humphry Davy, London 1840, Vol. 5, pp. 140/204, 185/91). – [5] J. Davy (The Collected Works of Sir Humphry Davy, Vol. 5, London 1840, p. 191 footnote).

[6] H. Davy (Phil. Trans. Roy. Soc. [London] **103** [1813] 263/79 from J. Davy, The Collected Works of Sir Humphry Davy, Vol. 5, London 1840, pp. 408/24). – [7] H. Davy (Phil. Trans. Roy. Soc. [London] **104** [1814] 62/73 from J. Davy, The Collected Works of Sir Humphry Davy, Vol. 5, London 1840, pp. 425/36). – [8] J. R. Partington (Manchester Lit. Phil. Soc. Mem. **67** No. 6 [1923] 73/87, 86). – [9] L. J. Gay-Lussac, L. J. Thenard (Ann. Chim. [Paris] **69** [1809] 204/20 from Ann. Physik **32** [1809] 1/15, 11/2; Rech. Phys. Chim. [Paris] **1** [1811] § 240, 313/4). – [10] L. J. Gay-Lussac, L. J. Thenard (Rech. Phys. Chim. [Paris] **2** [1811] §§ 314/9, 57/63, 261/2).

[11] Joubert (Ann. Chim. Phys. [6] **4** [1885] 5/13). – [12] A. M. Ampère (Ann. Chim. Phys. [2] **2** [1816] 5/32, 9, footnote 21/3). – [13] H. Davy (Phil. Trans. Roy. Soc. [London] **103** [1813] 263/79 from J. Davy, The Collected Works of Sir Humphry Davy, Vol. 5, London 1840, pp. 408/24, 410, 412). – [14] A. Ampère (Neues J. Pharm. Trommsdorff **1** II [1817] 322/432, 391). – [15] A. M. Ampère (Ann. Chim. Phys. [2] **1** [1816] 295/308, 373/94, **2** [1816] 1/32, 105/24, 118).

1.2.3 The First Preparation of Silicon by Berzelius

The significance of the discovery of elementary amorphous silicon by J. J. Berzelius in the spring of 1824 was pointed out 32 years later by Friedrich Wöhler (1800–1882) from his own experience, when he had been successful in improving the preparation of the element in a crystalline form [1]:

"Silicon unquestionably belongs to the most important elements of our planet as it is one of the main substances which have served in building up its structure. It is therefore worth the trouble to ascertain its properties as fully as possible. As is known it was first prepared in an isolated form by Berzelius in 1824 by decomposition of fluorosilicic gas or potassium fluorosilicate with potassium. I had the good fortune at the time when he was engaged in this instructive investigation to be his student (from November 1823 to September 1824 [5]) and to help him by preparing the necessary potassium. It is therefore a satisfying thought for me that, after the lapse of 31 years, it has been granted to me to be able to link up some observations to the work of the unforgettable fatherly friend which took place under my own eyes and which add a new interest to the substance discovered by him. As is known, Berzelius studied and described all the peculiar properties which are characteristic of silicon with a keenness and exactness which were his own. But he obtained it only in an amorphous form as a dull brown powder. He himself repeatedly said how interesting it would be to get to know this substance in a dense and crystalline state."

["Das Silicium gehört unstreitig zu den merkwürdigsten Elementen unseres Planeten, weil es eines der Hauptmaterialien ist, welche zum Bau desselben gedient haben; es ist daher wohl der Mühe werth, seine Eigenschaften möglichst vollständig kennen zu lernen. Es wurde bekanntlich 1824 zuerst von Berzelius in isolirter Form dargestellt durch Zersetzung von Fluorkieselgas oder Fluorkieselkalium mit Kalium. Ich hatte das Glück, zur Zeit, als er mit dieser lehrreichen Untersuchung beschäftigt war, sein Schüler zu sein (vom November 1823 bis September 1824 [5]) und ihm durch Darstellung des dazu erforderlichen Kaliums dabei Hülfe zu leisten. Es ist mir darum ein befriedigender Gedanke, daß es mir jetzt, nach Verlauf von 31 Jahren, noch vergönnt ist, an jene, unter meinen Augen entstandene Arbeit des unvergeßlichen väterlichen Freundes einige Beobachtungen anknüpfen zu können, durch welche der von ihm entdeckte Körper ein neues Interesse gewinnt. Mit der ihm eigenen Schärfe und Genauigkeit erforschte und beschrieb bekanntlich Berzelius alle die eigenthümlichen Eigenschaften, durch welche das Silicium characterisirt ist. Er erhielt es aber nur in amorpher Form, in Gestalt eines braunen, glanzlosen Pulvers. Er selbst äußerte wiederholt, wie interessant es sein müsse, diesen Körper im dichten und krystallinischen Zustand kennen zu lernen"].

To Berzelius himself the discovery appeared important enough and he wrote to P. L. Dulong in Paris [2] on 12th April, 1824: "In attempting to reduce hydrofluoric acid with potassium, I succeeded in reducing silica, zirconia, and the other earths, but I have been able to isolate only the silicon and the zirconium . . ." ["En voulant réduire l'acide fluorique moyennant le potassium, j'ai réussi à réduire la silice, la zirkone et les autres terres, mais je n'ai pu isoler que le silicium et le zirkonium. . . "] and commences the full description of his research with the words: "Nothing is easier than obtaining this substance . . ." ["Rien de plus facile que de se procurer cette substance . . ."]. A few days later, on 21st April, 1824, he also informed H. Davy in London of the discovery. This was somewhat briefer but contained the observation: "The substance that one thus obtains is the same as that which you know from your investigations of the nature of hydrofluoric acid" ["La masse qui en résulte est la même que celle que vous connaissez depuis vos recherches sur la nature de l'acide fluorique"]. A report on this letter was read before the Royal Society on 20th May, 1824 [3].

J. J. Berzelius [4] described his route leading to the discovery of the element much more precisely than in this letter under the title *"Decomposition of fluorosilicic acid by potassium"* [*"Zersetzung der Flußspaths[auren] Kieselerde durch Kalium"*]: "If one reads the description of the experiments carried out by the French chemists one cannot doubt that hydrofluoric acid has been decomposed under the conditions employed by them. Potassium burns in the gas and condenses it; a brown substance is formed which when boiled with water and dried burns in oxygen gas with expulsion of silica-containing hydrofluoric acid, leaving an earthy white final product. – When I set up the experiment I believed the reduction of the hydrofluoric acid with silica to be so certain that I judged that only a more exact determination of the composition of the reduced product to be necessary in order to be clear on this point. When I repeated Gay-Lussac and Thenard's experiment, I obtained the same result with the same phenomena as they describe, with the exception that the material burnt in oxygen was not white but had retained its original color without significant change. I expected that it would contain potassium fluorosilicate and therefore covered it with concentrated sulphuric acid. This, however, did not evolve any trace of hydrofluoric acid and could be evaporated off from the gray-brown material without changing it. It was attacked only by hydrofluoric acid among the acids, which abstracted silica and left behind a dark brown material which was insoluble in acids and did not burn in the flame. Was this the radical of hydrofluoric acid or that of silica or a compound of the two? To obtain this substance in greater amounts I took the following steps: In a glass retort of about 10 cubic inch capacity a small vessel of real porcelain was introduced on which a piece of potassium the size of a large hazle nut rested. The retort was rapidly made air free and then silica-containing hydrofluoric acid gas was admitted from a reservoir where it stood over mercury, and the section of the retort over which the vessel with potassium stood was heated with a spirit lamp. The potassium became white initially and then more and more dark in color and finally as black as coal; shortly afterwards it ignited and burned with a large dark red flame of low intensity while the mercury rose rapidly in the reservoir with which the retort was connected during the experiment. As soon as combustion had ended the retort was again pumped out in order to hinder formation of potassium fluorosilicate and then allowed to cool. The combustion product was a hard sintered porous mass, dark brown in color, but which did not change in the air but smelt of hydrogen gas if stirred with the finger or breathed on just as is characteristic of metallic manganese. Round the vessel which contained the potassium a loose light brown powder had collected in the retort which was kept on its own. The burnt material was thrown into water from which hydrogen gas was evolved vigorously at once. The water extracted a great deal of potassium fluoride and the brown material broke down with a slow gas evolution to a chestnut brown powder. The alkaline liquid was poured off and exchanged for fresh water. Gas evolution now became visibly less and, as this water was renewed after a time, ceased almost completely so that the material could be boiled up without further decomposition of the brown powder by water. The liquid obtained by boiling had a strongly acid reaction and the powder was therefore boiled with new portions of water as long as these still became acid. The water going through the filter was a saturated solution of potassium fluorosilicate. The brown powder was taken up on the filter and washed as long as the water passing through left a spot on evaporation. It was then dried and appeared as a loose chestnut-brown material which clearly contained brighter parts so that it appeared not to be of uniform in nature. The brown material which had settled on the glass of the retort during burning was much more uniform. With water it evolved no hydrogen gas but it became acid at once. This powder was washed with the same care as the previous one" ["Wenn man die Beschreibung der von den französischen Chemikern angestellten Versuche liest, so kann man nicht zweifeln, daß die Flußspathsäure unter den von ihnen angegebenen Umständen zersetzt wurde. Das Kalium brennt in dem Gase, und condensirt es; es wird eine braune Materie gebildet, die, mit Wasser ausgekocht und getrocknet, in Sauerstoffgas mit Ausstoßung von kieselhaltiger Flußspathsäure brennt, und eine weiße erdartige Materie hinterläßt. – Ich hielt, da ich den Versuch anstellte, die Reduction der Flußspathsäure mit der Kieselerde für so gewiß, daß ich nur noch eine nähere Bestimmung der Zusammensetzung des reducirten Productes für nöthig erachtete, um über diesen Punkt ins Reine zu kommen. Als ich Gay-Lussac's und Thenard's Versuch wiederholte, erhielt ich dasselbe Resultat unter gleichen Erscheinungen, wie sie beschrieben, nur mit der Ausnahme, daß die in Sauerstoffgas verbrannte Masse nicht weiß war, sondern ihre frühere Farbe ohne bedeutende Veränderung beibehalten hatte. Ich erwartete, daß das Verbrannte flußspathsaures Kieselkali enthalten werde, und übergoß es daher mit concentrirter Schwefelsäure; diese entwickelte aber keine Spur von Flußspathsäure, und konnte über der grau-braunen Materie abgedampft werden, ohne sie zu verändern. Unter den Säuren wurde dieselbe nur von der Flußspathsäure angegriffen, welche Kieselerde auszog und eine dunklere braune Materie hinterließ, die in Säuren unauflöslich und im Feuer unverbrennlich war. – War dieses das Radical der Flußspathsäure oder das der Kieselerde, oder eine Verbindung von beiden? – Um diesen Körper in größerer Menge zu erhalten, verfuhr ich folgender Maaßen: In eine Glasretorte von etwa 10 C[ubik] Z[oll] Inhalt, wurde ein kleines Gefäß von ächtem Porcellan gebracht, auf welchem ein Stück Kalium von der Größe einer großen Haselnuß lag; die Retorte wurde schnell luftleer gemacht; alsdann kieselhaltiges flußspathsaures Gas, aus einem über Quecksilber stehenden Reservoir hineingelassen, und darauf die Stelle der Retorte, über welcher sich das Gefäß mit Kalium befand, mit einer Spirituslampe erhitzt; das Kalium wurde anfänglich weiß, dann mehr und mehr dunkel gefärbt und endlich so schwarz wie Kohle; es entzündete sich kurz darauf, und verbrannte mit einer großen, dunkelrothen, aber nicht intensiven Flamme, während das Quecksilber schnell im Reservoir stieg, mit dem die Retorte während des Versuches in Verbindung stand. Sobald die Verbrennung beendigt war, wurde die Retorte wiederum ausgepumpt, um die Bildung von flußspathsaurem Kieselkali zu verhindern, und alsdann dem Erkalten überlassen. Das Product der Verbrennung war eine harte, zusammengebackne, poröse Masse, von dunkelbrauner Farbe, die sich zwar an der Luft nicht veränderte, aber wenn man sie mit dem Finger berührte oder anhauchte, nach Wasserstoffgas roch, wie es dem metallischen Mangan eigen ist. Rund um das Gefäß, welches das Kalium enthielt, hatte sich in der Retorte ein lockeres, hellbraunes Pulver gesammelt, das für sich bewahrt wurde. Die verbrannte Masse wurde in Wasser geworfen, aus welchem sie im ersten Augenblicke mit Heftigkeit Wasserstoffgas entwickelte. Das Wasser zog sehr viel flußspathsaures Kali aus, und die braune Masse zerfiel unter langsamer Gasentwickelung zu einem kastanienbraunen Pulver. Die alkalische Flüssigkeit wurde abgegossen und mit frischem Wasser vertauscht; die Gasentwickelung verringerte sich nun sichtlich und als dieses Wasser nach einer Weile erneut wurde, hörte sie fast ganz auf, so daß die Masse ausgekocht werden konnte, ohne daß das braune Pulver das Wasser ferner zersetzte. Die durch Kochen erhaltene Auflösung reagirte stark sauer, das Pulver wurde daher so lange mit neuen Portionen Wassers gekocht, als dieses noch sauer wurde. Das durchs Filtrum gehende Wasser war eine gesättigte Auflösung von flußspathsaurem Kieselkali. Das braune Pulver wurde aufs Filter genommen, und so lange ausgewaschen, als das durchgehende Wasser nach dem Verdunsten noch einen Fleck hinterließ. Es wurde hierauf getrocknet und stellte nun eine lose, kastanienbraune, pulverige Materie dar, die deutlich hellere Theile enthielt, so daß sie nicht von gleichförmiger Beschaffenheit zu seyn schien. Die braune Materie, welche sich während der Verbrennung auf dem Glase der Retorte abgesetzt hatte, war viel gleichartiger. Das Wasser entwickelte mit derselben kein Wasserstoffgas, wurde aber sogleich sauer. Auch dieses Pulver wurde mit derselben Vorsicht, wie das vorige ausgewaschen"].

After further purification of the brown substance he was convinced that "it now represents *silicon* in an isolated state" ["der Körper stellte nun das *Silicium* in isolirter Gestalt dar"]. After he had also shown that "siliceous hydrofluoric acid gas" ["flußspathsaures Kieselerdegas"] could also be decomposed by white-hot iron filings with formation of a thin brown film on the surface and after describing in detail experiments on the chemical behavior of the new element, he described a simple procedure for obtaining it. This he had already described in a letter on 12th July, 1824, to the Geneva chemist and academy rector C. G. de la Rive [6] as even "capable of being carried out in a lecture" ["in einer Vorlesung durchführbar"]:

"Method of preparing silicon. In order to burn potassium in gaseous siliceous hydrofluoric acid facilities are required which one often does not possess. On the other hand, the double salts which hydrofluoric acid forms with silica and potash or soda, provide an eminently suitable material for preparing silicon very readily . . . The salts are used in the following manner: the salt is rubbed down to a fine powder in case it was caked in drying, and heated as strongly as it will bear without decomposition in order to remove adhering water, i. e., far above 100°C. Then it is introduced in layers with potassium into a glass tube sealed at the bottom which has a capacity corresponding with the quantity of material and which is best selected so that the whole can be heated at once. If one wishes the potassium may be melted and mixed somewhat with the salt powder by a pure iron wire, whereupon the material is heated over a spirit lamp. Even before a red heat the silicon is reduced with a sizzling sound and flame. Nothing gaseous is evolved if the salt is properly dried. The mass is allowed to cool and is treated as described already. It must be put at once into a large quantity of water in order to render the alkaline liquid which is formed on oxidation of potassium in water as dilute as possible since the latter tends to oxidize and dissolve the silicon. Consequently the material must not be treated with warm water until the liquid ceases to be alkaline after repeatedly decanting fresh water. It is then first boiled and washed with hot water until liquid passing through no longer leaves a stain on evaporating a drop. For this a great deal of water and some days are usually required. Silicon prepared in this way now contains hydrogen but only in small amounts and perhaps in the same way as Davy considered our ordinary wood charcoal to be hydrogen-carbon. Additionally it contains silica which stems largely from the fact that potassium oxidizes somewhat previous to the reduction, and then a quantity of silica corresponding with the potash produced separates (the potash which forms after the reduction in water dissolves a part of the excess double salt without separating its silica). This silica must be removed with hydrofluoric acid; but, since silicon in this state dissolves in acid, it is first necessary to render it insoluble and noninflammable. If it is allowed to burn in air the unburnt part is obtained in this state after treatment with acid but one then normally loses ⅔ of the silicon, which burns away. This is avoided if the dried hydrogen-containing silicon is heated in an open crucible almost to red heat, held for some time at this temperature, the heating being then increased up to full redness. Should the silicon ignite the crucible is covered and the temperature reduced thus stopping the burning at once. After heating throughout the silicon is

noninflammable in air and is no longer attacked by acids, provided it contains no foreign metal, e.g., iron or manganese, in which case the alloy is dissolved completely by acids with evolution of hydrogen gas."

["*Darstellungs-Art des Siliciums.* Um Kalium im Gase von flußspathsaurer Kieselerde zu verbrennen, werden Anstalten erfordert, die man oft nicht besitzt. Die Doppelsalze dagegen, welche die Flußspathsäure mit Kieselerde und Kali oder Natron bildet, geben ganz vortreffliche Mittel ab, um Silicium auf eine sehr leichte Art darzustellen . . . Man bedient sich dieser Salze auf folgende Art: das Salz wird zu feinem Pulver gerieben, im Falle es beim Trocknen zusammengebacken war, und zur Verjagung der anhängenden Feuchtigkeit, so stark erhitzt als es ohne Zersetzung ertragen kann, d. h. weit über +100°C. Es wird dann schichtweise mit Kalium in eine unten zugeschmolzene Glasröhre gebracht, die eine Capacität hat, welche der Menge der Masse entspricht, und man am besten so wählt, daß das Ganze auf einmal erhitzt werden kann. Das Kalium kann, wenn man will, geschmolzen, und mit einem reinen Eisendrathe etwas mit dem Salzpulver gemengt werden; worauf man die Masse über der Spirituslampe erhitzt. Noch vor dem Glühen wird das Silicium mit einem zischenden Laut und einem schwachen Feuerphänomen reducirt. Nichts Gasförmiges wird entwickelt wenn das Salz gehörig getrocknet war. Man läßt die Masse erkalten, und behandelt sie wie vorhin angegeben wurde. Man muß dieselbe jedoch sogleich in eine große Menge Wassers bringen, um die alkalische Flüssigkeit, welche sich durch Oxydation des Kaliums im Wasser bildet, so verdünnt zu erhalten als möglich, weil letztere die Neigung hat das Silicium zu oxidiren und aufzulösen; deshalb muß man die Masse nicht eher mit warmem Wasser behandeln, als bis die Flüssigkeit nach mehrmaligem Aufgießen von frischem Wasser aufhört alkalisch zu seyn. Man kocht sie hierauf erst mit Wasser, und wäscht sie dann so lange mit heißem Wasser aus, bis das Durchlaufende, beim Verdunsten eines Tropfens, keinen Fleck mehr hinterläßt. Hierzu gehört gewöhnlich viel Wasser und einige Tage Zeit. Das auf diese Art dargestellte Silicium enthält nun Wasserstoff, jedoch nur in geringer Menge und vielleicht auf dieselbe Art, wie Davy unsere gewöhnliche Holzkohle als Wasserstoffkohle betrachtet. Es enthält außerdem Kieselerde, welche meist daher rührt, daß das Kalium vor der Reduction sich etwas oxydirt, und dann eine dem erzeugten Kali entsprechende Menge Kieselerde abscheidet; (dasjenige Kali, welches sich nach der Reduction im Wasser bildet, löst einen Theil des überschüssigen Doppelsalzes auf, ohne dessen Kieselerde abzuscheiden). Diese Kieselerde muß mit Flußspathsäure weggenommen werden; da sich aber das Silicium in diesem Zustande in Säure auflöst, so muß man es erst unauflöslich und unentzündlich machen. Läßt man es an der Luft verbrennen, so erhält man wohl den unverbrannten Theil nach der Behandlung mit Säure in diesem Zustande, aber man verliert dabei gewöhnlich ⅔ vom Silicium, was verbrennt. Man kommt diesem zuvor, wenn man das getrocknete, Wasserstoff haltende Silicium, in einem offnen Tiegel bis nahe zum Glühen erhitzt, einige Zeit lang so heiß erhält, und darauf die Hitze nach und nach bis zum vollen Glühen verstärkt. Sollte sich das Silicium entzünden, so bedeckt man den Tiegel und vermindert die Temperatur, wodurch die Verbrennung sogleich unterbrochen wird. Nach geschehener Durchglühung ist das Silicium unentzündlich an der Luft und wird nicht mehr von der Säure angegriffen, sofern es kein fremdes Metall enthält, z. B. Eisen oder Mangan, in welchem Falle die Legirung vollkommen mit Entwicklung von Wasserstoffgas aufgelöst wird"] [4].

It must be noted at this point that in 1897 E. Vigouroux [7] suggested on the grounds of differences in the strong hydrogen gas evolution from fresh silicon prepared by the two methods, that Berzelius had obtained two different modifications of silicon, one active and one passive (designated by Vigouroux as α and β), but no proof was given. In any case Berzelius emphasized the complete identity of the products obtained by the two methods after washing.

References:

[1] F. Wöhler (Liebigs Ann. Chem. **97** [1856] 266/70, 266). – [2] H. G. Söderbaum (Jac. Berzelius Lettres, Vol. 2, Pt. 4, Correspondance entre Berzelius et P. L. Dulong 1819–1837, Uppsala 1915, pp. 51/6). – [3] H. G. Söderbaum (Jac. Berzelius Lettres, Vol. 1, Pt. 2, Correspondance entre Berzelius et Sir Humphry Davy 1808–1825, Uppsala 1912, pp. 67/73, 68). – [4] J. J. Berzelius (Ann. Physik Chem. [2] **1** [1824] 169/230, 204/30). – [5] F. Wöhler (Ber. Deut. Chem. Ges. **8** [1875] 838/52).

[6] J. J. Berzelius (Bibliothéque Universelle Sci. Arts Genève [1] **26** [1824] 273/9, 274). – [7] E. Vigouroux (Compt. Rend. **120** [1895] 367/70).

1.2.4 The Year of the Discovery

In the preceding sections attempts to isolate the element which extended over more than three decades have been described in detail, beginning with speculation on the existence of metals in what at the time were designated as "earths", Lavoisier's substances which could not be broken down in 1789 and the definite preparation of silicon by Berzelius in 1824. From this it is understandable that diverse statements have been made in the literature as to the year of the discovery and the discoverer. Thus the discovery is often attributed to L. J. Gay-Lussac and L. J. Thenard in 1809 (as for example in [1] and [2]) but more frequently to J. J. Berzelius in 1824 (e.g., [3 to 5]). The publications on which this dating is based are quoted sufficiently on p. 22 and p. 29. According to E. Vigouroux [6] the year 1808 is often given as the year of the discovery with Berzelius as the discoverer with reference to his observation of an iron-silicon alloy formation, see p. 18, though the year is incorrect as the publication in question by Berzelius: "*Zerlegung der Kieselerde mit gewöhnlichen chemischen Mitteln*" [reduction of silica with ordinary chemical agents] first appeared in 1810 [7] describing the preparation of an iron with a 3% silicon content. To speak of these experiments as the time of the earliest preparation of "technical silicon" [technischem Silicium] as has happened recently [8], hardly accords with the facts. The true significance of this publication was probably recognized earlier by E. v. Meyer [9] when he showed that from then on Berzelius – and not only he, as is evident from what has been said – was convinced of the existence of a metallic substance as the "basis of silica" ["Basis der Kieselerde"]. When occasionally [10, 11] J. J. Berzelius and the year 1823 are given for the discovery of silicon then it can only be attributed to the fact it is overlooked that in Berzelius' publication [12] of his silicon preparation the main part is a translation from the Swedish Vetenskaps Academiens Handlingar for 1823 but the section containing the preparation of silicon "*Zersetzung der flußspaths.[auren] Kieselerde durch Kalium*" ["Decomposition of fluorosilicic acid by potassium"] was accompanied by the following footnote: "This last part is translated from [Berzelius'] manuscript by Mr. Dr. Wöhler and is not contained in the Stockholmer Denkschriften for 1823" ["Dieser letzte Theil ist aus der [Berzeliusschen] Handschrift von Hr. Dr. Wöhler übersetzt, und nicht mehr in den Stockholmer Denkschriften für 1823 enthalten"]. Furthermore, it seems that the letters mentioned on p. 30 in Berzelius hand with its conclusive 1824 dating remained unknown. – The year 1811 and the French investigators have been credited most recently by S. Engels, A. Nowak [21] for the discovery of silicon; here the year of appearance of a comprehensive paper by the two investigators has been taken for the discovery, see p. 22.

It cannot be easy to settle the time of the discovery of silicon given the unusual development of our knowledge of this element and it must depend on what meaning is given to the concept of "discovery". If one adheres to the definition given by J. R. Partington [13] in 1962: "The meaning of "discovery" is well known in chemistry. When a substance is prepared for the first time by a clearly described method, and sufficient details are given which distinguish it from other substances, it is said to be discovered" the question reduces, as E. Rancke-Madsen [14] critically noted, to the decision as to whether Berzelius has discovered the element in 1824 or the French researchers as early as 1809 and this has to be decided in favor of Gay-Lussac and Thenard. A few years after the publication of this definition, which is unquestionably adequate for the discovery of simple substances (e.g., of Glycerin, the sweet oil of C. W. Scheele in 1779 or of penicillin by A. Fleming in 1929) T. S. Kuhn [15] suggested that the discovery of an element implies a twofold recognition process: "Recognizing both that something is and what it is." With this condition we can only take 1824 as the discovery year with J. J. Berzelius responsible. This conclusion also applies if we adopted A. N. Meldrum's [16] idea of an "effective discovery of an element" which he introduced in 1930 in his studies on the discovery of oxygen. This entails not only the requirement of the recognition of the existence of the substance that has been discovered but also requires an announcement of the "discovery" by widespread scientific publication. This last requirement makes it difficult to name A. M. Ampère, who was the first to recognize in 1812 what researchers on both sides of the Channel seeking the radical of hydrofluoric acid actually had in their hands in the form of the brown substance, as the one who had made the discovery since he made his findings known to H. Davy only by letter, see p. 26. In his full discussion of the diverse ideas in the literature of "*the discovery of an element*" ["die Entdeckung eines Elementes"] E. Rancke-Madsen [14] would like to formulate the conditions under which one can speak of the "discoverer" and also the "discovery of an element" so as to satisfy modern requirements in the following way: "1. He has observed the existence of a new substance which is different from earlier described substances and this new substance is recognized by him or later by scientists as being elemental. The element in question may exist in chemical combination or as a free element without any demand of purity being made of this substance. 2. He must have published the discovery of the new substance in such a manner that it has been noticed by contemporaries outside the immediate circle of the discoverer, usually in a periodical or monograph, or he must have made it the subject of a lecture given to a professional group. Therefore the accepted date of discovery must always be some time (shorter or longer) *after* the performance of the experiments which resulted in discovery of the element."

With these stipulations still further possibilities open up for dating the discovery of silicon. In the first place we may name 1810 with Berzelius' postulated existence of an alloy-forming substance in silica, see above and p. 18, or the "discovery of silicon" may be associated with the time when it was recognized that silica, which previously was considered an element, was in fact a compound of an element, which had not been isolated, with oxygen. The year 1789 is named by E. Rancke-Madsen [14, p. 308] for this event when A. Lavoisier's famous *Traité Élémentaire de Chimie* appeared. There Lavoisier [17] placed together in his table of elements (Tableau des substances simples) in a group of "substances simples salifiables terreuses" ["earth elements which form neutral salts"] silica together with lime, magnesia (MgO), heavy oxide [BaO] and alumina thus treating them equally as oxygen compounds of unknown metals. In fact Lavoisier spoke in this way of the "earths" as a supposition or hypothesis as has been shown above, see p. 8, but it must be borne in mind here that, in discussing the table [17, pp. 164, 172/3] cited above, he no longer mentioned silica at the time. Why he omitted it, S. F. Hermbstädt, the translator of Lavoisier's works into German, believed to be because: "that Mr. Lavoisier could not include silica here is self-evident, it does not mix (form a compound) with any other acid apart from hydrofluoric acid and even with this forms no true neutral salt" ["daß Herr Lavoisier die Kieselerde hier nicht mit aufstellen konnte, versteht sich von selbst, da sie, die Flußsäure ausgenommen, mit keiner andern Säure in Mischung [Verbindung] geht, auch selbst mit dieser kein wahres Neutralsalz bildet"] [18]. For further evidence for his supposition E. Rancke-Madsen cites a quotation which according to his view comes from Lavoisier's work [17, p. 229] on "*Radical boracique, Silice et Alumine*": ". . . they are completely unknown. One only knows that these radicals combine with oxygen; that they form hydrochloric, hydrofluoric, and boric acids . . ." [". . . elles sont toutes absolument inconnues. On sait seulement que ces radicaux s'oxygènent; qu'ils forment les acides muriatique, fluorique et boracique . . ."]. This quotation is, however, taken from a section of Lavoisier's work under the heading "Sur les Radicaux muriatique, fluorique et boracique, et sur leur combinaisons" ["Concerning the hydrochloric, hydrofluoric, and boric acids radicals and their compounds"] and contains Lavoisier's opinion that both hydrochloric and hydrofluoric acid are formed by combination of their "radicals" with oxygen, while neither silica nor alumina nor the other "earths" are mentioned in this section. One is thus hardly able to identify Lavoisier as the researcher who recognized silica as the oxide of a "metal". It must also be mentioned here that Lavoisier's interest in silica can not have been exactly large: it is missing in the alphabetical index and is mentioned only twice in the text, once in the discussion of hydrofluoric acid [17, p. 263], which volatilizes silica, and once in connection with the infusibility of quartz [17, p. 555], which is pure silica. Otherwise, Lavoisier does not appear to have known about the acidic character of silica, which Otto Tachenius (d. 1699) had already made known in the year 1669 [19]; nevertheless, any mention of this among the acid groups is missing.

A further possibility for assigning a "discovery" according to his previously cited conditions was pointed out by E. Rancke-Madsen [14, pp. 308/10] with his dating of the discovery of aluminum. He thereby passed over the dating of the "discovery" by the recognition of the oxide character of alumina by Lavoisier and went for a point in time still further back to the differentiation of alumina from lime by Marggraf and his preparation of several aluminum compounds in the year 1754 [20], and assigned to H. C. Oersted alone in the year 1824 the "initial preparation", see "Aluminium" A 1, 1934, p. 2. He proceeds quite similarly for several rare earth elements. If this method of settling the "discovery" is also used for silicon, it is necessary to go back to have one of the early investigators named at the beginning, see p. 7, J. H. Pott or J. J. Becher for the "discoverer" and to name J. J. Berzelius as the "first to prepare" it in 1824.

References:

[1] N. Friend (Man and Chemical Elements, Griffin, London 1951, pp. 69/71). – [2] T. M. Lowry (Historical Introduction to Chemistry, London 1915, p. 288). – [3] M. E. Weeks, H. M. Leicester (Discovery of the Elements, 7th Ed., Easton, Pa., 1968, pp. 556, 806). – [4] S. Neufeld (Chronologie Chemie 1800–1970, Verlag Chem., Weinheim/Bergstr. 1977, p. 16). – [5] R. Partington (A History of Chemistry, Vol. 4, St. Martin's Press, New York 1964, p. 152).

[6] E. Vigouroux (Ann. Chim. Phys. [7] **12** [1897] 5/74, 7, footnote 1). – [7] J. J. Berzelius (Ann. Physik **36** [1810] 89/102, 94/5; Afhandl. Fys. Kemi Mineral. Hisinger **3** [1810] 117/28). – [8] W. Dietze (Metall **34** [1980] 676/7). – [9] E. v. Meyer (Geschichte der Chemie von den ältesten Zeiten bis zur Gegenwart, 3rd Ed., Leipzig 1905, pp. 361/2). – [10] E. Pilgrim (Entdeckung der Elemente mit Biographien der Entdecker, Mundus, Stuttgart 1950, pp. 205/7).

[11] O. Dammer (Kurzes Chemisches Handwörterbuch, Berlin 1873, p. 373). – [12] J. J. Berzelius (Ann. Physik Chem. [2] **1** [1824] 169, 251, 204/5, footnote). – [13] J. R. Partington (J. Chem. Educ. **39** [1962] 123/5). – [14] E. Rancke-Madsen (Centaurus **19** [1976] 299/313, 308). – [15] T. S. Kuhn (The Structure of the Scientific Revolution, 2nd Ed., Chicago 1970, p. 55).

[16] A. N. Meldrum (The Eighteenth Century Revolution in Science; The First Phase, Longmans Green, Bombay 1930, p. 47 from [15, p. 300]). – [17] A. Lavoisier (Traité Élémentaire de Chimie, Vol. 1, Masson, Paris 1789, pp. 192, 172). – [18] S. F. Hermbstädt (in: A. L. Lavoisier, System der antiphlogistischen Chemie [German translation by S. F. Hermbstädt], 2nd Ed., Berlin – Stettin 1803, p. 204 footnote). – [19] O. Tachenius (Antiquissimae Hippocraticae

Medicinae Claris Manuali Experientia in Naturae Fontibus, Elaborata, Venedig 1699, p. 169). – [20] J. R. Partington (A History of Chemistry, Vol. 2, St. Martin's Press, New York 1961, p. 727).

[21] S. Engels, A. Nowak (Auf den Spuren der Elemente, 2nd Ed., VEB Verl. Grundstoffe, Leipzig 1977, pp. 129, 270).

1.3 Silicon in Ancient Bronzes

Even as early as 1809 H. Davy [1] was able to say in his Bakerian Lecture on 16th November: "Copper, M. Berzelius informs me, is hardened by silicium". This information must have come from a private preliminary report, which has now been lost, relating to J. J. Berzelius' [2] experiments on the "Decomposition of silica by ordinary chemical agents" ["Zerlegung der Kieselerde durch gewöhnliche chemische Mittel"], which was published a year later. In this work the Swedish investigator had been able to obtain silicon-copper alloys quite readily by melting together copper turnings with silica and carbon, though he said nothing about their properties. He also omitted this five years later when, in describing the properties of the (amorphous) element which he had discovered [3], he mentioned its strikingly easy combination with copper. The hardness of silicon-copper alloys first became known in 1857 when H. Sainte-Claire Deville, H. Caron [4] investigated such alloys somewhat more systematically. The hardening effect of silicon on copper appeared to them to be so surprising and great that they spoke of the alloys as "copper steel" ["acier de cuivre"]. According to F. Gautier [8] the first modern silicon bronzes were prepared and used a little later in France. However, this hardening agent for copper, which it would also have been easy for early metal workers to obtain, was not detected in ancient bronzes, perhaps because it was not looked for, until 1979, when silicon was found in such bronzes by using the nondestructive method of spectroscopic microanalysis.

Although this all was known it was surprising that in 1979 J. Pelleg, J. Baram, E. Oren [5] announced that they had been able to detect 0.3, 1.0, and 3.7% Si as an alloy component in two bronze arrow heads and a half-moon shaped bugel of the same material which had been discovered in the course of excavations by the North Sinai expedition of Ben Gurion University on the north coast of the Sinai peninsula between the Suez canal and the Gaza strip. These belonged to the Egyptian Säite period (7th to 6th century B. C.). Of particular interest and consequently more fully studied was the flake-shaped arrow head with the highest of the silicon contents quoted since, equally surprisingly, this also contained 4.6% magnesium. It came to light at an excavation site designated as T 21 near the ancient town of Pelusium in the ruins of a fortress destroyed by the Persian king Kambyses in his conquest of Egypt in 525 B. C. Judging from the finds of copper ore, crucibles, and slags it must be concluded that bronze was made there on a considerable scale, though presumably only for local use. Mean values for the composition of the metal freed from the corrosion layer are in wt%: 66.8 Cu, 22.9 Sn, 4.6 Mg, 3.7 Si, 0.4 Fe, and 1.2 Cl, the absence of lead being notable. Spectroscopic analysis of the dendritic component of the alloy gave: 86.6 Cu, 6.8 Sn, 3.7 Mg, 1.9 Si, 0.4 Fe, and 0.6 to 1.4 Cl while for the interdendritic zone the values were: 40.6 Cu, 42.2 Sn, 10.4 Mg, 4.6 Si, 0.4 Fe, 1.2 to 6.5 Cl. As far as the surprising silicon content of the alloy is concerned (it is not possible to go into the magnesium content and its probable origin here), the micro methods of investigation used exclude the possibility that it has been confused with particles of silica embedded in the metal and such particles were also not once present in the corrosion layers. Furthermore, it is highly unlikely that such quantities of (the two) impurities could have been taken up from material of the molds and crucible used in working up the alloy. It is however probable that the process of smelting the ore was carried out in such a way that foreign oxides from the walls of the hearth were partially reduced and remained dissolved in the crude metal. In favor of this assumption is the fact that a piece of metallic copper which was largely coated with slag and which originated at the same excavation showed 1.3% Si on analysis. Whether such a method of carrying out the process was consciously aimed at obtaining hardening of the bell metal-like alloy and the good cold- and hot workability of such Si-containing bronzes must remain uncertain [5]. Further examination of bronze finds provides no clarification of this point. Other arrow heads from T 21, the excavation site described, proved to contain Mg and Pb but a silicon content of 1% occurred in only two (of six) samples. The composition was (in wt%) 82.0 Cu, 4.0 Sn, 10.4 Pb, 2.6 Mg, 1.0 Si in one case and 78.0 Cu, 8.8 Sn, 8.7 Pb, 3.6 Mg, and 1.0 Si in the other, while bronze objects other than weapons found at the same site (beads, weights, and one brass ornament) contained no magnesium although 0.4 to 0.9% Si was present. Arrow heads from other sites showed varying amounts of silicon [6], as in the case of ancient Daphnae (with 71.4 Cu, 5.4 Sn, 11.3 Pb, 1.0 Si, and 0.6 Fe) and Naucratis (with 66.6 Cu, 23.3 Sn, 6.0 Pb, 3.3 Si, and 0.7 Fe).

It must be borne in mind that silicon alloys can form very readily under the simplest conditions. J. J. Berzelius [3] drew attention to the fact that J.-B. Boussingault [7] from the École des Mineurs in Saint-Étienne had observed in 1821 that even the SiO_2 contained in coal ashes sufficed to produce silicon-platinum alloys during melting experiments in platinum and even brings about the uptake of silicon in steels.

References:

[1] H. Davy (Phil. Trans. Roy. Soc. [London] **100** [1810] 16/74 in J. Davy, The Collected Works of Sir Humphry Davy, Vol. 5, London 1840, pp. 225/83, 275). – [2] J. J. Berzelius (Afhandl. Fys. Kemi Mineral. **3** [1810] 117/28 from Ann. Physik **36** [1810] 89/102, 100/1). – [3] J. J. Berzelius (Ann. Physik Chem. [2] **1** [1824] 109/236, 220). – [4] H. Sainte-Claire Deville, H. Caron (Compt. Rend. **45** [1857] 163/7). – [5] J. Pelleg, J. Baram, E. Oren (Metallography **12** [1979] 313/24).

[6] J. Baram, J. Pelleg (Metallography **14** [1981] 347/58). – [7] J.-B. Boussingault (Ann. Chim. Phys. [2] **16** [1821] 5/16). – [8] F. Gautier (in: R. A. Hadfield, J. Iron Steel Inst. [London] **1889** II 222/55, 243).

1.4 Amorphous Silicon

Initially the method developed by J. J. Berzelius which was described earlier (see p. 29) was the only way in which the element could be prepared but, nine years after its publication, J. J. Berzelius [1] gave another method in the third edition of his *Textbook*. In it "chlorkiesel" [that is $SiCl_4$] was passed as vapor over warm metallic potassium which "burned at the expense of the silicon tetrachloride vapor" ["auf Kosten des Chlorkieselgases verbrennt"]. The resulting material was dissolved in water, thus freeing silicon from the potassium chloride produced. In 1855 H. Sainte-Claire Deville [2] replaced potassium by metallic sodium in this procedure, when it was found to be necessary to raise the temperature to a red heat to initiate reaction. W. Hempel, v. Haasy [3] found that when silicon tetrafluoride was used in place of the tetrachloride this was the only method "suitable for preparing amorphous silicon in larger quantities". They therefore described it quite fully, having first established that none of Berzelius' methods could be used for this purpose.

In 1857 H. Buff, F. Wöhler [4] attempted to improve the preparation from the double salts of fluorosilicic acid by introducing a mixture of sodium silicofluoride, sodium chloride, and sodium cut into small pieces into a red hot clay crucible, when reaction set in at once. When it had ended and the mass was quite cool it was washed out and the product purified with acids.

According to W. Hempel, v. Haasy [3] this method suffers from the disadvantage that sodium does not mix sufficiently with the salt mixture and a considerable part of the silicon usually burns when the reaction is finished. The earliest attempts to replace alkali metals, which are difficult to handle, by magnesium, which is easier to work with, and at the same time get away from gaseous silicon halides was made in 1864 by T. L. Phipson [5]. He reported that quartz powder is reduced to amorphous silicon by magnesium in a violent reaction carried out in an open crucible, alloys of the two elements being formed at the same time. Subsequently most investigators who studied this reaction were concerned with the latter, as it was possible to obtain silicon hydrides from the magnesium silicides, see p. 75. Among them were J. Parkinson [6] who, three years later, also observed magnesium silicate among the reaction products, then, in 1889, L. Gattermann [7] and shortly afterwards C. A. Winkler [8], who attempted to prepare well defined magnesium silicides and also reported that fusion of pure amorphous silicon could be observed under suitable conditions. It was not until 1895 that E. Vigouroux [9] succeeded in conducting the reaction in such a way that only amorphous silicon resulted. This he did by adding magnesium oxide to the mixture of quartz powder and magnesium powder. On completion of the reaction, magnesium oxide was removed by hydrochloric and sulphuric acids, when no trace of silicon hydrides was observed, and he thus obtained brown amorphous silicon of 99.6% purity. H. N. Warren's experiments [10] on the reduction of silicon tetrafluoride by magnesium, published in 1888, were undertaken with the primary aim of the formation of magnesium silicide rather than of the element. According to W. Hempel, v. Haasy [3], this method, which is adversely affected by the poor fusibility of the magnesium fluoride produced, gives only a poor yield of the element. Shortly after his experiments described above E. Vigouroux [21] replaced magnesium by aluminum and, when the mixture of quartz and aluminum powder was heated to 800°C, he obtained a bright chestnut-brown colored amorphous powder in a vigorous reaction which was accompanied by a bright glow. This yielded brown amorphous silicon on treatment with acids.

Melting experiments were undertaken in 1849 by C. Despretz [11] with amorphous silicon which Laroque, the former preparator of the École de Pharmacie, had prepared by Berzelius' method. He used a vessel described as an "electric egg" ["oeuf électrique"] which could be evacuated and was heated by 496 Bunsen cells arranged in series in four groups. The results were described as follows: "The silicon which I have exposed to the action of the electrical furnace melted with ease: it contracted immediately into a small sphere with a slightly glassy surface. The fracture surface of the silicon converted into a small sphere is in several places dull and little different from that of coal; at others it is similar to the glassy surface of certain anthracites. By rubbing with emery powder this surface can take on the luster of a deep black glass. The color of the silicon powder has not completely vanished, one finds it again on part of the surface. This silicon thus melted scratches glass. It can include no more silicic acid. It has been treated very often with hydrofluoric acid" ["Le silicium que j'ai soumis à l'action du feu électrique s'est fondu avec facilité; il s'est immédiatement rassemblé en un globule, un peu vitreux à la surface. Dans quelques points, la cassure de ce silicium réduit en globule est matte et peu différente de celle du charbon; dans d'autres, c'est à peu près la cassure vitreuse de certaines anthracites. Par le frottement avec le poudre de l'émeri, cette surface peut acquérir le poli d'un verre noir trèsfoncé. La couleur du silicium en poudre n'a pas disparu complétement, on la retrouve sur une partie de la surface. Ce silicium ainsi fondu raye le verre. Il ne devait pas renfermer d'acide silicique. Il avait été traité un grand nombre de fois par l'acide hydrofluorique"]. Six years later H. Sainte-Claire Deville [2] also observed melting of the amorphous element during his experiments on its preparation, though he came to quite other results than C. Despretz, which he described somewhat later [12]. He had carried out his experiments in a platinum crucible which was well coated with lime, in a forced draft furnace with a "lampe-forge" constructed by him and found that: ". . . the melting point of the silicon is a little high, it lies about in the middle between the melting point of cast iron and that of steel" [". . . le point de fusion du silicium est peu élevé, il est intermédiaire entre les points de fusion de la fonte et celui de l'acier"]. He also observed that the element always occurs in a crystalline form on solidification.

As early as 1825 H. C. Oersted (1777–1851) had referred to the possibility of obtaining silicon, then known only as an amorphous powder, in a compact form when he published his new method for preparing aluminum and silicon chlorides [13]. He had undertaken an experiment aimed at obtaining silicon from its chloride by means of an alkali metal amalgam, a route which had been used for many metals. He writes, however: "The volatility [of the silicon chloride] appears to interfere with its decomposition by potassium amalgam but, should it succeed, one may hope to see the combustible base of silica, which Berzelius has already prepared as a powder, in a coherent state or even perhaps in a form with a metallic luster" ["Die Flüchtigkeit desselben [des Siliciumchlorids] scheint seine Zersetzung durch Kaliumamalgam zu verhindern, aber wenn sie gelingen sollte, darf man auch hoffen, den brennbaren Grundstoff des Kiesels, der bereits von Berzelius als ein Pulver hergestellt ist, in zusammenhängender, vielleicht sogar metallglänzender Gestalt erzeugt zu sehen"]. Although C. A. Winkler [14] had established in 1864 that there was no way of bringing about alloy formation between silicon and mercury, A. Stock, F. Zeidler [15] repeated Oersted's experiment in 1923, though probably with other objectives. After 40 hours shaking of sodium amalgam and silicon tetrachloride the amalgam was found to have reacted and had assumed a golden-coppery appearance, while nothing volatile at room temperature was present any more in the tube. A more detailed investigation of this reaction appears not to have been published.

W. Manchot [16] was able to show in 1922 that "amorphous" silicon obtained by the preparative methods under consideration was actually structurally identical with crystalline specimens obtained by methods described below. X-ray diagrams which Debye and Frauenfelder had taken on silicon specimens which W. Manchot [17] had prepared for this purpose by the old method, and which appeared visually to be amorphous under 1000-fold magnification, all showed a diamond lattice. The higher reactivity of many forms of so called amorphous silicon, which is governed by the method of preparation, was thought by F. Roll [18] in 1926 to be explicable on the basis that silicon dioxide dissolves in elementary silicon.

In 1952 R. Schwarz, A. Köster [19] were successful in preparing silicon which was amorphous and in providing proof of this when they investigated the product of the thermal decomposition of unsaturated silicon chlorides. They describe the new form of the element as having a metallic glance, a dark gray color and as appearing outwardly to be crystalline. Under 60-fold magnification the glistening round particles showed themselves as fragments from the breakdown of small bubbles, which yielded a brown powder when ground in a mortar. This material dissolved slowly in hot caustic soda and in hydrofluoric acid but did not ignite in a flame. "After tempering for four days at 800°C the silicon atoms start to become ordered and the interference lines of normal silicon with a tetrahedral structure appear, though they are still very indistinct" [19].

So-called amorphous silicon appears to have become of no significance in connection with the preparation of crystalline silicon, which has now been obtained in the highest state of purity by many methods, see below. Recently, however, it has been found that solar cells for the direct transformation of solar to electrical energy, and which are prepared from silicon crystals of the highest purity, may be prepared more efficiently and cheaply from amorphous silicon, which is obtained by decomposing silane, SiH_4, in a glow discharge [20].

References:

[1] J. J. Berzelius (Lehrbuch der Chemie, German translation by F. Wöhler, 3 Ed., Vol. I, Dresden – Leipzig 1833, pp. 327/8). – [2] H. Sainte-Claire Deville (Compt. Rend. **40** [1855] 1034/6). – [3] W. Hempel, v. Haasy (Z. Anorg. Allgem. Chem. **23** [1900] 32/42). – [4] H. Buff, F. Wöhler (Liebigs Ann. Chem. **104** [1857] 94/109, 107 footnote). – [5] T. L. Phipson (Proc. Roy. Soc. [London] **13** [1864] 217/8).

[6] J. Parkinson (J. Chem. Soc. **20** [1867] 117/31, 129). – [7] L. Gattermann (Ber. Deut. Chem. Ges. **22** [1889] 186/97, 186/7). – [8] C. A. Winkler (Ber. Deut. Chem. Ges. **23** [1890] 2642/68, 2652/7). – [9] E. Vigouroux (Compt. Rend. **120** [1895] 94/6). – [10] H. N. Warren (Chem. News **58** [1888] 215/6).

[11] C. Despretz (Compt. Rend. **29** [1849] 545/8, 545). – [12] H. Sainte-Claire Deville (Ann. Chim. Phys. [3] **49** [1857] 62/78, 65/6). – [13] H. C. Oersted (Kgl. Danske Videnskab. Selskab. Oversigt **1824/25** 13/6 from K. Meyer, Scientific Papers, Vol. 2, Kopenhagen 1920, pp. 464/6). – [14] C. A. Winkler (J. Prakt. Chem. **91** [1864] 193/208, 202/8). – [15] A. Stock, F. Zeidler (Ber. Deut. Chem. Ges. **56** [1923] 986/97, 989).

[16] W. Manchot (Z. Anorg. Allgem. Chem. **124** [1922] 333/4). – [17] W. Manchot (Z. Anorg. Allgem. Chem. **120** [1922] 277/99). – [18] F. Roll (Z. Anorg. Allgem. Chem. **158** [1926] 343/8, 344). – [19] R. Schwarz, A. Köster (Z. Anorg. Allgem. Chem. **270** [1952] 2/15, 14). – [20] Anonymous (Physik Unserer Zeit **12** No. 5 [1981] A 45).

[21] E. Vigouroux (Compt. Rend. **120** [1895] 1161/4).

1.5 Crystalline Silicon

1.5.1 Discovery and Preparation

The earliest observation of a crystalline form of elementary silicon appears to have been made as early as 1843 by Friedrich Wöhler (1800–1882). On 4th August of that year he wrote to J. Liebig [1]: "It will hardly interest you now [as you are busy editing several books] that the chlorides of tungsten, molybdenum, and silicon in the vapor form when brought to a red heat with hydrogen give metallic tungsten and molybdenum as thick black crusts and silicon in the form of black crystals" ["Es wird Dich [weil mit der Herausgabe mehrerer Bücher überbeschäftigt] jetzt kaum interessieren, daß die Chloride von Wolfram, Molybdän und Silicium, dampfförmig mit Wasserstoff geglüht, metallisches Wolfram und Molybdän in dichten schwarzen Rinden und Silicium in schwarzen Kristallen geben"]. A study of the *Catalogue of Scientific Papers* of the Royal Society of London, other reference works and a comprehensive biography [2] shows that he clearly did not publish anything else on this. When J. N. Pring, W. Fielding [3] took up such simple experiments again in 1909 they actually obtained beautiful crystalline silicon, perhaps because they worked with higher temperatures than Wöhler was able to reach. As in the case of vanadium, see "Vanadium" A 1, 1968, p. 5, F. Wöhler missed the discovery by not following up an observation further. So credit for having prepared the first crystalline silicon belongs to Henri Étienne Sainte-Claire Deville (1818–1881) [4], who found during his experiments on the preparation of aluminum, made in 1854, that this metal is sometimes contaminated with iron but always contains silicon in quantities up to 10%, particularly if it is dark gray in color and brittle. He was successful in isolating it by treating his aluminum melts with hot hydrochloric acid and it was in the new and hitherto unknown form of small laminae with a metallic luster which were somewhat like platinum filings. In this form the element had a high resistance to chemical agents and had the same chemical properties as the so-called brown amorphous silicon which Berzelius had discovered. H. Sainte-Claire Deville characterized it as follow: "However, I do not believe that silicon is a true metal; on the contrary, I think that this new form of silicon is to ordinary silicon as graphite is to carbon" ["Cependant je ne crois pas que le silicium soit un véritable métal; je pense, au contraire, que cette nouvelle form de silicium est au silicium ordinaire ce qu'est graphite au carbon"].

Two years later F. Wöhler [5], without going into his earlier experience with crystalline silicon mentioned above, published an improved method for its preparation. This he stumbled on when he wished to prepare aluminum from cryolite with metallic sodium by Rose's method [6] and, in so doing, changed from the iron crucible recommended to a Hessian clay crucible. He often obtained, in addition to small reguli of aluminum, "brittle beads which were interpenetrated with a black crystalline material" ["öfters spröde Kugeln, die mit einer schwarzen, krystallisierten Substanz durchwachsen waren"], which he recognized as crystalline silicon. Consideration of how his experiments could have given rise to silicon led him to assume the intermediate formation of salts of fluoroacids from the crucible material and induced him to melt aluminum with 20 to 40 times the amount of sodium fluorosilicate. He obtained in this way a dark iron-black regulus from which, after removal of an aluminum-silicon alloy by hydrochloric acid and then treating with hydrofluoric acid, 65 to 75% of crystalline silicon (referred to the aluminum used) was obtained as "larger or smaller crystalline flakes" ["größere oder kleinere Krystallblätter"], the density of which was less than that of quartz, which yielded a dark brown powder when ground. In the following year F. Wöhler simplified the process by replacing the sodium fluorosilicate by a mixture of cryolite and powdered fused water glass [7]. This change to cheaper starting materials also had the advantage that it was very much simpler to remove excess aluminum. He describes his observations thus: "It is strange that these aluminum-containing substances cannot be remelted at the temperature at which they form. Such a pellet may be kept in an open crucible at a strong heat for a long time without melting or oxidizing. If it is thrown into water while still glowing about half of the volume of aluminum separates as a coherent mass which continues to glow under water for some minutes. Silicon, however, remains behind as a globule of jagged crystalline material filled with cavities. In preparing silicon it is therefore advisable to treat the regulus in this way before solution in hydrochloric acid as it is then possible to recover much of the aluminum which, even if it contains silicon, can be used in new preparations" ["Es ist sonderbar, daß diese aluminiumhaltigen Siliciummassen nicht wieder schmelzbar sind bei der Temperatur, bei der sie sich gebildet haben. Eine solche Kugel läßt sich in einem offenen Tiegel lange in starker Glühhitze erhalten, ohne zu schmelzen und ohne sich zu oxydiren. Wirft man sie glühend in kaltes Wasser, so fließt ungefähr das halbe Volumen Aluminium als eine zusammenhängende Masse aus, die minutenlang unter dem Wasser glühend bleibt; das Silicium aber bleibt in der Form der Kugel als eine mit leeren Räumen erfüllte, zackig krystallinische Masse zurück. Bei der Darstellung des Siliciums ist es daher zweckmäßig, den Regulus vor der Auflösung in Salzsäure auf diese Weise zu behandeln, indem man dadurch viel Aluminium, wenn auch siliciumhaltig, wieder gewinnt und zu neuen Darstellungen anwenden kann"].

Another preparative method which avoided the use of aluminum, which was then expensive, and offered the same advantage as Wöhler's method had that excess of reducing agent could readily be removed while also giving well formed crystals, was published in 1861 by H. Caron [8, 9] when, jointly with H. Sainte-Claire Deville he sent specimens of silicon prepared by different methods to V. Regnault for a determination of the specific heat of the element. Caron used a mixture of sodium metal and zinc as the reducing agent. H. Sainte-Claire Deville [10] had already used the latter to obtain silicon and was content to dissolve all of the excess reducing agent in acids. Four years later, when aluminum had already become quite cheap, E. Vigouroux [11] recommended that sodium be replaced in Caron's procedure by aluminum turnings. In this process, where aluminum serves simultaneously as a reducing agent and solvent, the presence of fluorine-containing compounds appeared always to be necessary. The light metal, however, is also able to reduce silicate melts to form crystalline silicon as has been found very recently; glassworks experienced difficulties in treating old broken glass which is not free of aluminum foil and caps [31, 32].

On 14th January, 1856, the day on which J. B. Dumas told the Paris Academy of a letter sent to him by F. Wöhler [5] about his first improvement in the preparation of crystalline silicon, H. Sainte-Claire Deville [12] published a possible new way of obtaining crystalline silicon which brings to mind Wöhler's earlier observations which had not been followed up. The French investigator had been able to obtain rhombohedral silicon crystals by passing silicon chloride vapor in a current of hydrogen over red hot aluminum.

After H. N. Warren [13] had succeeded in 1891 in producing the element in the form of crystals which were ½ inch and later 1½ inch big and "as perfect as a crystal of alum" by dissolving amorphous silicon in an aluminum-tin melt [14], it is not surprising that the transformation of amorphous silicon to a crystalline form by solution in other metal melts was attempted. H. Moissan, F. Siemens [15] investigated zinc- and lead melts and, by slow cooling, obtained from lead melts very thin transparent yellow-brown lamellae, while rapid cooling resulted in octahedra which were partly free and partly linked together in chains, together with free and linked tetrahedra and hexagonal plates. The so-called HF-soluble modification (see below) was obtained by P. Lebeau [16] from copper melts and also by H. Moissan, F. Siemens [15] from silver. Investigations on the crystallization of silicon which has been dissolved in melts of metals which do not form silicides were undertaken in 1951 by H. v. Wartenberg [30], though it is true that they were then concerned with the question of the degree of purity that could be attained. He found that the crystals always contained up to 0.15% of the solvent and concluded that "it is not possible to obtain pure silicon by this method".

Since it was well known to iron metallurgists that, like carbon, silicon in varying small amounts always accompanies iron in all its technical forms, it is not surprising, that – clearly unaware of the unsuccessful experiments by J. J. Berzelius, H. Davy, and F. Stromeyer made almost fifty years earlier in trying to isolate silicon in its elementary form from iron, see pp. 18 and 19 –, M. Buchner in Graz in 1857 shortly after the discovery of crystalline silicon, which he wrongly attributed to F. Wöhler, expressed what was certainly the conviction of many of his colleagues [17] that "from Wöhler's discovery" it was "more than probable" that when crude iron is dissolved in acids, as is done in determining carbon in iron, crystalline silicon, like graphite, must also be present in the residues. This (in addition to other factors which interfere) must impair the accuracy of the iron analysis. With a new analytical procedure he then believed he had found a maximum of 1.62% of crystalline silicon in some few of the 18 sorts of iron which he investigated. As A. E. Jordan, T. Turner reported [18], Buchner's view was supported ten years later by the English metallurgist J. A. Phillips [19] in his *Metallurgy* when he wrote: "Cast iron often contains silicon in the graphitic state" (however, elsewhere Phillips states that this was not by any means the unanimous opinion of English iron metallurgists on the form in which silicon occurs in iron), while Dr. Percy [20] expressed himself somewhat more cautiously in discussing the residues obtained on dissolving iron in acids: "There is reason to believe that crystallized silicon may also not unfrequently be present", though he then adds: "Mr. T. H. Henry informed me that he found crystallized silicon among the graphitic scales obtained by the action of hydrochloric acid on pig iron." In 1862 Professor Robert Richter [21] at Leoben claimed to have obtained graphite-like plates with a metallic luster in the same way. These remained unchanged when heated in oxygen and first yielded some iron oxide and silica on melting with a mixture of saltpeter and soda. The earliest information on crystalline silicon in iron was thought by A. E. Jordan, T. Turner [18] to be found in the *Wagnersche Jahresberichte über die Fortschritte der chemischen Technologie für 1857* [22]. They maintained that there was "a reference. . . to some analyses by Wöhler, who is stated to have found crystallized silicon in cast iron". This rests on a complete misunderstanding of the reference cited, which reports Buchner's publication [17] factually quite correctly, quoting his analytical results and, as has already been indicated, mentions Wöhler's name, but says nothing further about the occurrence of crystalline silicon. In 1859 K. E. Schafhäutl [23] referred to the fact that he had already drawn attention to the occurrence of "metallic silicon" in the processing of iron in 1846 [24] at a time when its crystalline form was not yet known. He had then written: "In our refining processes even silicon, for example, behaves as if in its metallic state while iron and manganese burn to oxides. I will mention only one of several cases which I observed. In the *Tividale* ironworks in England I noticed a bubble of considerable size on the upper part of a billet which had just come from the reheating furnace to be converted to railway rails in the rolling mill. As the billet passed between the rollers a small explosion occurred and a burst of fire like burning iron filings was emitted into the air from the billet. The star-like sparks were transformed to web-like flocks resembling Lana philosophica [latin: philosophical wool, an old name [25] for light flocculent zinc oxide formed in burning molten zinc in air], which not only floated around in the air but also completely covered the jacket of W. Knight, the rolling superintendent, on one side with this web so that it could readily be collected. On examination under the microscope the flocks were found to be composed of extremely fine lustrous silky threads made up of small spherules joined together in strings, which were fairly transparent. No trace of crystalline structure was apparent in polarized light. The whole behaved chemically like silica with which a few percent of alumina was admixed, though it resembled amorphous silica since it dissolved quite easily in caustic lye, which is never the case with crystalline silica" ["Es erhält sich z. B. sogar das Silicium in unseren metallischen Frischprozessen häufig in seinem metallischen Zustande, während Eisen und Mangan zu Oxyd verbrennen. Von mehreren Fällen, die unter meine Augen kamen, will ich nur eines erwähnen. Ich bemerkte in den *Tividaler* Eisenwerken in England auf dem oberen Theile eines Paquets von Eisenschienen, welches eben aus dem Schweißofen kam, um durch das Walzwerk in eine Eisenbahnschiene verwandelt zu werden, eine Blase von ziemlicher Ausdehnung. Als das Paquet zwischen den Walzen durchpassierte, entstand eine kleine Explosion und ein Feuerstrom wie von verbrennender Eisenfeile sprühte aus dem Paquet in die Luft. Die sternartigen Funken verwandelten sich in Spinnweben-artige Flocken gleich der Lana philosophica [lateinisch: philosophische Wolle, alte Bezeichnung [25] für das bei der Verbrennung von geschmolzenem Zink an der Luft sich bildende, leichte, flockige Zinkoxid], die nicht allein in der Luft umherschwammen, sondern das Koller des Walzmeisters W. Knight noch überdies auf einer Seite ganz mit diesem Gewebe überdeckten, so daß es leicht gesammelt werden konnte. Bei der Untersuchung mit dem Mikroskop fanden sich die Flocken aus äußerst feinen Seiden-glänzenden Fäden und diese aus aneinandergereihten Kügelchen von ziemlicher Durchsichtigkeit zusammengesetzt. Auch im polarisirten Lichte waren durchaus keine Spuren von krystallinischer Struktur zu erkennen. Das Ganze verhielt sich chemisch wie Kieselerde, welcher ein paar Prozent Thonerde beigemengt wären – jedoch wie amorphe Kieselerde, denn es löste sich ganz leicht in Ätz-Lauge, was mit krystallinischer Kieselerde nie der Fall ist"]. He then mentioned a similar occurrence described still earlier in the literature and the many reports on the occurrence of such silica threads which L. Gmelin [26] had collected in the fourth edition of his *Handbook*. That the substance held by Schafhäutl to be metallic silicon was actually silicon sulphide was established unequivocally in 1864 by H. Hahn [27] in his *"Chemische Untersuchung der beim Lösen des Roheisens entstehenden Produkte"*. He added that "the other view was disproved by the reported inflammability in air" ["Die andere Meinung werde allein schon durch die Angabe der Verbrennbarkeit durch Luft widerlegt"]. He also rejected other earlier claims that elementary silicon occurred in iron and suggested that the silicides Fe_2Si, FeSi, and $FeSi_2$, all of which he believed he had found, had been confused with the element. Even in 1874, however, E. H. Morton [28] was able to speak of a theory then current, according to which silicon occurs, "being intimately mixed with pig iron" though he was able to show that this is by no means the case, but that it "exists combined with a portion of the iron as a silicide of iron. . ." and may readily be separated out as silica by dissolution. The question of whether free silicon is obtainable from an iron melt was first clarified in 1913 by N. Kurnakow, G. Urasow [29] by a study of the iron-silicon system, according to which up to a content of 22% Si there is a solid solution of the element in iron. Between the limits 22.3 to 55.18% Si formation of FeSi takes place primarily and free silicon is present only with a content in excess of 61% Si.

References:

[1] A. W. Hofmann (from Justus Liebig's und Friedrich Wöhler's Briefwechsel in den Jahren 1829/73, Vol. I, Braunschweig 1888, pp. 230/1). – [2] J. Valentin (Friedrich Wöhler, Stuttgart 1949, pp. 158/78). – [3] J. N. Pring, W. Fielding (J. Chem. Soc. **95** [1909] 1497/506, 1501/2). – [4] H. Sainte-Claire Deville (Compt. Rend. **39** [1854] 321/6, 323; Ann. Chim. Phys. [3] **43** [1855] 1/33, 31/2). – [5] F. Wöhler (Compt. Rend. **42** [1856] 48/9; Liebigs Ann. Chem. **97** [1856] 266/70).

[6] H. Rose (Ann. Phys. Chem. [2] **96** [1855] 152/63). – [7] F. Wöhler (Liebigs Ann. Chem. **102** [1857] 382/3). – [8] H. Caron (Ann. Chim. Phys. [3] **63** [1861] 1/38, 26/9 footnote). – [9] H. Sainte-Claire Deville, H. Caron (Ann. Chim. Phys. [3] **67** [1863] 435/43, 437). – [10] H. Sainte-Claire Deville (Compt. Rend. **45** [1857] 163/7, 164).

[11] E. Vigouroux (Ann. Chim. Phys. [7] **12** [1897] 1/74, 55/6). – [12] H. Sainte-Claire Deville (Compt. Rend. **42** [1856] 49/52). – [13] H. N. Warren (Chem. News **63** [1891] 46). – [14] H. N. Warren (Chem. News **67** [1893] 136/7). – [15] H. Moissan, F. Siemens (Compt. Rend. **138** [1904] 657/61; Ber. Deut. Chem. Ges. **37** [1904] 2086/9).

[16] P. Lebeau (Compt. Rend. **142** [1906] 154/7). – [17] M. Buchner (Sitz.-Ber. Akad. Wiss. Wien Math. Naturw. Klasse **25** [1857] 231/5; J. Prakt. Chem. **72** [1857] 364/9, 365). – [18] A. E. Jordan, T. Turner (J. Chem. Soc. **49** [1886] 215/22). – [19] J. A. Phillips (Metallurgy, p. 115 from [2]: probably: J. A. Phillips, Elements of Metallurgy, London 1874, here: p. 117). – [20] Dr. Percy (Iron and Steel, p. 145 from [2] perhaps in: K. Styffe, translated from the Swedish by C. P. Sandberg, with a preface by J. Percy, London 1869).

[21] R. Richter (Berg- Hüttenmänn. Jahrb. Bergakad. Pribram Leoben Kgl. Ung. Bergakad. Schlemnitz **12** [1862] 289/90). – [22] J. R. Wagner (Jahresber. Chem. Technol. **1858** 8). – [23] K. E. Schafhäutl (J. Prakt. Chem. **76** [1859] 257/310, 266/7). – [24] K. E. Schafhäutl (Neues Jahrb. Mineral. Geol. Paläontol. **1846** 641/95, 689/91 footnote). – [25] H. Kolbe (Handwörterbuch der reinen und angewandten Chemie, Vol. 4, Braunschweig 1849, p. 770).

[26] L. Gmelin (Handbuch der Chemie, 4th Ed., Vol. 2, Heidelberg 1844, p. 340). – [27] H. Hahn (Liebigs Ann. Chem. **129** [1864] 57/76). – [28] E. H. Morton (Chem. News **29** [1874] 107/9). – [29] K. Kurnakow, G. Urasow (Z. Anorg. Allgem. Chem. **123** [1922] 89/131, 105). – [30] H. v. Wartenberg (Z. Anorg. Allgem. Chem. **265** [1951] 188/200, 191/3).

[31] H. Petermöller (Glastech. Ber. **48** [1975] R 131 No. 75 R 1133). – [32] L. Schumacher (Glastech. Ber. **48** [1975] R 131 No. 75 R 1134).

1.5.2 Further Preparative Methods

1.5.2.1 Experiments on Electrolytic Preparation

From Aqueous Solution. Following the rather limited success which had attended early attempts to prepare silicon electrolytically, see p. 12, it is not surprising that experiments to obtain the element by this method were not renewed until the middle of the 19th century, when the galvanic technique had undoubtedly reached a peak [1]. A start was made by L.-N. Junot de Bussy [2] in a report to the Institut de France on 21st March, 1853, in which he stated that "tungsten, molybdenum, silicon, and titanium had been obtained as coatings on other metals by electrolytic reduction. These were silver white and had a marked metallic luster. Reduction of these metals was successful only from their nonacidic compounds. . . For the reduction of silicon use was made of a liquid prepared by saturation of boiling aqueous sodium carbonate with gelatinous silica and addition of potassium cyanide to the liquid, which had been decanted from the sodium carbonate that had crystallized and diluted, boiling and filtering" ["Wolfram, Molybdän, Silicium und Titan durch Electrolyse ihrer Verbindungen, als Ueberzug anderer Metalle, reducirt erhalten zu haben; sie seien silberweiß und stark metallglänzend. Die Reduction dieser Metalle gelinge nur aus ihren nichtsauren Verbindungen. . . Zur Reduction des Siliciums diene eine Flüssigkeit, bereitet durch Sättigen von kochendem, wässrigem kohlens[aurem] Natron mit Kieselgallerte, Zusatz von Cyankalium zu der vom auskristallisirenden kohlens[auren] Natron abgegossenen und verdünnnten Flüssigkeit, Kochen und Filtriren"]. In the French original the color of the silicon precipitate is described as "couleur isabelle, légèrement irisée", which the German reporter passes over in silence. Clearly doubts about Junot's findings had also arisen and a commission was set up to check the product and report to the Academy. As a result only the title of Junot's work, "Mémoire sur la Réduction et l'Application électrochimique du Tungstène, du Molybdène, du Titane et du Silicium", appeared in the corresponding report in the *Comptes Rendues* of the Paris Academy. At the following meeting the announcement was made that Junot de Bussy had withdrawn his work as the committee to which it had been referred, and which comprised C. S. M. Pouillet, H. V. Regnault, and A. J. Balard, had drawn his attention to a serious error which they had found while checking the results [3]. Two years later A. J. Balard, the discoverer of bromine, announced [4], following further work initiated by the Academy in the interim, according to which aqueous electrolysis of salts of the elements mentioned above appeared to be possible, that the white deposits on such experiments, which had also been obtained by Junot de Bussy, did not consist of the metals in question. At about the same time similar experiments were undertaken in England. On 24th February, 1854, the electrochemist G. Gore [5] wrote to the editor of the *Philosophical Magazine* to say that he had been able to deposit silicon on copper from a dilute solution of waterglass: "When the [Smee] battery was operating very mildly the metal deposited was even whiter than aluminum and almost as white as silver" ["Bei sehr schwacher Wirkung der [Smee'schen] Batterie war das abgesetzte Metall noch weißer wie Aluminium, fast so weiß wie Silber"]. He enclosed a specimen with his letter, though confirmation or rejection of his claim appears not to have ensued. It was not until eighty years later, in 1934, that a proposal for obtaining silicon electrolytically from aqueous solution was again made. According to the Russian M. J. Dshems-Levy [6] this is successfully accomplished if a silicon-containing starting material is attached to the anode, the cathode being metallic magnesium and the electrolyte an alkali solution. The latter should also contain ammonium fluoride or ammonium carbonate.

For the electrolysis of alkali silicate solutions with an alternating current see p. 68.

From Melts. The earliest observation of the occurrence of elementary silicon on the electrolysis of a melt was made in 1854 by H. Sainte-Claire Deville [7] when he electrolysed molten aluminum trichloride and discovered crystalline silicon in the resulting metallic aluminum. Initially he expressed the view that the element had only been produced indirectly through hydrogen which is evolved during the operation: "And actually in this latter process one removes by means of the hydrogen the silicon, the sulfur, and even the iron, which passes over at the operating temperature as the unchangeable ferrous chloride; while now all these impurities remain in the liquid, which one decomposes with the electrical battery, they are removed with the first part of the reduced metal" ["Et, en effet, dans ce dernier procédé, on enlève, au moyen de l'hydrogène, le silicium, le soufre et même le fer, qui passe à l'état de protochlorure fixe à la température on l'on opère, tandis que toutes ces impuretés restent dans le liquide que l'on décompose par la pile, et sont enlevées avec les premières portions de métal réduit"]. He seems, however, soon to have changed his mind for, three years later, he was able to publish a method for obtaining silicon from a melt [8]: "Finally one can also set free the silicon of the silica by means of the electrical battery, in a very simple procedure, that can be used for the electrolysis of all substances which dissolve in hot alkali fluorides, and these substances are very numerous. One prepares a mixture of approximately equal parts of sodium and potassium fluorides, which one melts down in front of the flame additionally strengthened by a blowpipe, of a torch with a double air duct [constructed by the author] operated with turpentine oil. As soon as the mass is completely liquid, one adds the calcined silica, which dissolves very rapidly. If one then dips in the carbon and platinum pole of a battery of four Bunsen elements, one sees silicon deposit on the negative electrode, while on the positive pole oxygen is freed. Mostly I was content in this study of the decomposition of silica by the battery to determine that I transformed a negative electrode platinum wire, which was changed thereby into a surprisingly low melting silicide." ["Enfin on peut encore extraire le silicium de la silice au moyen de la pile, par un procédé fort simple, qui peut être utilisé pour l'électrolyse de toutes les substances qui se dissolvent à chaud dans les fluorures alcalins, et ces substances sont fort nombreuses. On fait un mélange, de fluorure de sodium et de fluorure de potassium, à peu près à parties égales, que l'on fond ensemble au-dessus de la flamme d'une lampe à double courant, alimentée d'alcool térébenthiné et activée par un chalumeau. Lorsque la matière est en pleine fusion, on y introduit de la silice calcinée qui se dissout très-vite. Alors, en plongeant dans le creuset les pôles platine et charbon d'une pile de quatre éléments de M. Bunsen, on voit se déposer du silicium au pôle négatif, et se dégager de l'oxygène au pôle positif. Le plus souvent je me suis contenté de constater, par cette expérience, la décomposition de la silice par la pile, en employant, comme électrode négative un fil de platine qui se transforme en siliciure fusible avec une excessive facilité"].

A similar method was reported in 1866 by F. Ullik [9] from Vienna. Some time previously, while still in Bunsen's laboratory in Heidelberg, he had attempted to deposit rare earth metals electrolytically from melts of their fluorides with potassium fluoride as a flux and had also obtained silicides of these metals, the silicon having been dissolved from the porcelain crucible walls. This observation provided an incentive for Ullik to undertake experiments with a KF-K_2SiF_6 melt, from which (with 8 Bunsen cells) he obtained, in addition to potassium, a brown material which proved to be the so-called amorphous silicon. In 1939, M. Dodero [10] was able to show, by cooling the cathode, that in such electrolyses metallic potassium is first produced and that this then liberates silicon from the double salt by the reaction already recommended by Berzelius, see p. 32. Electrolysis of potassium silicofluoride was taken up again in 1951 by H. v. Wartenberg [11] in his search for ways of preparing silicon of high purity, though he had to conclude that this method did not offer advantages of any kind, especially as the element was obtained as thin brown needles, which were attacked by hydrofluoric acid, even at a concentration of one percent, with bubble formation.

In 1888, W. Hampe [12], from his experience in investigating the electrolysis of borax melts, had expressed his conviction that it should be possible to obtain silicon in an analogous manner from sodium silicate melts. This speculation was changed to reality in 1934 by L. Andrieux, M. Dodero [13] using lithium silicate melts stabilized with lithium halides. As early as 1890 A. Grätzel von Grätz [14] had patented a process for the electrolytic production of light metals, among which he also included silicon, in which the melt was made up by adding to the oxide of the light metal it was desired to obtain the chloride of a strongly electropositive metal and its oxide. For the preparation of silicon the bath is composed of fused silica, strontium chloride, and strontium oxide. In the course of a systematic investigation, in which the suggested chloride had been replaced by the corresponding fluoride, M. Dodero [15] was able to obtain free silicon (together with calcium silicide) only with salts of calcium among the possible alkaline earths that could be added. In a report summarizing his work M. Dodero [16] maintains that, on the electrolysis of alkali silicates (with a suitable fluoride as flux), so-called amorphous silicon is obtained whereas it is crystalline if aluminum- or magnesium silicate is electrolysed (also with the fluorides as fluxes). According to a 1965 patent [17], pure silicon is produced on the electrolytic decomposition of silica if it is dissolved in a cryolite melt, contact with foreign substances being reduced by having the melt in a container coated with pure silica.

References:

[1] O. P. Krämer, R. Weiner, M. Fett (Die Geschichte der Galvanotechnik und die Entwicklung der galvanischen Metallüberzüge bis zur Neuzeit, Saulgau 1959, pp. 27/59). – [2] L.-N. Junot de Bussy (Inst. Sci. Math. Phys. Nat. Sect. I **21** [1853] 97/8; Jahresber. Fortschr. Chem. **1853** 335/6). – [3] L.-N. Junot de Bussy (Compt. Rend. **36** [1853] 540, 602). – [4] A. J. Balard (Compt. Rend. **41** [1855] 1069/71). – [5] G. Gore (Phil. Mag. [4] **7** [1854] 227/8; J. Prakt. Chem. **61** [1854] 447; Arch. Pharm. **130** [1854] 296).

[6] M. J. Dshems-Levy (U.S.S.R 39088 [1934] from C. **1935** II 2993). – [7] H. Sainte-Claire Deville (Compt. Rend. **39** [1854] 321/6, 326). – [8] Sainte-Claire Deville (Ann. Chim. Phys. [3] **49** [1857] 62/78, 69). – [9] F. Ullik (Sitz.-Ber. Akad. Wiss. Wien Math. Naturw. Klasse II **52** [1866] 115/7). – [10] M. Dodero (Compt. Rend. **208** [1939] 799/801).

[11] H. v. Wartenberg (Z. Anorg. Allgem. Chem. **265** [1951] 186/200, 198/9). – [12] W. Hampe (Chemiker-Ztg. **12** [1888] 841). – [13] L. Andrieux, M. Dodero (Compt. Rend. **148** [1934] 753/5). – [14] A. Grätzel v. Grätz (Ger. 58600 [1890] from Dinglers Polytech. J. **283** [1892] 124). – [15] M. Dodero (Bull. Soc. Chim. France [5] **6** [1939] 209/18).

[16] M. Dodero (Silicon Sulphur Phosphates Colloq., Münster 1954 [1955], pp. 15/20, 16). – [17] General Trustee Co. Inc. (Neth. Appl. 65-12726 [1966]; Swiss Appl. [1965] from C.A. **67** [1967] No. 17403).

1.5.2.2 Silicon from Quartz and Carbon

Alloys of iron with silicon, which are referred to as ferrosilicon and contain up to 20% Si, may be obtained in a blast furnace by the same method that J. J. Berzelius used in his research in 1810, see p. 18, the raw materials being quartz, siliceous iron ore, coke, and above all aluminiferous material for slag formation. Because of the relatively great difficulty in reducing silica, large additions of coke and high blast temperatures, that is hot operating conditions in the furnace, are necessary [1]. The earliest preparation of such alloys was carried out by Porrcel in the blast furnaces of Terre-Noire (near St. Étienne, France) in 1872 [2], but it is not possible to discuss the technically very important subject of ferrosilicon here.

Silicon-iron alloys with a 25 to 90% Si content, which were also called ferrosilicon for short, may be smelted in open electric arc furnaces, the charge consisting of quartz and coke, the quantity of which is very important for the furnace operation and voltage, together with scrap iron in the form of turnings [1]. Such alloys are thought to have been first produced in the U.S.A. by the Wilson Aluminium Company. They were prepared in Europe somewhat later by the firm of Bozel in furnaces which were patented in 1899 [2, 3] by Walther Rathenau who, having in the meantime moved from technology to politics, was assassinated in 1922. Silicon powder was obtained from such alloys, as for example in a patent by de Chalmot [4], by grinding and treatment with acids. It was found that, with a total 69% Si in the ferrosilicon, 39% of the free element was present. If crude silicon with 96 to 98% Si is to be smelted in the electric furnace addition of iron turnings to the charge is omitted and, in order to reduce the iron and aluminum content, one goes over to heating with wood charcoal or retort carbon as well as the use of quartz which is as pure as possible. Since high vaporization losses both of silica and also of metal have to be reckoned with in this process it can be operated successfully only in furnaces which are very advanced technically. In Germany a three phase 75000 kW furnace with amorphous carbon electrodes of 900 mm diameter in a triangular arrangement has proved to be particularly good [1]. It is therefore understandable that the development of silicon production from quartz and carbon is described almost exclusively in a large number of patents which cannot be discussed in detail here; see in this connection "Silicon" B, 1959, p. 11.

The first observation on the reduction of silica by carbon at high temperatures appears to have been made in 1895 by H. Moissan [5] during his experiments on the vaporization of quartz in his electric furnace. He found that, in addition to flocculent silica and solid colorless layers of silicon carbide, there were black lustrous crystals on the cover of the graphite crucible, which had been charged with quartz and carbon powder. These were mixed here and there with fused beads and he was able to identify this material as silicon. Only a little later in America E. G. Acheson, who discovered carborundum, see p. 153, observed that metallic silicon occurred in carborundum furnaces the operation of which had been suspended for technical reasons. It was thought to have been formed either through direct reduction of silica by carbon or by reduction of quartz by silicon carbide [6]. F. M. Becket [7] described what F. J. Tone of the Carborundum Company had shown to him at the beginning of the century as several hundred pounds of a substance "which consisted of a fused mixture of silicon and fire bricks". Trials to obtain the element on this way were finally initiated in 1898 and it was found that the latter reaction proceeded more smoothly and was more readily controlled [6] though it was probably more expensive than the former by which, as E. G. Acheson [8] was able to say even in 1911 "silicon is made in thousands of tons". This process remains the only one for the large-scale production of crude silicon and has since operated unchanged from the turn of the century. Very recently a purity of 99.99% Si has been achieved by using pure quartz and pure lampblack briquetted with cane sugar [9].

References:

[1] H. Hoff, O. Heerhaber, O. Meyer (in: K. Winnacker, E. Weingaertner, Chemische Technologie Metallurgie Allgemeines: Das Eisen, Hauser, München 1953, pp. 457/9). – [2] R. Fichte, H.-J. Retelsdorf (in: K. Winnacker, L. Küchler, Chemische Technologie, 3rd Ed., Vol. 6, Stahlveredler und hochschmelzende Metalle, München 1973, p. 489). – [3] W. Rathenau (Ger. 99232 [1897] from C. **1899** I 316). – [4] de Chalmot (U.S. 589415 [1897]) from F. J. Tone (Ind. Eng. Chem. **23** [1931] 1312/6, 1313). – [5] H. Moissan (Bull. Soc. Chim. France **13** [1895] 972/3).

[6] F. J. Tone (Ind. Eng. Chem. **23** [1931] 1312/6). – [7] F. M. Becket (J. Ind. Eng. Chem. **16** [1924] 147/205, 199). – [8] E. G. Acheson (Trans. Faraday Soc. **7** [1911] 217/20). – [9] J. Dietl, D. Helmreich, E. Sirtl (in: J. Grabmaier, Silicon: "Solar"-Silicon, Springer, Berlin 1981, pp. 43/107, 57/8).

1.5.2.3 Aluminothermic Preparation of Silicon

In 1897, when Hans Goldschmidt (1861–1923), the originator of the "thermite process", published this and his other aluminothermic experiments, in working out what Claude Vautin from London had encouraged him after he had abandoned his own attempts, he did not include silicon among the twenty elements for whose extraction he believed the new method to be suitable [1]. He was familiar with the earlier experiments of H. Sainte-Claire Deville [2] with aluminum. The latter had been the first to prepare this metal in large amounts and had also made alloys of it with a high silicon content. Since, however, he had also read the publications of the former assistants of the French investigator, the brothers Charles and Alexandre Tissier [3], in which they reported that alkali silicate melts coated the surface of metallic aluminum with a layer of silicon, but that in no case was the change complete, his silence is understandable, especially as in the reaction of quartz with aluminum further prerequisites which, in his opinion, determine if a "thermite reaction" has actually occurred, appeared to be lacking. He later defined such a reaction as follows: "A thermite reaction is one in which one or several metals or metallic alloys with a reducing action act on a metallic compound in such a way that the mixture, once it has been ignited at one place, continues to burn spontaneously so that the active metal is completely oxidized and a liquid slag forms, the reduced metal separating as a homogeneous regulus. If excess of the reducing metal is taken, the reduced metal should be free or practically free of the metal employed for the reduction" ["Eine Thermitreaktion ist eine solche, bei der ein oder mehrere reduzierend wirkende Metalle oder Metallegierungen auf eine Metallverbindung derartig einwirken, daß das Gemisch, an einer Stelle zur Entzündung gebracht, von selbst weiter brennt, so daß sich unter völliger Oxydation des aktiven Elementes eine flüssige Schlacke bildet und das reduzierte Metall sich als einheitlicher Regulus abscheidet; falls ein Überschuß des zu reduzierenden Metalles genommen wird, ist das reduzierte Metall frei oder praktisch frei von dem zur Reduktion verwendeten Metalle"] [4]. In the following year it became apparent that aluminum had been closely investigated as a reducing agent elsewhere. L. Franck [5] had reported his extensive investigations in which he had worked with sand and silicates and obtained silicon powder, promising to report in detail on this case, though it did not happen. In 1902 K. A. Kühn [6] was able to demonstrate how the "firework reaction" – it was thus that M. Trautz, J. D. Holtz [7] named the thermite reaction – could be brought more closely into line with the concept of its discoverer "by introducing sulfur, an aid which is not currently employed in aluminothermie" ["unter Beiziehung eines sonst in der Aluminothermie nicht gebräuchlichen Helfers, des Schwefels"]. This addition not only induces more ready combustion of the reaction mixture (though even in 1907 F. M. Perkin [8] complained that with a mixture of silica and aluminum powder it is "a very difficult matter to get this mixture to react") but also results in the formation of a more fluid slag, so that a homogeneous regulus was able to form. In 1908 A. F. Holleman [9] and H. Biltz [10] confirmed the effectiveness of sulfur additions and improved the process. They also verified the good yields and complained that (after removal of excess aluminum) only a small amount of well formed crystals was obtained. Since the thermite reaction depends on the basicity of aluminum and, as M. Trautz, J. D. Holtz [7] showed, the metalloids silicon and boron represent the limit of the elements which can be reduced by the metal, reduction is, in contrast to other thermite reactions, not quantitative but fairly incomplete. It was hardly to be expected therefore that technical production of silicon by this process would be successful. It is, however, operated on a large technical scale in the Wacker process in which special fluxing agents are added to the sand or silicates used [11].

References:

[1] H. Goldschmidt (Z. Elektrochem. **4** [1897/98] 494/9; Liebigs Ann. Chem. **301** [1898] 19/28, 19/20). – [2] H. Sainte-Claire Deville (Compt. Rend. **44** [1857] 20/1; J. Prakt. Chem. **71** [1857] 368/71). – [3] C. Tissier, A. Tissier (Compt. Rend. **43** [1856] 1187; J. Prakt Chem. **71** [1857] 76/7). – [4] H. Goldschmidt (Z. Elektrochem. **14** [1908] 558/64, 559). – [5] L. Franck (Chemiker-Ztg. **22** [1898] 236/45, 245).

[6] K. A. Kühn (Ger. 147871 [1902] from C. **1904** I 64). – [7] M. Trautz, J. D. Holtz (J. Prakt. Chem. [2] **148** [1937] 225/65, 224, 227). – [8] F. M. Perkin (Trans. Faraday Soc. **3** [1907] 115/8). –

[9] A. F. Holleman (Rec. Trav. Chim. **23** [1908] 380/3). – [10] H. Biltz (Ber. Deut. Chem. Ges. **41** [1908] 2634/45, 2635).

[11] J. Dietl, D. Helmreich, E. Sirtl (in: J. Grabmaier, Silicon: "Solar"-Silicon, Springer, Berlin 1981, pp. 43/107, 58).

1.5.3 Purification and Preparation of the Pure Element

Silicon "metal" of commerce (crude silicon) contains 93 to 98% Si and must therefore be submitted to exhaustive purification if silicon with semiconductor properties is required. This should contain as the most 10^{-3} at.% of electrically active impurities [1]. The cost of achieving this is reflected in the price. In the mid-fifties more than 2300 times the price of the raw material had to be paid for silicon of high purity [2]. The commercial grades of silicon and the limits for their impurity contents are given by J. Dietl and co-workers [1]. The prices of the raw material have fallen sharply as a result of increasing production; whereas more than 100 $ per ounce had to be paid for silicon in 1900, when it was a chemical curiosity, according to J. W. Richards [3], 10 cents a pound for silicon "metal" was a good market price in 1916. Since purification is also associated with loss of material, methods of production are sought which yield the element in a purer form. Therefore it is not surprising that finally in 1966 it was possible to state that: "Silicon is probably the purest and most perfect crystalline element to be prepared up to 1965. The impetus for the effort required came from the semiconductor industry" [4].

References:

[1] J. Dietl, D. Helmreich, E. Sirtl (in: J. Grabmaier, Silicon: "Solar"-Silicon, Springer, Berlin 1981, pp. 43/107, 48). – [2] D. W. Lyon (in: C. A. Hampel, Rare Metals Handbook, Reinhold, New York 1954, p. 380). – [3] J. W. Richards (Met. Chem. Eng. **15** [1916] 26/31, 26). – [4] W. G. Pfann (Zone Melting, 2nd Ed., Wiley, New York 1966, p. 141).

1.5.3.1 Methods of Purification

1.5.3.1.1 Acid Purification. Flotation

Particularly if larger amounts of impurities are to be removed, but also in other circumstances, purification of silicon may still be carried out today in the same way as was used in 1824 by J. J. Berzelius [1], its discoverer, namely by treatment with acids of varying concentrations including aqua regia and hydrofluoric acid. This method, which had been used by most of those who prepared the element, appears first to have been investigated in detail and quantitatively by N. P. Tucker [2] in 1927. He established that impurities are segregated in the main at grain boundaries and that under favorable circumstances a product with 99.94% Si may be obtained from it. The good results with the method which Tucker had developed were confirmed in 1939 by A. B. Kinzel, T. R. Cunningham [3]. F. M. Becket [4] had already pointed out in 1919 that during acid purification the treatment with hydrofluoric acid had to be undertaken with sufficient care if larger silicon losses were to be avoided. How purification on the basis of the foregoing investigations was carried out in industry (British General Electric Company) in 1943 was described by H. Guyford Stever [5] with the results of a test run with 7 to 8 lb. of silicon. Eight stages are required: "1. Commercial silicon powder of about 98 percent purity is crushed so that it will pass through a 200-mesh screen. Then 750 g of the powder are covered with water and HCl is added. – 2. After the first violent reaction, an excess of HCl is added together with some HNO_3. The mixture is digested for 24 h. – 3. The mixture is diluted, filtered, and washed thoroughly with distilled water. – 4. The powder is placed in a platinum pail and a liter of distilled water is added. – 5. To the mixture is added about 300 cc of H_2SO_4 and about 200 cc of concentrated HF (about 40 percent). The HF is added slowly. The mixture is then warmed for 4 h, and further similar quantities of H_2SO_4 and concentrated HF are added. – 6. The mixture is evaporated to fuming; the residue is washed thoroughly with distilled water and then filtered. – 7. The material is treated with 700 cc of concentrated HF and left standing for 12 h. – 8. It is then filtered, washed thoroughly and dried. After undergoing this purification process, the silicon is analyzed and found to be almost spectroscopically pure".

Tucker's observation of the layer-like occurrence of the impurities in crude silicon suggests the idea of flotation tests with the finely divided material, H. P. Hood [6] being the first to propose this in 1937. Starting from a raw material with 40% Si (50% Al and 5% each of Fe and Ti) M. M. Striplin [7] obtained an enrichment to 96.7% Si when he ground the material so that half passed a sieve with 100 meshes per cm^2. The coarser material was rich in aluminum, the finer being separated in a mixture of tetrabromoethane and nitrobenzene with D = 2.43.

References:

[1] J. J. Berzelius (Ann. Physik. Chem. [2] **1** [1824] 160/230, 209/11). – [2] N. P. Tucker (J. Iron Steel Inst. [London] **115** [1927] 412/6). – [3] A. B. Kinzel, T. R. Cunningham (Trans. AIME Eng. **137** [1940] 425/9). – [4] F. M. Becket (U.S. 1386227 [1919/21] from C.A. **1921** 1131). – [5] H. G. Stever (in: H. C. Torrey, C. A. Whitmer, Crystal Rectifiers, 1st Ed., 3rd Reprint, New York – London 1948, p. 303).

[6] H. P. Hood (Fr. 48995 [1937/38] from C. **1939** 1029; supplement to Fr. 805111 [1936]; C.A. **1937** 3656). – [7] M. M. Striplin (U.S. 2469418 [1946/49] from C. **1950** I 1136).

1.5.3.1.2 Purification by Distillation

Purification of 99% technical silicon by high vacuum distillation (from beryllium oxide crucibles) was attempted by W. Kroll [1] in 1934. He found great difficulties in this, in the first place on account of the high gas content of the silicon and the considerable volatility of silicon dioxide, which was always present as an impurity and, secondly, because the element was able to absorb the high frequency flux on which heating depended only after preheating to about 1000°C. Part of the condensate occurred with the most diverse color tones, ranging from ocher-yellow to deep brown and blue-black. Distilled silicon in the dendritic form had a blue-violet color. Furthermore, "it was not possible to carry out a chemical analysis of the deposit as, in a very short time, it took up large amounts of water like silica gel. A metallographic study of the residue showed a high enrichment in iron".

Twenty years earlier E. Tiede, E. Birnbräuer [2] had already undertaken experiments on the distillation of the element, starting with amorphous silicon. In their experiments with carbon boats they obtained a transparent deposit on the bell jar even at a bright red heat and, with rising temperature, this became opaque and assumed a graphitic luster. After melting, the element reacted with the material of the boat. The silicon carbide formed decomposed at a still higher temperature, giving a steel-like condensate which "could be converted to SiO_2 by igniting over the blowpipe". They also observed distillation in melting tests in an electron beam furnace in which a crucible was not employed. The deposits, the ready oxidizability of which was striking, had a beautiful graphitic luster. In spite of the poor success of these experiments, E. Enk [3] stated that fractional distillation leads to an increase in purity without, however, quoting literature to substantiate this.

Apparently, therefore, successful purification by distillation is not easily achieved in the case of silicon. A. Smakula [4], however, pointed out that when silicon is heated or melted the impurities evaporate more readily than the element, and this had also been noticed already by E. Tiede, E. Birnbräuer [2]. This effect may perhaps be significant in the preparation of single crystals by the zone melting process. S. E. Bradshaw, A. I. Mlavsky [5] have proposed an equation for the "concentration decrease per second" for numerous impurities in such cases.

References:

[1] W. Kroll (Metallwirtsch. Metallwiss. Metalltech. **13** [1934] 725/31, 729; Metal-Ind. [London] **47** [1935] 3/6). – [2] E. Tiede, E. Birnbräuer (Z. Anorg. Allgem. Chem. **87** [1914] 129/68, 144/5). – [3] E. Enk (Ullmanns Encykl. Tech. Chem. 3rd Ed. **15** [1964] 688). – [4] A. Smakula (Einkristalle, Berlin – Göttingen – Heidelberg – München 1962, p. 288). – [5] S. E. Bradshaw, A. I. Mlavsky (J. Electron. **2** [1956/57] 134/44).

1.5.3.1.3 Purification by a Melting Process

Simple Melting. Purification of crude silicon by simply remelting under a layer of silicate, which should free the element from iron-, aluminum-, and calcium compounds, was attempted as early as 1912 by C. J. Brockbank [1], but 30 years later F. Seitz [2] pointed out that, in the experience of many experts, the danger of taking up fresh impurities from the crucible material should never be underestimated especially with prolonged melting. In 1951 H. v. Wartenberg [3] confirmed, on the basis of his own experiments, that all attempts to purify silicon by remelting were futile for this reason. He even asserted that there was only one way to obtain pure silicon by melting, namely to fuse a larger quantity of pure silicon powder by means of a burning glass or mirror in an evacuated quartz flask in such a way that the regulus comes into contact only with silicon powder, and the wall, which remains cold, does not give off any gas. This had been done in 1909 by Alfred Stock, H. Heynemann [4] with a burning glass of 40 cm diameter and a small regulus had been obtained, but nothing was said about its purity. In spite of the fact that success seems unattainable, J. H. Scaff [5] was granted a patent in 1941 according to which even simple melting in a quartz crucible, followed by heating above the melting point and very slow cooling in an oxygen-free atmosphere, enabled him to make very pure material with special electrical properties from 99% silicon. J. H. Scaff later reported fully on the latter [6], dealing with the discovery of the fact that silicon can occur as a p- and as an n-semiconductor – two terms used for the first time in connection with this discovery – and with the observation of the photovoltaic effect in the case of this element. According to C. E. Ransley, J. W. Ryde, S. V. Williams [24], the disadvantages which can accompany the use of quartz crucibles may be avoided if they are replaced by a beryllium oxide crucible, or one made of metallic tantalum which has previously been heated to redness in an atmosphere of naphthalene to form a layer of tantalum carbide, which is chemically very resistant. The authors say nothing in their patent about the degree of purity attained.

The Zone Melting Process. The technique of zone melting, which was worked out by W. G. Pfann [7] and published in 1952, consists in melting a narrow region in a polycrystalline metal rod which is at right angles to its length and then conducting the "zone" along the rod sufficiently slowly for recrystallization to occur, when single crystals mostly result and impurities are transferred from the melt to the end of the rod. After repeating this procedure a number of times single crystals of high purity may be obtained. According to K. Reuschel [8], W. G. Pfann arrived at his process independently of early experiments on growing single crystals with a moving hot zone which were made by P. Kapitza [9] in 1928 and by E. N. da C. Andrade, R. Roscoe [10] in 1937. It may be mentioned at this point that the first observations on the peculiarities associated with the incorporation of impurities in single crystals, later termed the Coring effect, were made by P. W. Bridgman [11] in 1925, but it was not until more than twenty years later that they were investigated more closely by R. H. McFee [12].

The technique, which had been tested on lower melting elements, was applied to silicon simultaneously in various places in the second half of 1952, without, however, using a crucible, because of the element's great reactivity. According to K. Reuschel [8], patent priority lies with A. C. Theurer [13] and publication priority with P. H. Keck, M. J. E. Golay [14]. The first description of unexceptionable rod-like crystal was provided by R. Emeis [15]. A detailed description of the apparatus is given, for instance, by P. H. Keck and his co-workers [16]. According to this a silicon rod without a sheath (obtained by casting or sintering in vacuum) is held vertically in an inert atmosphere (He or Ar) and submitted to zone melting, the melting zone being kept so narrow that the (quite high) surface tension of the liquid phase holds the rod together. After several passages of the "zone" the result was very satisfactory. In spite of this, zone melting in very thin-walled quartz vessels has been investigated in industrial laboratories. The advantage of this lies not only in the greater simplicity of the apparatus, but several "zones" can be passed simultaneously through the rod. This was carried out successfully [17] and H. Schildknecht [18] found that wetting of the walls of the vessel could be avoided if they were coated with a fluoride mixture. W. Schreiter [19] recommended silicon nitride boats for this operation, as they are not wetted by molten silicon. Using quartz vessels, single crystals may be prepared which, while free from numerous impurities, such as aluminum, antimony, and phosphorus, for example, always still contain oxygen [20]. Even more troublesome was the fact that the element boron is not removed from the silicon at all by simple zone melting and its use in the preparation of diodes, transistors, rectifiers, and solar batteries is therefore not possible [21]. H. C. Theurer was successful in oxidizing boron in the molten zone in an atmosphere of moist hydrogen, and the oxidation product, which was insoluble in molten silicon, either vaporized off or was transferred to the end of the rod with the moving zone [22]. E. Buehler [23] described a suitable apparatus with which silicon with less than 1 ppb of electrically active impurities is obtainable, which corresponds with a specific electrical resistance of 16000 Ω ·cm. To achieve this 67 zone passes were necessary.

References:

[1] C. J. Brockbank (U.S. 1180968 [1912/16] from C.A. **1916** 1698). – [2] F. Seitz (PB-2628 [1942] 1/10). – [3] H. v. Wartenberg (Z. Anorg. Allgem. Chem. **265** [1951] 186/200, 191/3). – [4] A. Stock, H. Heynemann (Ber. Deut. Chem. Ges. **42** [1909] 2863/6). – [5] J. H. Scaff (U.S. 2402582 [1942/46] from C. **1947** I 376).

[6] J. H. Scaff (Met. Trans. **1** [1970] 561/73). – [7] W. G. Pfann (Trans. AIME **194** [1952] 747/53). – [8] K. Reuschel (Siemens-Z. **41** [1967] 669/74, 670, footnote 1). – [9] P. Kapitza (Proc. Roy. Soc. [London] A **119** [1928] 358/86, 360). – [10] E. N. da C. Andrade, R. Roscoe (Proc. Phys. Soc. **49** [1937] 152/77, 154).

[11] P. W. Bridgman (Proc. Am. Acad. Sci. **60** [1925] 305/83). – [12] R. H. McFee (J. Chem. Phys. **15** [1947] 856/61). – [13] A. C. Theurer (Ger. 1014332 [1952] from C. **1958** 13603). – [14] P. H. Keck, M. J. E. Golay (Phys. Rev. [2] **89** [1953] 1297). – [15] R. Emeis (Z. Naturforsch. **9a** [1954] 67).

[16] P. H. Keck, W. van Horn, J. Soled, A. MacDonald (Rev. Sci. Instr. **25** [1954] 331/4). – [17] D. Hartman, P. L. Ostapkovich (Metal Progr. **70** No. 4 [1956] 100/3). – [18] H. Schildknecht (Zonenschmelzen, Weinheim 1964, pp. 146/8). – [19] W. Schreiter (Seltene Metalle, Vol. 2, Leipzig 1961, p. 412). – [20] E. A. Taft, F. H. Horn (J. Am. Electrochem. Soc. **105** [1958] 81/3).

[21] R. S. Aries, A. P. Sachs (Dechema Monograph. No. 477/502 [1959] 83/8, 84). – [22] H. C. Theurer (Trans. AIME **206** [1956] 1316 from [18]). – [23] E. Buehler (Rev. Sci. Instr. **28** [1957] 453/60). – [24] C. E. Ransley, J. W. Ryde, S. V. Williams (U.S. 2419966 [1942/47] from C. **1947** I 1106).

1.5.3.2 Methods of Preparation

1.5.3.2.1 Silicon Tetraiodide as Starting Material

Many high-melting metals may be successfully prepared in a pure state by the "growth method", which was described by A. E. van Arkel and J. H. de Boer [1] in 1925. In it, formation of the element is affected by decomposition of gaseous compounds on an incandescent wire, the corresponding iodides being usually employed because of their ready decomposition. The process is normally operated in such a way that circulation of the required metal is achieved through continuous regeneration of the volatile iodide from impure metal inserted in the same apparatus, impurities remaining behind. This is therefore often spoken of as a "chemical transport reaction".

It was obvious to the discoverers that their new method might also be applied to silicon, but A. E. van Arkel [2] encountered difficulties, as silicon is not a metal and thin silicon rods obtained on the hot wires of foreign material are not ductile, and cannot therefore be used in making hot filaments for the next experiment. As H. v. Wartenberg [3] showed, a carbon filament is completely ruled out for the initial run as silicon carbide is apparently formed at once. Furthermore, silicon tetraiodide cannot be used for this process, as its regeneration from the impure silicon which has been introduced proceeds too slowly. In 1954 F. B. Litton, H. C. Andersen [4] showed, however, that by dispensing with the regeneration of the iodide in the same vessel and going over to an intermittent operation in which silicon tetraiodide, which has previously been very carefully fractionated, is carried in argon, high purity silicon may be obtained in good yield. They also found, however, that elements of the III and V group of the periodic system, which occur as impurities and are critical for the semiconductor properties of the product, are likewise carried over, so that purification of the iodide employed, is of the highest importance. The influence of impurities was investigated five years later by L. V. McCarty [5] with tetraiodosilane which had been purified very carefully by the most varied methods. In this work he went over from a hot tantalum wire, which was used initially, to a thin quartz tube heated externally and obtained products with a specific electrical resistance, which serves as a measure of the purity, which was three powers of ten higher than appeared to have been attained before. E. Enk [6] considered this form of the growth method to be suitable only for operating on a small scale, and emphasized that the silicon crystals obtained must be submitted to further purification with hydrofluoric acid before their conversion to single crystals.

The mechanism of this "transport reaction", in which a gaseous silicon diiodide plays a major rôle, was investigated very fully by H. Schäfer, B. Morcher [7] in 1957 and they were able to clarify why the first attempts to apply this method to silicon had run into difficulties. In the same year G. Szekely [8] changed the process, so as to be able to operate at atmospheric pressure, by replacing the purely thermal decomposition of silicon tetraiodide by its reduction with hydrogen at a hot surface according to the equation $SiI_4 + 2H_2 \rightarrow Si + 4HI$. He passed the corresponding mixture of high purity gases at normal pressure through a graphite tube held at temperatures between 900 and 1000°C, the tube being lined with tantalum foil in later experiments. It was possible in this way to obtain the element in compact crystalline layers with the degree of purity necessary for semiconductor purposes.

References:

[1] A. E. van Arkel, J. H. de Boer (Z. Anorg. Allgem. Chem. **148** [1925] 345/50). – [2] A. E. van Arkel (Reine Metalle: Herstellung – Eigenschaften – Verwendung, Springer, Berlin 1939, p. 479). – [3] H. v. Wartenberg (Z. Anorg. Allgem. Chem. **265** [1951] 186/200, 193). – [4] F. B. Litton, H. C. Andersen (J. Am. Electrochem. Soc. **101** [1954] 287/92). – [5] L. V. McCarty (J. Am. Electrochem. Soc. **106** [1959] 1036/42).

[6] E. Enk (Ullmanns Encykl. Tech. Chem. 3rd Ed. **15** [1964] 689). – [7] H. Schäfer, B. Morcher (Z. Anorg. Allgem. Chem. **290** [1957] 279/91). – [8] G. Szekely (J. Am. Electrochem. Soc. **104** [1957] 663/7).

1.5.3.2.2 Silicon Tetrabromide as Starting Material

"Transport phenomena" for silicon in the thermal decomposition of tetrabromosilane were studied by H. Schäfer, B. Morcher [1] in 1957 in the same way as has been described for the iodo compound in the previous section, and analogous results were obtained. The reducing pyrolysis of high purity silicon tetrabromide in an atmosphere of pure hydrogen was used in the same year by R. C. Sangster, E. F. Maverick, M. L. Croutch [2] for obtaining compact silicon. They employed thin needle-shaped silicon crystals obtained in another way as the filaments on which growth occurred.

Another process suitable for the "transport" of silicon, which can be represented by the equation $Si + 3SiBr_4 + 2H_2 \rightleftharpoons 4SiHBr_3$, was proposed in 1978 by L. M. Woerner, E. B. Moore [3] (of the firm J. C. Schumacher Co.). Powdered crude silicon is treated in a fluidized bed reactor with a mixture of the two highly purified gases at 650°C. The reaction product passing over is purified, and then submitted to disproportionation at 800°C, which yields highly pure silicon. The gaseous products are transferred back to the cycle (Schumacher Process) [4].

References:

[1] H. Schäfer, B. Morcher (Z. Anorg. Allgem. Chem. **290** [1957] 279/91). – [2] R. C. Sangster, E. F. Maverick, M. L. Croutch (J. Am. Electrochem. Soc. **104** [1957] 317/9). – [3] L. M. Woerner, E. B. Moore (Ger. Offen. 2919086 [1980] from C.A. **92** [1980] No. 218730). – [4] J. Dietl, D. Helmreich, E. Sirtl (in: J. Grabmaier, Silicon, Springer, Berlin 1981, pp. 43/107, 54).

1.5.3.2.3 Silicon Tetrachloride as Starting Material

The fact that the preparation of pure silicon by the growth method using silicon tetrachloride alone offers no prospect of success because the thermal cleavage is very small, even at 1600°C, and the resulting formation of dichloride interferes, was recognized not only by H. Schäfer [1] in 1954: in 1909 J. N. Pring, W. Fielding [2] had already been unsuccessful in such experiments. When, however, they added hydrogen – as F. Wöhler had already done in 1843 when he observed crystalline silicon, though he did not publish his results, see p. 41, – they first obtained a thin layer of silicon carbide on the very hot carbon filament, and a deposit of hard crystalline silicon then formed on this. At 1055°C this was transformed completely into carbide. In 1927, R. Hölbing [3] worked with a hydrogen-silicon tetrachloride vapor mixture at low filament temperatures and obtained from both carbon and tungsten filaments thin compact silicon rods, which, however, it was impossible to use further as filaments in the same process because they could not be worked mechanically. The rods reached a thickness of about 4 mm with a purity of 99.92% Si. Greater thickness could not be obtained as decomposition of the gaseous mixture requires a rod surface temperature of at least 900°C, while the internal temperature must not reach 1200°C, which is the formation temperature for silicon carbide. When a tungsten filament was employed, a silicide was formed at once and the purity of the rod reached only 93.5% Si. In 1951, H. v. Wartenberg [4] found that this gas mixture was unsuitable for preparing high-purity silicon, not only on the grounds mentioned already, but particularly because there are stringent requirements for the purity of the hydrogen, which are met only with difficulty. In 1953, H. Schäfer, J. Nickl [6] were able to show that transport of silicon in the pyrolysis of tetrachlorosilane can take place without the presence of hydrogen in accordance with the equation $Si+SiCl_4 \rightleftharpoons 2SiCl_2$, as indeed L. Troost, P. Hautefeuille [5] had stated in 1876. After a detailed study of the reaction, they pointed out its possible significance for preparing very pure silicon. Three years later, H. Schäfer, H. Jacob, K. Etzel [7] investigated the reaction again and showed the limitations of the method.

While the thermal decomposition of tetrachlorosilane in presence of hydrogen has not led to technically useful production of the pure element, P. C. van der Linden, J. de Jonge [8] were able to show in 1959 that pyrolysis of silicochloroform, $SiHCl_3$, in a hydrogen atmosphere on a hot tantalum filament can yield polycrystalline silicon rods with diameters up to 15 mm. The reaction that occurs is certainly more complicated than that represented by the equation $2SiHCl_3 \rightarrow Si+SiCl_4+2HCl$, which describes its end products. Single crystals of silicon may be obtained from the polycrystalline rods by zone melting and these, from their electrical resistance, are estimated to contain only 10^{13} foreign atoms per unit volume. It must not, however, be forgotten that the earliest reference to the possibility of splitting out silicon from silicon-halogen-hydrogen compounds is to be found in a patent application by the Bosnische Elektrizitäts-A.-G. in Vienna for 1915 [9]. The claim was there put forward of being able to obtain "thick coatings of silicon or silicon alloys" on hot metal objects from gaseous silicon compounds named, "if necessary in presence of suitable substances which bring about or promote the separation of silicon". It was in this way that the technical usefulness of the thermal decompositon, which H. Buff, F. Wöhler [10] had observed in 1857 when they discovered the compound, was first established. The discoverers found the decomposition products to be tetrachloride and hydrogen chloride, in addition to elementary silicon whereas, almost 50 years later, O. Ruff, K. Albert [11], in checking the work, were able to detect only the tetrachloride and hydrogen as gaseous products. In 1923, A. Stock, F. Zeidler [12], in studying the temperature dependence of the reaction, were able to show that the apparently contradictory results were in fact both correct. The earlier observations were made at about 400°C, whereas the more recent ones were made at a bright red heat.

There was a patent application in 1954 by F. Bischoff [13] for the preparation of high-purity silicon by thermal decomposition of silicochloroform and what was known as the Siemens-C process based on it was operated on an industrial scale. The silicochloroform was obtained from technical silicon and hydrogen chloride and was carefully purified by fractional distillation [14].

The methods for preparing pure silicon which have been described are based on the thermal decomposition of silicon chlorides, sometimes with addition of hydrogen, and secure a high degree of purity. There are, however, two further processes which have become even more significant and in which silicon tetrachloride is reduced by metallic elements on the basis of earlier observations. Thus, in 1949, D. W. Lyon, C. M. Olson, E. D. Lewis [15] reverted to a method described thus almost ninety years previously by N. Beketoff [16]: "I allowed silicon chloride vapor, which was carried in a stream of hydrogen, to act on zinc vapor in a porcelain tube. Even the first experiment was completely successful. The inner wall of the tube became covered with lustrous crystals of silicon, and in the cooler parts of the tube a quantity of zinc separated which was mixed with flat graphite-like crystals of silicon which were several millimeters long" ["Ich ließ Chlorsiliciumdampf, welchen ein Wasserstoffstrom zuführte, auf

Zinkdampf in einer Porcellanröhre einwirken; schon der erste Versuch gelang vollkommen. Die inneren Wandungen der Röhre bedeckten sich mit glänzenden Krystallen von Silicium, und gegen den kälteren Theil der Röhre hin schied sich eine Zinkmasse ab, welche von flachen, mehrere Millimeter langen Krystallen von graphitartigem Silicium durchsetzt war"]. The process was worked out and somewhat modified by American investigators, presumably during the second world war. Starting with 86 kg of the purest tetrachlorosilane and 65 to 68 kg of 99.99% zinc in a quartz apparatus at 950°C, with pure nitrogen as transport gas, and a time for the run of 30 h, they obtained 7 kg of very pure silicon [15]. The operation is carried out today under the name of the Batelle process on a technical scale. The gas phase reaction takes place in a fluidized bed reactor in which the temperature does not fall below 918°C because of the possibility of zinc condensation. It should also not exceed 927°C, as otherwise formation of $SiCl_2$ will increase markedly. The size of the Si particles obtained lies between 500 and 1000 μm [14].

The second method for preparing pure silicon, in which a metal is introduced as the reducing agent, is the so-called Westinghouse process [14]. Silicon chloride and sodium vapor are blown into the plasma of an arc burning in a hydrogen-argon atmosphere at about 2000 K. The temperature of the furnace is controlled so that the silicon can be separated as a liquid in a centrifugal separator from the sodium chloride produced, which is gaseous.

References:

[1] H. Schäfer (Silicon Sulphur Phosphates Colloq., Münster 1954 [1955], pp. 24/8, 25). – [2] J. N. Pring, W. Fielding (J. Chem. Soc. **95** [1909] 1497/506, 1501/2). – [3] R. Hölbing (Z. Angew. Chem. **40** [1927] 655/9). – [4] H. v. Wartenberg (Z. Anorg. Allgem. Chem. **265** [1951] 186/200, 193). – [5] L. Troost, P. Hautefeuille (Ann. Chim. Phys. [5] **7** [1876] 452/76).

[6] H. Schäfer, J. Nickl (Z. Anorg. Allgem. Chem. **274** [1953] 250/64). – [7] H. Schäfer, H. Jacob, K. Etzel (Z. Anorg. Allgem. Chem. **286** [1956] 27/41). – [8] P. C. van der Linden, J. de Jonge (Rec. Trav. Chim. **78** [1959] 962/6). – [9] Bosnische Elektrizitäts-A.-G. (Ger. 302305 [1915/17] from C. **1918** I 323). – [10] H. Buff, F. Wöhler (Liebigs Ann. Chem. **104** [1857] 94/109, 94/9).

[11] O. Ruff, K. Albert (Ber. Deut. Chem. Ges. **38** [1905] 2222/43, 2230/1). – [12] A. Stock, F. Zeidler (Ber. Deut. Chem. Ges. **56** [1923] 986/97, 988). – [13] F. Bischoff (Ger. 1102117 [1954] from C.A. **56** [1962] 6901). – [14] J. Dietl, D. Helmreich, E. Sirtl (in: J. Grabmaier, Silicon, Springer, Berlin 1981, pp. 43/107, 49/54). – [15] D. W. Lyon, C. M. Olson, E. D. Lewis (Trans. Am. Electrochem. Soc. **96** [1949] 359/63).

[16] N. Beketoff (Liebigs Ann. Chem. **110** [1859] 374/6; Bull. Soc. Chim. Paris Bull. Seances **1858/62** 22/4).

1.5.3.2.4 Silicon Fluoride as Starting Material

As gaseous silicon tetrafluoride occurs as an unwanted byproduct of the preparation of superphosphates from natural phosphates, it was of interest to render this useful for the production of silicon. A. Sanjurjo, L. Nanis, K. Sancier, B. Bartlett, V. Kapur [1] at the Stanford Research Institute attempted this by first preparing a solution of fluorosilicic acid, H_2SiF_6, precipitating the sodium salt, decomposing it thermally to regenerate the tetrafluoride and reducing the latter at 650°C by sodium metal in the form of fine droplets in a suitable reactor. The heat of formation of silicon suffices to maintain the temperature constant. The reaction product is separated by dissolving sodium fluoride in acidified water, the element being left as a brown crystalline powder with grain sizes up to 150 μm. Separation may also be effected by melting the reaction product [2], when two immiscible liquids result which are readily separated, and by a sort of liquid-liquid extraction unwanted impurities may also be removed, so that very pure silicon may be obtained. A further method, known as the Moturola process, uses the interaction of silicon tetrafluoride, which has previously been purified, and impure silicon at 1100°C, which results in the formation of gaseous SiF_2. This separates in the cold as solid polymeric $(SiF_2)_n$, the final step consisting of the decomposition of the polymer at 250°C, when very pure silicon in the powder form and silicon tetrafluoride result. The latter is recycled [2].

References:

[1] A. Sanjurjo, L. Nanis, K. Sancier, B. Bartlett, V. Kapur (J. Am. Electrochem. Soc. **128** [1981] 179/84). – [2] J. Dietl, D. Helmreich, E. Sirtl (in: J. Grabmaier, Silicon, Springer, Berlin 1981, pp. 43/107, 51/4).

1.5.3.2.5 Silicon Hydride as Starting Material

Monosilane has also been used for preparing very pure silicon. In a method proposed by A. Böttcher [1], a gas discharge is used for "separating out elements with a metallic character" (silicon being one of them) from their volatile compounds. Glow discharges at low pressures were employed, the metals resulting in the form of coatings on the electrodes. Later, E. Amberger, E. Wiberg [2] studied the decomposition of monosilane-hydrogen mixtures not only in the glow discharge but also in arc discharges and particularly in what they termed the "flame arc discharge" (Flammenbogenentladung). They found the type of discharge to be of critical significance for the result. Brown amorphous silicon was obtained with a glow discharge while, with a pure arc discharge, the molten phase resulted, though there was a risk of dripping from the electrode and of vaporization. With the flame arc discharge controllable growth occurred, leading to the formation of polycrystalline rods with a specific electrical conductivity of 10^{-2} to $10^{-3}\,\Omega^{-1}\cdot cm^{-1}$. The homogeneous gas phase decomposition of monosilane was applied for the production of high-purity silicon in the so-called Union Carbide process, which was claimed to be cheaper (than other methods). The very pure monosilane required was obtained by disproportionation of dichloromonosilane, SiH_2Cl_2, on a catalyst at 60°C and 3 atm pressure. The tetrachlorosilane produced at the same time reacted with silicon obtained metallurgically, hydrogen and a catalyst in a fluidized bed at 500°C and 30 atm pressure to give silicochloroform. From this, dichloromonosilane was obtained by disproportionation in presence of a tertiary amine resin, at 55°C and 3 atm pressure, the chlorosilane obtained at the same time being returned to the process. To get 1 kg of very pure silicon, about one thousand times this weight of $SiCl_4$ must be in circulation. As the process yields silicon in the form of quite small particles, these may be used as seed for further decomposition in a fluidized bed until the required size is reached, unless it is thought preferable to compact the fine material by sintering or fusion [3].

References:

[1] A. Böttcher (Ger. 863997 [1951/53] from C. **1953** 6349). – [2] E. Amberger, E. Wiberg (Diss. Univ. München 1955, pp. 1/82). – [3] J. Dietl, D. Helmreich, E. Sirtl (in: J. Grabmaier, Silicon, Springer, Berlin 1981, pp. 43/107, 51/4).

1.5.3.2.6 Quartz as Starting Material

Recently attempts have been made to prepare pure silicon by the direct reduction of quartz by carbon, a process which appears much simpler than those described above. According to a patent by L. P. Hunt, V. D. Dosay [1] of the Dow Corning Corporation use may be made of the fact that natural quartz deposits which are frequently very extensive, are also often sufficiently pure to serve as a source of raw material. Using a pure soot, which has been briquetted by means of sucrose, and operating in an electric arc furnace, silicon with a purity of 99.99% may be prepared. The chief impurities, aluminum and iron, occur in amounts of 50 to 100 ppm and less than 10 ppm of each of the remaining impurities, such as boron, phosphorus and certain metals is present [2].

Reduction of quartz with aluminum, as was shown on p. 49, does not lead to pure silicon. As E. Vigouroux [3] found in 1899, the reducing agent present in the product, which is always quite large in amount, may be removed by chlorination, as reaction between chlorine and silicon sets in at a higher temperature than that with aluminum. Surprisingly, some patents have been granted for this process [2].

References:

[1] L. P. Hunt, V. D. Dosay (Belg. 879715 [1980] from C.A. **94** [1981] No. 124648). – [2] J. Dietl, D. Helmreich, E. Sirtl (in: J. Grabmaier, Silicon, Springer, Berlin 1981, pp. 43/107, 58). – [3] E. Vigouroux (Compt. Rend. **129** [1899] 334/5).

1.5.4 Crystallography. Modifications

Because of the quite different appearance of elementary silicon, depending on the method of preparation, it was believed for a long time that the occurrence of various modifications had also to be inferred in the case of this element. Thus in 1844 J. J. Berzelius [1] in his major publication entitled "*Allotropie einfacher Körper*" assumed the existence of two different modifications of the element, one amorphous and one crystalline: "Kiesel [that is the element silicon] displays the closest similarity to carbon with respect to both of the first allotropic states of the latter [these were, according to Berzelius, C_α = wood charcoal, which was held to be amorphous and C_β = graphite]. Si_α results when kiesel is prepared by reduction with potassium . . . while Si_β is produced if the first one is subjected to an intense red heat, kiesel needing a less intense red heat for this transformation than carbon" ["Der Kiesel [das ist das Element Silicium] bietet die nächste Ähnlichkeit mit dem Kohlenstoff in Rücksicht auf die beiden ersten allotropischen Zustände des letztgenannten dar [das sind nach Berzelius C_α = die für amorph gehaltene Holzkohle, und C_β = der Graphit]. Si_α entsteht, wenn man den Kiesel durch Reduktion mit Kalium darstellt . . . Si_β entsteht, wenn man den ersteren einer starken Glühhitze aussetzt, wobei der Kiesel einer weniger starken Glühhitze zu diesem Übergang bedarf als die Kohle"]. It must be pointed out that in 1904 H. Moissan, F. Siemens [2] still believed that they could confirm Berzelius' ideas with a "silicon modification soluble in hydrofluoric acid", see p. 61.

Eleven years after Berzelius' differentiation of the two forms, H. Sainte-Claire Deville [3] put forward the view that the similarity between the two elements extended still further and that silicon, like carbon, occurred in three modifications. The first was the "silicon of Berzelius" ["Silicium nach Berzelius"], corresponding with ordinary carbon. Secondly there was a "graphitic silicon" ["graphitoides Silicium"], the counterpart of graphite, which could be obtained under the same conditions as artificial graphite. Finally came a "crystalline silicon" ["kristallisiertes Silicium"] which he himself had discovered and which was the analog of diamond, as crystallographic investigations by H. de Sénarmont [4] showed. What exactly he understood by "graphitic silicon" he first made clear two years later [5]. Initially he observed this modification of the element during his electrolytic preparation of aluminum in 1855 [6], when silicon derived from the carbon electrodes and which formed hexagonal lamellae exactly like graphite was found on dissolving the metal in acids. The best way of obtaining them is during the preparation of crystalline silicon as described by F. Wöhler [7], see p. 42, where it occurs as well. H. Sainte-Claire Deville's view appeared to be confirmed by Richter's observations [8], in which the element was obtained in graphite-like platelets on dissolving cast iron in dilute hydrochloric acid, and also by T. L. Phipson's claim [9] that silicon occurred in the free as well as in the bound state in iron and could be isolated from this material. C. A. Winkler [10] also reported that diamond-like silicon could be obtained from silicon-rich aluminum after exposing it to an intense white heat and subsequently cooling the melt rapidly and removing the aluminum. Finally, H. Kopp [11] observed somewhat different values for the specific heat of graphitic and crystalline silicon. In 1866, however, the mineralogist W. H. Miller [12] questioned the existence of two different crystalline modifications of the element after he had made precise measurements on crystals of the two types. H. Sainte-Claire Deville never reverted to the idea of a "graphitic modification" subsequently and only occasionally was the readily obtainable black flaky form spoken of as "graphitic" by other authors [13].

In 1910 claims appeared in the literature that massive fused silicon possessed allotropic modifications. L. Baraduc-Muller [14] concluded on the basis of anomalies in the coefficient of thermal expansion that there was a stable modification between 700 and 1000°C capable of existing in presence of small amounts of foreign metals, while J. Königsberger, K. Schilling [15] found from measurements of the temperature dependence of electrical resistance that there were two regions of discontinuity, one at 250 and the other at 430°C, which they attributed to polymeric transitions. Two years earlier F. G. Wick [16] had been unable to obtain any indication of a transition in the course of measurements of various electrical properties of the element and J. Königsberger [17], when he later repeated the resistance measurements, was likewise unable to detect the lower lying of the two discontinuities again. Later, more accurate work of this sort, which included thermal analysis, failed to confirm the existence of a transition point up to 1300°C [18] so that in 1936 M. C. Neuburger [19] came to the conclusion in his report on *Die Allotropie der chemischen Elemente* that "no allotropic modifications of silicon exist". Confirmation of this statement based on measurements of coefficients of thermal expansion made on a compact 99.85% silicon bar was published in 1939 by Gonser, Seybolt [20].

H. Stöhr, W. Klemm [21] referred to the fact that elementary silicon has good crystallizing properties and should therefore form relatively large crystals. The habit and size of the crystals, which belong to the cubic system, do, however, depend on the method of preparation and rate of cooling, while other properties, such as chemical reactivity may also vary widely. In 1904 H. Moissan, F. Siemens [22] believed they had obtained a crystalline "modification" of silicon, which was readily soluble in hydrofluoric acid, by dissolving "amorphous" silicon in a silver melt. After crystallization of the regulus and removal of the silver, finely crystallized silicon resulted which underwent an extraordinarily high loss in weight when treated with the acid mentioned. A few years later P. Lebeau [23] and M. Philips [24] were able to confirm these observations by etching experiments on polished sections of copper-silicon melts. In 1912 W. Manchot [25] in collaboration with Heinrich observed that the criterion for forming this hydrofluoric acid-soluble "modification" was the quenching of the melt. They also noted that this form of silicon could even be obtained from solutions in aluminum. In the following year W. Manchot, H. Funk [26] confirmed that the element isolated from Cu-Si melts also exhibited the same phenomenon and were able to show at the same time that solubility in hydrofluoric acid was a general property of crystalline silicon, which was more or less readily observable, depending on the crystal size, so that one could not speak of a special modification.

Silicon crystals obtained from silicon-containing metal melts and by subsequent removal of the alloying metal by suitable means all consist of the normal cubic modification, though they differ in shape and appearance according to the solvent. From *aluminum melts* crystallization takes place in the form of small platelets with a metallic luster which resemble platinum filings

[27] while fine needles crystallize from *zinc melts* together with dendritic crystals resembling α-iron [28]. *Silver melts* yield minute lamellae which are very thin and occasionally even transparent, when they are yellow [29]. If silicon-containing silver melts are very rapidly cooled, however, the modification described previously as a product from *lead and copper melts* is also produced in this case. Silicon crystals produced by other methods also belong to the same system. This is so for the long needles which are produced in the pyrolytic reduction of silicon tetrachloride by zinc vapor [30]. They were very pure and could also be obtained as single crystals at higher working temperatures [31].

A modification of the element consisting of two-dimensional silicon networks was produced in 1953 by H. Kautsky, L. Haase [32] by decomposing calcium silicide, $CaSi_2$, which F. Wöhler had discovered in 1863 [33] and which investigations by J. Böhm, O. Hassel [34] had shown to contain networks of six-membered silicon rings in parallel layers, by means of a suspension of antimony trichloride in o-dichlorobenzene. Kautsky and Haase described the product as dark brown laminae, which bleached in the air and had pyrophoric properties. A simpler route to this active lepidoidal silicon [from the Greek λεπίδιον (lepidion) – a flake] was described in 1961 by E. Bonitz [35] and involved careful chlorination of a suspension of the calcium silicide in carbon tetrachloride with chlorine gas.

In the autumn of 1946 F. Heyd, F. Khol, A. Kochanovská [36] announced at the 20th Congress of Industrial Chemistry in Paris that they had been able to prepare what was probably a metastable hexagonal modification of silicon by vacuum distillation of the element. It could be detected in a more or less pure state only immediately after the distillation and best at an elevated temperature, since it changed spontaneously to the stable modification on keeping. The same hexagonal modification is formed in compressed silicon briquets, to which 20% of calcium fluoride and 1% of sodium fluoride have been added to prevent grain growth, when these are heated for 2 h at 900°C, and it can still be identified in quenched specimens after a long period. Annealing at 400°C, however, results in transition to the normal diamond-like modification, even with these stabilized specimens.

Single Crystals. Single crystals of silicon may be prepared in a hydrogen atmosphere from a melt using a method first described in 1917 by J. Czochralski [37] and later named after him. One can achieve, in theory at least, a uniform distribution of impurities in the crystal by suitable control of the drawing rate, the melt surface and the temperature of the melt and also, through the latter, the rate of vaporization of impurities, which varies within wide limits depending on the pressure of the inert gas as well as on the melt temperature [38]. The usual diameter of single crystals of silicon is 2 to 3 cm, though 15 cm has been attained [39], the length being up to ten times the diameter. The rate of drawing lies between 2 and 50 cm per hour. The specific resistance (which is a measure of the degree of purity) is, in the case of crystals prepared by this method, only about $100\,\Omega\cdot cm$ while ideally silicon should have a value of $230000\,\Omega\cdot cm$ [38]. H. Kleinknecht [40] appears to have been the first to use this technique with silicon in 1952. In his experiments with commercial powdered silicon fused in a quartz crucible the [110]-direction was that of preferred growth and a significant segregation of impurities took place. According to W. Dietze, W. Keller, A. Mühlbauer [41] in 1981 about 80% of Si single crystals for semiconductor purposes were still prepared by this method though contamination by impurities from the crucible walls could not be prevented because of the high reactivity of the element. When quartz was used as the material for the crucible the single crystals contained about 10^{12} oxygen atoms per cm^3 as W. Kaiser, P. H. Keck, C. F. Lange [42] and H. J. Hrostowski, R. H. Kaiser [43] were able to determine from the infrared absorption. This stems from the reaction $Si + SiO_2 \rightarrow 2SiO$ between the element and the crucible walls, which was observed by G. Hass [44].

In 1954, P. H. Keck, M. J. E. Golay [45] and R. Emeis [46] discovered independently how to avoid difficulties arising from the crucible walls during the preparation of single crystals by modifying another method, namely the zone melting technique published by W. G. Pfann [47] in 1952. The melting zone was passed vertically through a silicon rod, prepared either by sintering or casting, which was free standing (in vacuum) and clamped at both ends. The narrow liquid zone was maintained as drops between the two ends of the rod through the relatively high surface tension of the molten silicon. Mention may be made of the fact that H. C. Theurer [48] had already applied for a patent for a similar process in 1952 which, however, was not granted until ten years later. For further details of the zone melting process, see p. 53.

References:

[1] J. J. Berzelius (Liebigs Ann. Chem. **49** [1844] 247/64, 251). – [2] H. Moissan, F. Siemens (Ber. Deut. Chem. Ges. **37** [1904] 2540/4). – [3] H. Sainte-Claire Deville (Compt. Rend. **40** [1855] 1034/6). – [4] H. de Sénarmont (Ann. Chim. Phys. [3] **47** [1856] 169/72). – [5] H. Sainte-Claire Deville (Ann. Chim. Phys. [3] **49** [1857] 62/78, 70/2).

[6] H. Sainte-Claire Deville (Ann. Chim. Phys. [3] **43** [1855] 1/33, 21/2). – [7] F. Wöhler (Ann. Physik Chem. **97** [1856] 484/8). – [8] H. Richter (Berg- Hüttenmänn. Jahrb. Bergakad. Leoben **11** [1862] 289/90). – [9] T. L. Phipson (Compt. Rend. **60** [1865] 1030/2). – [10] C. A. Winkler (J. Prakt Chem. **91** [1864] 193/208, 198/9).

[11] H. Kopp (Liebigs Ann. Chem. Suppl. **3** [1864/65] 1/126, 72). – [12] W. H. Miller (Phil. Mag. [4] **31** [1866] 397/9). – [13] F. S. Hyde (J. Am. Chem. Soc. **21** [1899] 603/5). – [14] L. Baraduc-Muller (Rev. Met. **7** [1910] 657/834, 774, 827). – [15] J. Königsberger, K. Schilling (Ann. Physik [4] **32** [1910] 179/230, 189/92).

[16] F. G. Wick (Phys. Rev. **27** [1908] 76/86). – [17] J. Königsberger (Physik. Z. **14** [1913] 658/9). – [18] A. Schulze (Z. Tech. Physik **11** [1930] 443/52). – [19] M. C. Neuburger (Die Allotropie der chemischen Elemente und die Ergebnisse der Röntgenographie, Stuttgart 1936, p. 40). – [20] Gonser, Seybolt (from A. Kinzel, T. R. Cunningham, Am. Inst. Mining Met. Eng. Tech. Publ. No. 1138 [1939] 1/5, 3).

[21] H. Stöhr, W. Klemm (Z. Anorg. Allgem. Chem. **241** [1939] 305/23, 316). – [22] H. Moissan, F. Siemens (Ber. Deut. Chem. Ges. **37** [1904] 2540/4). – [23] P. Lebeau (Compt. Rend. **142** [1906] 154/7). – [24] M. Philips (Metallurgie [Paris] **4** [1907] 587/92, 591). – [25] W. Manchot (Ber. Deut. Chem. Ges. **54** [1921] 3107/11, 3110).

[26] W. Manchot, H. Funk (Z. Anorg. Allgem. Chem. **122** [1922] 22/6). – [27] H. Sainte-Claire Deville (Compt. Rend. **39** [1854] 321/6; Ann. Chim. Phys. [3] **43** [1855] 5/33, 31/2). – [28] A. B. Albro (Trans. Am. Electrochem. Soc. **7** [1905] 251/72, 253). – [29] H. Moissan, F. Siemens (Compt. Rend. **138** [1904] 1299/303). – [30] G. L. Pearson, J. B. Bardeen (Phys. Rev. [2] **75** [1949] 865/83, 867).

[31] D. W. Lyon, C. M. Olson, E. D. Lewis (Trans. Am. Electrochem. Soc. **96** [1949] 359/63). – [32] H. Kautsky, L. Haase (Ber. Deut. Chem. Ges. **86** [1953] 1226/34). – [33] F. Wöhler (Liebigs Ann. Chem. **127** [1863] 257/74, 262). – [34] J. Böhm, O. Hassel (Z. Anorg. Allgem. Chem. **160** [1927] 152/64). – [35] E. Bonitz (Ber. Deut. Chem. Ges. **94** [1961] 220/3).

[36] F. Heyd, F. Khol, A. Kochanovská (Collection Czech. Chem. Commun. **12** [1947] 502/9). – [37] J. Czochralski (Z. Physik. Chem. **92** [1917] 219/21). – [38] A. Smakula (Einkristalle, Wachstum, Herstellung und Anwendung, Berlin – Göttingen – Heidelberg – München 1962, p. 289). – [39] R. Rost (Silicium als Halbleiter, Stuttgart 1966, p. 29, Fig. 56, p. 37). – [40] H. Kleinknecht (Naturwissenschaften **39** [1952] 400/1).

[41] W. Dietze, W. Keller, A. Mühlbauer (in: J. Grasmeier, Silicon, Springer, Berlin 1981, p. 2). – [42] W. Kaiser, P. H. Keck, C. F. Lange (Phys. Rev. [2] **101** [1956] 1264/8). – [43] H. J. Hrostowski, R. H. Kaiser (Bull. Am. Phys. Soc. [2] **1** [1956] 295). – [44] G. Hass (J. Am. Ceram. Soc. **33** [1950] 353/60). – [45] P. H. Keck, M. J. E. Golay (Phys. Rev. [2] **89** [1953] 1207).

[46] R. Emeis (Z. Naturforsch. **9a** [1954] 67). – [47] W. G. Pfann (Trans. AIME **194** [1952] 747/53). – [48] H. C. Theurer (U.S. 3060123 [1952/62] from C.A. **58** [1963] 136).

1.6 Atomic Weight. Isotopes

The Chemical Atomic Weight. In the second part of his *New System of Chemical Philosophy,* which appeared in 1810, John Dalton (1766–1844), who originated the concept of atomic weights, considered that not only the metals and some nonmetals and gases were "simple substances", which were elementary in character, but also the "earths". He said of "elementary silica" that: "It is much more difficult to find the weight of an atom of silica than that of an atom of any other earth... Nevertheless I have been successful in obtaininig this weight approximately by studying its relationship to potash, lime, and baryta." He concluded: "When I take everything into consideration, I am inclined to assume that the weight of an atom of silica is 45, that of hydrogen being taken equal to 1 [and that of oxygen equal to 7]" ["Es ist ungleich schwieriger, das Gewicht eines Atomen Kieselerde, als den [Wert] eines Atomen irgend einer anderen Erde zu finden... Es ist mir jedoch gelungen, dieses Gewicht dadurch, daß ich ihre Verhältnisse zu dem Kali, der Kalkerde und Baryterde untersuchte, auszumitteln. Wenn ich alles in Erwägung ziehe, so bin ich geneigt, anzunehmen, daß das Gewicht eines Atomen Kieselerde 45 ist, das des Wasserstoffs gleich 1 [und das des Sauerstoffs gleich 7] gesetzt"] [1].

In the year when Dalton's work appeared, that is 14 years before J. J. Berzelius isolated the element, see p. 29, the latter, being convinced that silica was the oxide of a metal, undertook the first experiments to determine its constitution [2]. He prepared alloys of the unknown metal with iron and copper by melting together turnings of these metals with silica and powdered carbon. He then analyzed the alloys and attempted to determine the quantity of oxygen necessary to convert the element to silica in the decomposition. As he was unable to determine the carbon present in the silicon alloys, the results were not exact, as he himself emphasized. He found the following values for the composition of silica: 38 parts of "basis" and 62 of oxygen, 51.5 "basis" and 49.5 oxygen or 52.25 "basis" and 47.75 oxygen. From this, B. Brauner [3] calculated Si = 19.6, 26.3, or 29, referred to O = 16.

This experiment was repeated at once by F. Stromeyer [4] in Göttingen, with particular reference to its significance in iron metallurgy, which had already been emphasized by Berzelius. He found for the composition of silica "47.37 parts of metallic basis or silicium and 52.63 parts of oxygen" from which, according to B. Brauner [3], a value of Si = 27.25 may be calculated. Three years later, J. J. Berzelius [5] tried to avoid the difficulties which had arisen in his first experiments because of the carbon content of the sample used. He utilized J. Davy's investigations [6] on the compound of silica with hydrofluoric acid and ammonia to determine the oxygen content of silica, finding a value of 49.46%, which W. Becker, J. Meyer [7] considered to be the first useful value. When, however, J. J. Berzelius [8] repeated Davy's work in 1818 he stated: "... I was not in a position [with reference to the proportions in the composition of siliceous hydrofluoric acid (SiF_4)] to obtain reasonably reliable results" ["... ich war [in bezug auf das Verhältnis der Zusammensetzung der kieselerdehaltigen Flußsäure (SiF_4)] nicht im Stande, einigermaßen zuverlässige Resultate zu erhalten"] – they varied between 128 and 147 parts of silica to 100 parts of hydrofluoric acid. For this reason, and also because of some other experimental difficulties, he analyzed a synthetic aluminosilicate, from which he deduced a silicon content (in SiO_2) of 49.641 and 49.716, leading to Si = 31.62

according to B. Brauner [3]. In 1826 J. J. Berzelius [9] determined the proportions of the components in silica by burning elementary amorphous silicon, and arrived at an oxygen content of 51.23 to 51.12%, from which W. Becker, J. Meyer [7] deduced values of the atomic weight of silicon between 29.5 and 30.4. According to B. Brauner [3], Berzelius preferred a recalculated value of Si = 29.63. Since, however, he found that he was wrong in his picture of the composition of silicic acid, he wrote: "It is consequently most likely . . . that the atom of "kiesel" [that is elementary silicon] weighs 277.8 [referred to O = 100], or 44.49 times as much as that of hydrogen" ["Es ist folglich am wahrscheinlichsten . . ., daß das Atom des Kiesels [das ist: elementares Silicium] wiegt 277.8 [bezogen auf O = 100], oder 44.49 so viel, als das des Wasserstoffs"]. He adopted this value in his *Lehrbuch* [10] and it was still there, almost unchanged, in the fourth edition of 1835 [11].

It was not until 1845, that new experiments to determine the atomic weight chemically were made by J. Pelouze [12], using what O. Hönigschmid, M. Steinheil [13] later termed the "classical standard method" (for the experiments to determine the atomic weight of silicon by physical methods which were made in the intervening years, see below). Pelouze tried to determine the equivalent weight of silicon from the composition of silicon chloride, the chlorine content of which he determined by titration with silver nitrate after decomposing the compound in water, and obtained the value 88.94 (Pelouze calculated 92.43 from Berzelius' results). According to B. Brauner [3] this gives Si = 28.39, while W. Becker, J. Meyer [7] gave Si = 28.46. The latter also pointed out that the number obtained is the mean of only two values, which are appreciably different and that Pelouze had worked with quite small quantities of a preparation which was probably only moderately pure. Fourteen years later, J. Dumas [14] employed the same method with somewhat greater precautions and obtained an equivalent weight from which B. Brauner [3] derived Si = 28.09. W. Becker, J. Meyer [7] pointed out, however, that the two values determined by Dumas differ from one another too much for the mean to be useful. In 1861 J. Schiel [15] found Si = 28.01 by the same method, in agreement with the results from vapor density measurements with silicon chloride and silicon fluoride, see below. A recalculation by B. Brauner [3] led to the very low value of Si = 27.96.

Because of the critical remarks made by F. W. Clarke [16] in 1882 in his *Recalculations of the Atomic Weights* – he found the values all to be too high and considered silicon chloride to be little suited to such determinations – T. E. Thorpe, J. W. Young [17] started with silicon tetrabromide specially prepared from very pure starting materials. This they also decomposed in water and used the precipitated SiO_2 to determine the atomic weight, obtaining Si = 28.332. Using the same method, W. Becker, J. Meyer [7] determined a value of Si = 28.21 in 1905, though later authors questioned the purity of their starting material [18]. Fifteen years later, G. P. Baxter, P. F. Weatherill, E. O. Holmes [19] reverted to the procedure which J. Dumas [14] had used in 1859, and which they considered to be the best, and determined the value Si = 28.111 from four analyses, which was appreciable lower than the value Si = 28.3 in general use until then. As they developed this investigation further in 1923, G. P. Baxter, P. F. Weatherill, E. W. Scripture [18] arrived at a value Si = 28.063 from their results with silicon chloride and bromide, and this value, Si = 28.06, was adopted in 1925 by the International Commission for Atomic Weights in their tables [20]. A year earlier, however, O. Hönigschmid, M. Steinheil [13] had found the higher value of Si = 28.105 as the result of their work with silicon tetrachloride prepared and very carefully purified by Alfred Stock (1876–1946). The latter were quite unable to explain the discrepancy between their value and that of the American workers. It was not until 1932 that new measurements of the atomic weight of silicon were published, when P. F. Weatherill, P. S. Brundage [21] derived a value Si = 28.103 from the transformation of silicon tetrachloride to silica. An even greater value, Si = 28.15, had been found by A. Stock, E. Kuss [22] in 1923 when they decomposed monosilane with caustic soda, and calculated the atomic weight from measurements of the volume of hydrogen evolved.

Physical Methods. The first attempt to determine the atomic weight of the element by vapor density measurements was made in 1821 by Amedeo Avogadro [di Quaregna] (1776–1856) in his third paper on the determination of the mass of molecules entitled: "Nouvelles Considérations sur la Théorie des Proportions Déterminées dans les Combinaisons, et sur la Détermination des Masses des Molecules des Corps". This he read before the Academy of Sciences in Turin on 14th February of the year in question [23]. From the values measured by J. Davy [6] for silicon tetrafluoride – he was the first to formulate it as SiF_4 – and the values given by J. J. Berzelius [2] for the ratio of oxygen to silicon in silica (where Avogadro again used a valency of four), he calculated the atomic weight Si = 31.6, referred to H = 1. The next experiments were made by J. Dumas [24] in 1826 with silicon tetrachloride, but his measured values were too high because of insufficient purification of his material. Incorrect assumptions were also made as to the valency of silicon and other errors occurred, but B. Brauner [3] recalculated his data to obtain a value Si = 29.7. Thirty five years then passed before such determinations were taken up once more. J. Schiel [15], who stressed the tetravalency of the element, obtained the value Si = 28.01 from the vapor densities of silicon fluoride and chloride. He had also obtained this value by chemical methods, see above. He conjectured that the same value could certainly be calculated from the (then unknown) specific heat of the element. In the following decades other liter weight determinations with gaseous silicon compounds were published, but in the view of A. F. O. Germann, H. S. Booth [25], they were carried out by unrefined methods and cannot therefore be used as a basis for atomic weight determinations. Only the vapor density measurements on silicon fluoride made in 1913 by A. Jaquerod, M. Tourpaian [26] appeared suitable, and from these a value Si = 28.496 was calculated later by P. L. Robinson, H. C. Smith [27]. A somewhat lower value, namely Si = 28.306, was also calculated by them from the measurements of A. F. O. Germann, H. S. Booth [25]. Finally, in 1938, E. Moles [28] obtained the value Si = 28.105 from the ratio of the limiting densities of silicon tetrafluoride and oxygen, which was in good agreement with the chemical values then known.

In 1947 the "pykno-X-ray method", which was first used by C. A. Hutchinson, H. L. Johnston [29] to determine the atomic weight of fluorine and which is based on a comparison of the X-ray data for the substance under investigation wi'.h those of calcite or sodium chloride, was applied to α-quartz by T. Batuecas [30], the values calculated being Si = 28.075 or 28.076. A year later G. Schmid [31], using the comparison with sodium chloride, obtained Si = 28.103.

Mass-Spectroscopic and Similar Data for Natural Isotopes. In his first mass-spectroscopic study with silicon tetrafluoride in August 1920, Francis William Aston (1877–1945), who had discovered this technique, was able to detect with certainty the occurrence of only two isotopes, ^{28}Si and ^{29}Si [32]. It remained doubtful whether an atom of mass 30 existed because of the possibility of the existence of distinct hydrogen compounds with this mass. In 1924, R. S. Mulliken [33] analyzed the band spectrum attributed to the diatomic molecule SiN, which W. Jevons [34] had photographed ten years before, and was able to detect the occurrence of the isotope ^{30}Si and to show that it was only a little less abundant than ^{29}Si. Shortly afterwards, F. W. Aston [35] also identified the heaviest of the natural isotopes mass-spectroscopically.

The earliest calculation of the abundance of the separate isotopes was made in 1934 by A. McKellar [36] on the basis of his reinvestigation of the SiN spectrum. He found $^{28}Si : ^{29}Si : ^{30}Si = 89.6 : 6.2 : 4.2$. This relationship was first investigated mass-spectroscopically in 1946 simultaneously at three different locations in the United States: in Charlottesville, Virginia [37], in Los Alamos, New Mexico [38], and in Chicago, Illinois [39], the results differing little from one another. The mean value was $^{28}Si : ^{29}Si : ^{30}Si = 92.28 : 4.68 : 3.06$. In later years a whole series of such studies was carried out, all of which gave results quite similar to that quoted [40].

F. W Aston made a more precise determination of the isotope masses in 1936, using a new mass spectroscope [41] and obtaining $^{28}Si = 27.9863 \pm 0.0007$. By using the so-called doublet method it was possible to increase the accuracy of these values substantially and, with its aid, H. Ewald [42] in 1951 obtained $^{28}Si = 27.985792 \pm 0.000032$. In the following year C. W. Li [43] was able to measure the isotopic masses for all the light elements referred to ^{16}O from nuclear decomposition data and obtained the values: $^{28}Si = 27.985767 \pm 0.000032$, $^{29}Si = 28.985650$ (± 0.000034), $^{30}Si = 29.983237$ (± 0.000036), and $^{31}Si = 30.985140$ (± 0.000039). From such data and the abundance values quoted, A. O. Nier [44] in 1954 calculated the *physical atomic weight* Si = 28.086. This confirmed the value Si = 28.09 adopted as the *chemical atomic weight* by the International Atomic Weight Commission in 1952 [45]. For the radioactive isotope ^{32}Si (see later) A. Turkevich, A. Tompkins [74] in 1953 established on the basis of nuclear stability considerations that it occurs naturally. By radiochemical methods they calculated its abundance to be 4×10^{-6}%.

The first experiments to examine the question of whether the mass ratio of the isotopic components of cosmic and terrestrial silicon of different origins was constant, or whether it varied, was undertaken in 1921 to 1923 by F. M. Jaeger, D. M. Dykstra [46]. They determined as accurately as possible the densities of 12 samples of tetraethylsilane – [$Si(C_2H_5)_4$] – prepared from Si samples of different origins (6 stony meteorites and 6 rock samples), using a pycnometric method. They found the atomic weight of silicon derived from such measurements to be constant to within 0.019 atomic weight unit. F. W. Aston [47] in the *Annual Reports* described this investigation as a "conclusive paper" which proves "that the ratio of isotopes of silicon is constant within the limits of the highest accuracy attainable at present". H. V. A. Briscoe, P. L. Robinson [48] at once disputed this favorable view, calculating that the accuracy of atomic weights derived from Jaeger and Dykstra's results was about 30% worse than that from the standard methods. It was found, however, from similar measurements by P. L. Robinson, H. C. Smith [27] with silicon tetrachloride, the accuracy of which was said to be four times greater than that in the Jaeger-Dykstra study, that the constancy of the isotope ratio for silicon of different origins was confirmed.

Radioactive and Artificial Isotopes. The first observation indicating the possibility of forming an artificially radioactive isotope of silicon was made in 1934 by I. Curie, P. Joliot [49] when they bombarded magnesium with α-rays from polonium and were able to observe a new radioactivity involving the emission of electrons and positrons. By analogy with the radioactive intermediates obtained from boron and aluminum under the same conditions (which had been identified chemically) they concluded that (chemically unidentified) *radiosilicon* had been formed in accordance with the reaction $^{24}_{12}Mg + ^{4}_{2}He \rightarrow ^{27}_{14}Si + ^{1}_{0}n$ (abbreviated as $^{24}Mg(He, n)^{27}Si$) and $^{27}_{14}Si \rightarrow ^{27}_{13}Al + e^{+}$. The half-life of this isotope was said to be about 3 min. Subsequently the occurrence of this reaction was repeatedly confirmed and the half-life was corrected [50, 51], though attempts in 1939 to obtain the isotope by bombarding aluminum with protons failed [52].

The next artificial silicon isotope, ^{31}Si, was discovered in 1935 by E. Fermi's research group in Rome when they irradiated phosphorus with slow neutrons, a half-life of 2.4 h being determined [53]. The same isotope was also obtained in 1935 by H. W. Newson [54] by bombarding ^{30}Si with neutrons.

The heaviest of the silicon isotopes, ^{32}Si, with a half-life in excess of 700 years [56] was obtained in 1953 by M. Lindner [55] on irradiating sodium chloride with protons in accordance with $^{37}Cl(p, \alpha 2p)^{32}Si$. The occurrence of this isotope was confirmed and its half-life corrected a year later by A. Turkevich, A. Samuels [57], who obtained it by the reaction $^{30}Si(n, \gamma)$ $^{31}Si(n, \gamma)^{32}Si$. Five years later D. Lal, E. D. Goldberg, M. Koide [58] were able to show that the isotope is also formed by spallation of atmospheric argon by cosmic rays and that it can be found in the skeletons of siliceous sponges in the oceans. They hoped that this would provide them with a method of dating a range of oceanographic phenomena [59]. In the autumn of

1961, K. Jantsch [60] and D. Geithoff [61] found almost simultaneously that the isotope may be produced by bombardment of silicon with tritium in the reaction $^{30}Si(t,p)^{32}Si$. A more detailed study of its radiation and a new estimate of its half-life were published in 1964 [62].

The short-lived isotope ^{26}Si, which emits protons, was first observed as a product of the reaction $^{27}Al(p,2n)^{26}Si$ by H. Tyrén, P.-A. Tove [63] and then obtained six years later by E. L. Robinson, O. E. Johnson [64], using the reaction $^{24}Mg(^{3}He,n)^{26}Si$. The older findings on this isotope were confirmed three years later [65].

The lightest nuclide, ^{25}Si, was observed in 1963 by a Canadian group in Montreal [66]. It was a short-lived positron emitter formed on bombarding aluminum with 96 MeV protons. Two years later it was investigated in greater detail by R. McPherson, J. C. Hardy [67].

No further comment need be made here on the fact that the known stable isotopes may also result in some of the nuclear processes described by using other isotopes as targets. Other quite expensive reactions have been found in which nuclides of silicon are formed. That they can also be obtained quite cheaply was claimed in 1965 in a French patent [68] entitled: "Fabrication d'aciers spéciaux par transmutation à faible énergie". According to this ^{28}Si may be synthesized from carbon and oxygen at low energies, iron and nickel being also formed. The idea that carbon, among other things, is a "constituent of silicon" was put forward 60 years earlier by Th. Gross [69, 70] of the Technische Hochschule Charlottenburg on the basis of electrolytic experiments with alkali silicate solutions, using an alternating current. Contemporaries [71] considered that the main result of this experiment was the fact that the silver crucible used melted. The notion that substances which are generally called elements are made up of larger units, as was assumed in Prout's hypothesis, was, however, widespread not only in the 19th century, as W. V. Farrar [72] showed for silicon, but survived into the present century. Even in 1924, H. Collins [73] believed he could explain the properties of silicon correctly if he thought of it as composed of four parts of unequal size, the largest of which was ^{23}Na.

References:

[1] J. Dalton (A New System of Chemical Philosophy, Pt. 2, Manchester 1810, pp. 536, 547 from German translation by F. Wolff, Ein neues System des chemischen Theiles der Naturwissenschaft, Vol. 2, Berlin 1813, pp. 368, 371). – [2] J. J. Berzelius (Ann. Physik **36** [1810] 89/102). – [3] B. Brauner (in: R. Abegg, F. Auerbach, Handbuch der anorganischen Chemie, Vol. 3, Pt. 2, Leipzig 1909, pp. 280/7). – [4] F. Stromeyer (Ann. Physik **37** [1811] 335/9, **38** [1811] 321/30). – [5] J.J. Berzelius (Schweiggers J. Chem. Physik **11** [1814] 193/233, 215/6 footnote).

[6] J. Davy (Phil. Trans. Roy. Soc. [London] **102** [1812] 352/69). – [7] W. Becker, J. Meyer (Z. Anorg. Allgem. Chem. **43** [1905] 260/6, 254/5). – [8] J. J. Berzelius (Schweiggers J. Chem. Physik **23** [1818] 277/99, 277/85). – [9] J. J. Berzelius (Ann. Physik Chem. [2] **8** [1826] 1/24, 20/1). – [10] J. J. Berzelius (Lehrbuch der Chemie [German translation by F. Wöhler], Vol. 3, Pt. 1, Dresden 1827, p. 118).

[11] J. J. Berzelius (Lehrbuch der Chemie [German translation by F. Wöhler], 4th Ed., Vol. 1, Dresden-Leipzig 1835, p. 330). – [12] J. Pelouze (Compt. Rend. **20** [1845] 1047/55, 1052/3; J. Prakt. Chem. **35** [1845] 73/82, 79/80). – [13] O. Hönigschmid, M. Steinheil (Z. Anorg. Allgem. Chem. **141** [1924] 101/8). – [14] J. Dumas (Ann. Chim. Phys. [3] **55** [1859] 129/210, 183/4). – [15] J. Schiel (Liebigs Ann. Chem. **120** [1861] 44/99).

[16] F. W. Clarke (A Recalculation of the Atomic Weights, Washington 1882 from Chem. News **48** [1883] 231/2). – [17] T. E. Thorpe, J. W. Young (J. Chem. Soc. **51** [1887] 576/9). – [18] G. P. Baxter, P. F. Weatherill, E. W. Scripture (Proc. Am. Acad. Sci. **58** [1922/23] 245/68). – [19] G. P. Baxter, P. F. Weatherill, E. O. Holmes (J. Am. Chem. Soc. **42** [1920] 1194/7). – [20] Office of the General Secretary IUPAC (J. Am. Chem. Soc. **47** [1925] 597/601).

[21] P. F. Weatherill, P. S. Brundage (J. Am. Chem. Soc. **54** [1932] 3932/8). – [22] A. Stock, E. Kuss (Ber. Deut. Chem. Ges. **56** [1923] 314/6 footnote 2). – [23] A. Avogadro (Mem. Reale Accad. Sci. Torino **26** [1821] 1/162, 105/6, 121/5). – [24] J. Dumas (Ann. Chim. Phys. [Paris] **33** [1826] 337/91, 367/8). – [25] A. F. O. Germann, H. S. Booth (J. Phys. Chem. **21** [1917] 81/100).

[26] A. Jaquerod, M. Tourpaian (J. Chim. Phys. **11** [1913] 269/74). – [27] P. L. Robinson, H. C. Smith (J. Chem. Soc. **1926** 1262/82, 1264). – [28] E. Moles (Arch. Sci. Phys. Nat. [5] **20** [1938] 59/65, 65). – [29] C. A. Hutchinson, H. L. Johnston (J. Am. Chem. Soc. **63** [1941] 1580/2). – [30] T. Batuecas (Nature **159** [1947] 705).

[31] G. Schmidt (Diss. Darmstadt T.H. 1948, p. 64). – [32] F. W. Aston (Phil. Mag. [6] **40** [1920] 628/34, 630, 633). – [33] R. S. Mulliken (Nature **113** [1924] 423/4). – [34]. W. Jevons (Proc. Roy. Soc. [London] A **89** [1914] 187/97, 192). – [35] F. W. Aston (Phil. Mag. [6] **49** [1925] 1191/201, 1198; Nature **114** [1924] 273).

[36] A. McKellar (Phys. Rev. [2] **45** [1934] 761). – [37] E. P. Ney, J. H. McQueen (Phys. Rev. [2] **69** [1946] 41). – [38] D. Williams, P. Yuster (Phys. Rev. [2] **69** [1946] 556/67, 561). – [39] M. G. Ingram (Phys. Rev. [2] **70** [1946] 653/60, 659). – [40] J. F. Norton, P. D. Zemany (J. Chem. Phys. **20** [1952] 525).

[41] F. W. Aston (Nature **138** [1936] 1094). – [42] H. Ewald (Z. Naturforsch. **6a** [1951] 293/302, 300). – [43] C. W. Li (Phys. Rev. [2] **88** [1952] 1038/42). – [44] A. O. Nier (Z. Elektrochem. **58** [1954] 559/67, 565). – [45] E. Wichers (J. Am. Chem. Soc. **74** [1952] 2447/50).

[46] F. M. Jaeger, D. M. Dykstra (Proc. Koninkl. Ned. Akad. Wetenschap. **27** [1924] 393/406, 402; Z. Anorg. Allgem. Chem. **143** [1925] 233/58, 252). – [47] F. W. Aston (Ann. Rept. Progr. Chem. [Chem. Soc. London] **21** [1925] 238/58, 243). – [48] H. V. A. Briscoe, P. L. Robinson (Nature **117** [1926] 377/8). – [49] I. Curie, P. Joliot (J. Phys. Radium [7] **5** [1934] 153/6). – [50] H. Fahlenbrach (Z. Physik **96** [1935] 503/11).

[51] M. L. Pool, J. M. Cork, R. L. Thornton (Phys. Rev. [2] **52** [1937] 239/40). – [52] G. Kuerti, S. N. van Voorhis (Phys. Rev. [2] **56** [1939] 614/5). – [53] E. Amaldi, O. D'Agostino, E. Fermi, P. Pontecorvo, F. Rosetti, E. Segrè (Proc. Roy. Soc. [London] A **149** [1935] 522/58, 544). – [54] H. W. Newson (Phys. Rev. [2] **48** [1935] 482). – [55] M. Lindner (Phys. Rev. [2] **89** [1953] 1150).

[56] W. Lindner (Phys. Rev. [2] **91** [1953] 642/4). – [57] A. Turkevich, A. Samuels (Phys. Rev. [2] **94** [1954] 364). – [58] D. Lal, E. D. Goldberg, M. Koide (Phys. Rev. Letters **3** [1959] 380). – [59] D. Lal, E. D. Goldberg, M. Koide (Science **131** [1960] 332/7). – [60] K. Jantsch (Kernenergie **4** [1961] 846).

[61] D. Geithoff (Radiochim. Acta **1** [1962] 3/6). – [62] R. L. Brodzinski, J. R. Finkel, D. C. Conway (J. Inorg. Nucl. Chem. **26** [1964] 677/81). – [63] H. Tyrén, P.-A. Tove (Phys. Rev. [2] **96** [1954] 773/4). – [4] E. L. Robinson, O. E. Johnson (Phys. Rev. [2] **120** [1960] 1321/7). – [65] G. Frick, A. Gallmann, D. E. Alburger, D. H. Wilkinson, J. P. Coffin (Phys. Rev. [2] **132** [1963] 2164/84, 2178/9).

[66] R. Barton, R. McPherson, R. E. Bell, W. R. Frisken, W. T. Link, B. Moore (Can. J. Phys. **41** [1963] 2007/25). – [67] R. McPherson, J. C. Hardy (Can. J. Phys. **43** [1965] 1/11). – [68] Y. Sakurazawa-Osawa (Fr. 1427109 [1964/66]; C.A. **65** [1966] 14751). – [69] Th. Gross (Elektrochem. Z. **12** [1905] 48/50). – [70] Th. Gross (Elektrochem. Z. **10** [1903] 70/1).

[71] H. Dannell (Jahrb. Elektrochem. **9** [1902/4] 493). – [72] W. V. Farrar (Brit. J. Hist. Sci. **2** [1965] 297/323, 319). – [73] H. Collins (Chem. News **129** [1924] 47/9). – [74] A. Turkevich, A. Tompkins (Phys. Rev. [2] **90** [1953] 247).

1.7 Uses

In 1916, J. W. Richards [1] related an experience he had had 14 years earlier which, at the time, had appeared to him as surprising. F. J. Tone, the manager of the carborundum plant at Niagara Falls had shown him "a vessel with about 100 lb of silicon" which had been obtained as an unwanted byproduct from the preparation of silicon carbide in the electric furnace, and had asked: "What can we use this material for?" In fact, as F. J. Tone [2] records in looking back in the early thirties, silicon was, at the turn of the century, a rarity encountered only in scientific laboratories, for which 100 dollars an ounce had to be paid and with which little could be done. At the time of this reminiscence [1931], production of the element had risen so much as the result of applications discovered in the meantime that the price had fallen to 10 cents a pound. Initially the element was introduced as an additive in iron metallurgy. The favorable mechanical and magnetic properties [11] resulting from the presence of the silicon were soon recognized, leading to its use for preparing a whole series of silicon steels with further alloying elements. Cast iron alloyed with a high proportion of silicon has become an important technical material under various names, particularly on account of the acid resistance brought about by the presence of silicon. Silicon is also introduced into other metals as an alloying component. Thus Si-Cu alloys, the so-called silicon bronzes, are widely used, their favorable properties being enhanced by additions of nickel and manganese. A special rôle is played by Al-Si alloys, which have great strength and a good resistance to corrosion. Furthermore they can be improved by thermal treatment, though usually after the addition of further alloy components. For example, a special aluminum alloy with 0.6% Si which can be hardened is much used as a piston alloy for automobiles. In practice, in this connection, silicon itself is replaced by silicocalcium, silico-manganese, silicochromium, or ferrozircosilicon, according to the subsequent use [3]. Silicon forms silicides with many metals, such as magnesium, which are often the starting materials for the preparation of silicon compounds, e.g., the large group of silicones (organosiloxanes). Recently, the silicides of the metals of groups IVa, Va, and VIa have increased in importance as heat-conducting materials ($MoSi_2$), and for wear-resistant, nonscaling and corrosion-resistant layers [4].

Some smaller scale and less known actual or suggested uses may also be mentioned here. Thus the element was used in the first world war for generating hydrogen for captive balloons by its reaction with caustic soda. Compared with gas generation by means of acids, this has the advantage of simpler operation and lower transport costs [2]. In the nineteen thirties two patents were granted for this [6, 7]. An interesting suggestion for the use of the element was made in 1905 by F. J. Tone [5], namely that it should replace aluminum in the Goldschmidt aluminothermic process. He was attracted to this not only because the element was more readily powdered, but as he calculated its use to be energetically more favorable. H. N. Potter [12] believed that priority for this proposal belonged to A. B. Albro, who had applied for a patent for such a mixture, which he called "Calorite", though the granting of this patent cannot be confirmed. As experiments had shown that the reaction between silicon powder and reagents such as nitrates and chlorates which furnish oxygen can occur explosively, F. J. Tone [5] believed that such mixtures offered the possibility of preparing new explosives. Many years later preparations of this sort were patented, in various countries for use in fireworks [8, 9, 10].

The high resistance of the element to corrosion resulted in a number of proposals for its use, such as that of C. Rossi [13] in 1913 for electrodes in gas reactions for obtaining nitrogen oxides, and the use of either the element or its alloys as protective films, these being obtained either by the decomposition of gaseous compounds or by cementation. The Bosnische Elektrizitätsgesellschaft in Wien [14] used the first route in 1915, see p. 57, and twenty years later a similar but somewhat modified proposal was put forward by J. Keller [15]. The formation of such surface layers had been described by A. Sansfourche [16] in 1926, and he stated that E. F. Fremy (1814–1894), professor of chemistry at the École Polytechnique in Paris, had made similar experiments in 1865. Formation of silicon-containing layers by cementation appears first to have been suggested in 1927 by L. H. Marshall [17]. The iron object to be treated was packed in silicon- or ferrosilicon powder, sand, and iron trichloride (or a similar compound which can vaporize at a higher temperature, partly dissociating and thus forming volatile silicon compounds). It was then heated to between 500 and 900°C. Later C. F. Lauenstein, P. F. Ulmer [18] recommended the replacement of iron chloride by common salt and E. Davidis [19] observed that the adhesion of the silicon coating was improved by intermediate layers of copper or nickel. According to G. Becker, K. Daeves, F. Steinberg [20] the carbon content of the iron should be below 0.1%, otherwise the surface must be decarbonized to a depth of at least 2 mm prior to the silicon treatment. This results in rapid diffusion, good properties of the diffusion layer und better adhesion to the base metal. The use of silicon carbide as a cementation agent for iron and steel was suggested by B. Manzini [21] in 1943.

The production of foils, which is impossible by the usual mechanical methods in the case of silicon, but which is highly desirable for modern applications of the element, was attempted in 1938 by H. Kamogawa [22], using the vaporization method. He evaporated silicon from a thorium oxide crucible on to a polished rock salt surface. He found, however, that the vaporized layer, separated by dissolving away the salt with water, gave electron diffraction rings attributable to silicon dioxide and not to elementary silicon. Five years later H. Richter [23] thought he had been able to vaporize amorphous silicon on to rock salt in much the same way, but this claim was later contested by E. Hass [24]. H. Richter [23] was able to obtain films on metal blocks. Initially these were amorphous, but, after heating for several hours at a red heat in high vacuum, they could be transformed through a fine crystalline structure to the normal crystalline state. E. Hass [24] was able to obtain well-crystallized layers by vaporizing silicon on to a hot rock salt surface (at above 600°C). In 1954, E. Wiberg, M. Schmidt [25] attempted to prepare strongly adhering silicon films on objects made up of silicates by treating them with the hydrides of the elements of the I and/or II group of the periodic system and then allowing the reduction reaction to take place in a hydrogen atmosphere or in vacuum. As H. Schulz [26] showed in 1936, such films are more simply produced by decomposing monosilane on hot glass surfaces, and may be used in preparing mirrors suitable for reflexion in the ultraviolet. Silicon films about 0.1 mm thick, which are polycrystalline with an average crystal size of 10 μ, were obtained on ceramic or carbon supports during the second world war by P. L. Günther, G. Rebentisch [27] by decomposing tetrachlorosilane with aluminum. In the reaction, which takes place at 800°C, the aluminum chloride escapes to the cooler parts of the apparatus. The aluminum content of the layers amounts to 0.1 to 1.0% [28].

In the course of the second world war a use of elementary silicon started to build up which, though, according to J. Dietl, D. Helmreich, E. Sirtl [29], it accounted for only about 1% of the total production of silicon, became the most significant and important, namely its application in semiconductor technology. It had the particular advantage over germanium, which had been chiefly used until then, that the temperature range in which such devices were capable of operating was substantially increased [30]. For the preparation of electronic units, rectifiers, transistors, etc., small thin highly polished strips are fabricated mechanically from ultra-pure single crystals of silicon, a process involving large material losses. From these the separate components needed for computers, for example, are prepared in further highly complicated operations [31]. The technical and economic significance which the element has acquired is seen from the fact that about 3000 related undertakings which have been established in the region between Palo Alto and San José to the south of San Francisco (USA) have led to this location being called "Silicon Valley" [32]. Two further possibilities for the use of high-purity silicon promise to acquire similar importance to that in semiconductor technology. These are the application of silicon in "solar batteries", in which the energy of sunlight is transformed directly into electrical energy [29, 31], and its use as a structural material in "micro mechanics", in which, by developing the techniques discovered in the semiconductor field, it becomes possible to construct out of silicon not only minute micro valves and very small microphones, but also complete micro-gas-chromatographs [33]. Less spectacular is the possibility of using silicon single crystals (and also polycrystalline material) for preparing lenses and windows for heat radiation, for which the element is transparent in the range 1 μm to 20 μm [34]. It may also be used in jewelry as a gemstone "with a beautiful dark metallic luster of the highest stability". This is due to "an iridescent silicon dioxide film produced by treatment in air at higher temperatures" [34].

References:

[1] J. W. Richards (Met. Chem. Eng. **15** [1916] 26/31, 26). – [2] F. J. Tone (Ind. Eng. Chem. **23** [1931] 1312/6). – [3] W. Schreiter (Seltene Metalle, Vol. 2, Verlag Grundstoffind., Leipzig 1961, pp. 408/11). – [4] R. Kieffer, G. Jangg, P. Ettmayer (Sondermetalle: Metallurgie Herstellung Anwendung, Springer, Wien 1971, p. 442). – [5] F. J. Tone (Trans. Am. Electrochem. Soc. **7** [1905] 243/9, 246/7).

[6] Elektrizitäts A.-G. vorm. Schuckert & Co. (Ger. 528498 [1930/31] from C. **1931** II 1738). – [7] G. F. Jaubert (Fr. 831329 [1937/38] from C. **1941** II 1985). – [8] Dr. G. Krebs, Geka-Werke Offenbach (Ger. 326761 [1917/20] from C. **1921** II 84). – [9] H. Zenftman, Imperial Chemical Industries Ltd. (Ger. 852360 [1948/52] from C. **1953** 2383). – [10] T. Matagi, Imperial Fire Works Co. (Japan. 179428 [1949] from C.A. **1951** 9863).

[11] P. H. Brace (Trans. Am. Electrochem. Soc. **70** [1936] 33/48, 43/5). – [12] H. N. Potter (from [5, Discussions, p. 249]). – [13] C. Rossi (Ger. 321287 [1913/20] from C. **1920** IV 245). – [14] Bosnische Elektrizitätsgesellschaft Wien (Ger. 302305 [1915/17] from C. **1918** I 323). – [15] J. Keller (Swiss 168455 [1933/34] from C. **1934** II 4018).

[16] A. Sansfourche (Compt. Rend. **183** [1926] 791/3). – [17] L. H. Marshall (U.S. 1853370 [1927/32] from C. **1932** II 1509). – [18] C. F. Lauenstein, P. F. Ulmer (U.S. 2105888 [1938] from C.A. **32** [1938] 2080). – [19] E. Davidis (Austrian 154362 [1938] from C.A. **1939** 1656). – [20] G. Becker, K. Daeves, F. Steinberg (Fr. 872392 [1938] from W. Machu, Metallische Überzüge, 3rd Ed., Geest & Portig, Leipzig 1948, pp. 624/5).

[21] B. Manzini (Met. Ital. **35** [1943] 247/8 from W. Machu, Metallische Überzüge, 3rd Ed., Geest & Portig, Leipzig 1948, pp. 624/5). – [22] H. Kamogawa (Phys. Rev. [2] **54** [1938] 91). – [23] H. Richter (Physik. Z. **44** [1943] 406/42, 411/2). – [24] E. Hass (Z. Anorg. Allgem. Chem. **257** [1948] 166/72). – [25] E. Wiberg, M. Schmidt, Siemens & Halske A.-G. (Ger. 975411 [1954/61] from C.A. **56** [1962] 12561).

[26] H. Schulz (Glashütte **66** [1936] 685/6). – [27] P. L. Günther, G. Rebentisch (in: G. Goubau, J. Zenneck, Fiat Review of German Sciences 1939–1946, Vol. 15, Pt. 1, Dieterich, Wiesbaden 1948, p. 279). – [28] K. Seiler (Z. Naturforsch. **5a** [1950] 393/7). – [29] J. Dietl, D. Helmreich, E. Sirtl (in: J. Grabmaier, Crystals, Vol. 5: Silicon, Springer, Heidelberg 1981, p. 45). – [30] D. W. Lyon (in: C. A. Hampel, Rare Metals Handbook, Reinhold, New York 1954, p. 385).

[31] R. S. Aries, A. P. Sachs (DECHEMA-Monogr. **33** No. 477/502 [1959] 83/8). – [32] Anonymous (Der Tagesspiegel [Berlin] Sep. 25, 1983, p. 12). – [33] H. Zettlev (FAZ [Frankfurter Allgemeine Zeitung] No. 39 [1984] 34). – [34] R. Rost (Silicium als Halbleiter, Berliner Union, Stuttgart 1966, pp. 18, 309/12).

2 History of Individual Silicon Compounds

2.1 Silicon and Hydrogen. Silicon Hydrides. Silicanes. Silanes

Introduction. Saturated silicon hydrides of the type Si_nH_{2n+2} are the analogs of the aliphatic hydrocarbons methane, ethane, propane etc. Whereas, however, the ability to form chains is enormous in the case of carbon – homologs of methane up to $C_{100}H_{202}$ are known [1] – they terminate in the case of silicon hydrides on reaching a six-membered chain according to Alfred Stock (1876–1946) in his comparative survey on *"Siliciumchemie und Kohlenstoffchemie"* [2] in which he coins the term "ligand", which is now an indispensable part of the language of chemistry. Longer chains can, however, be prepared in the case of the silicon chlorides, see p. 131. Ring formation, to form silicobenzene for example, has not so far been observed though the synthesis of "aromatic" silicon derivatives has been attempted, for example by R. Schwarz [3]. More recent gas-chromatographic investigations show [4] that chains with seven and eight members also occur with the usual methods for the preparation of saturated silicon hydrides, though in such small quantities that it was not possible to isolate them. The chromatograms also show that, as in the case of carbon, branched molecules occur for chains with four and more members. Studies conducted in the year 1971 [13] were able to identify still higher silanes, up to pentadecasilane $Si_{15}H_{32}$, but the chromatograms no longer allowed conclusions concerning the occurrence of isomers of the last members. It must also be noted that the great variety of carbon-hydrogen compounds which exist because of the ready formation of covalent multiple bonds by carbon is not encountered with silicon although its electronic structure is very similar to that of carbon. L. E. Gusel'nikov, N. S. Nametkin [5] summarize the numerous earlier and unsuccessful attempts to obtain such compounds. R. West, M. J. Fink, J. Michel [12] in 1981 were the first to demonstrate the existence of this type of compounds with the isolation of tetramesityl disilene.

In contrast to the hydrocarbons, which do not react at ordinary temperatures with oxygen or water, the chains are unusually sensitive in the case of the silicon hydrides, the silanes. Disilane burns spontaneously in air and all the homologous members are converted to silicic acid by alkali with hydrogen evolution. The name *Silane* for this series of compounds goes back to a proposal by A. Stock [6] in 1916. The somewhat older name *silicane* proposed by F. S. Kipping [7, 8] did not survive. Naming the unsaturated silicon-hydrogen compounds follows largely the usual rules for unsaturated hydrocarbons. Recently, but still before the existence of polar double bonds was observed in the case of silicon also, a proposal was made by G. Schott, W. Herrmann [9] to change the nomenclature of unsaturated silicon hydrides on the basis that these occur only as polymers, the monomers of which (conveniently called *silens* or *silins*) are unknown. Such compounds should be called polysilanes(n) where n denotes the "average number of bonds to foreign atoms for each silicon atom."

The similarity of the silicon hydrides and the carbon compounds struck F. Wöhler, who wrote to J. Liebig on 25th June, 1863 [10]: "...I live completely in the laboratory busy with the new silicon compound which results from silicon calcium [calcium silicide] and which is deep orange-yellow in the pure state. I become always more convinced that it is constituted in the same way as the organic compounds. All its behaviour is also analogous. It remains quite unchanged in the dark, when dry or in water, but in sunlight it evolves hydrogen gas as in a fermentation process and becomes snow white. This white coating behaves just like the "silicium-oxydul" which has already been described, but certainly cannot be a hydrated oxide of silicon. The yellow silicon, as I will call it, behaves on dry distillation like an organic compound. One obtains hydrogen gas, silicon hydride gas, brown amorphous silicon (corresponding with carbon) and silicic acid corresponding with carbonic acid" ["...Ich lebe ganz im Laboratorium, beschäftigt mit dem neuen Siliciumkörper, der aus dem Siliciumcalcium [Calciumsilicid] entsteht, und der in reinem Zustande tief orangegelb ist. Ich werde immer mehr davon überzeugt, daß er nach Art der organischen Körper zusammengesetzt ist; der Kohlenstoff ist hier durch Silicium vertreten. Auch sein ganzes Verhalten ist analog. Im Dunkeln, trocken oder in Wasser, bleibt er ganz unverändert, aber im Sonnenschein entwickelt er, wie bei einer Gährung, Wasserstoffgas und wird schneeweiß. Dieses weiße Zeug verhält sich dann vollkommen wie das früher beschriebene Siliciumoxydul, das aber gewiß kein Oxydulhydrat sein kann. Bei der trockenen Destillation verhält sich das gelbe Silicon, wie ich es nennen will, wie ein organischer Körper. Man erhält Wasserstoffgas, Siliciumwasserstoffgas, braunes amorphes Silicium (entsprechend der Kohle) und Kieselsäure, entsprechend der Kohlensäure"]. According to the analyses of Konrad Friedrich Beilstein (1838–1906), his assistant at the time, he had a compound $Si_4H_4O_3$ or $Si_6H_6O_4$ in his hands [11].

References:

[1] Beilstein's Handb. Org. Chem. 4th Ed. Suppl. Ser. III **1** Pt. 1 [1972] 604. – [2] A. Stock (Ber. Deut. Chem. Ges. **50** [1917] 170/82). – [3] R. Schwarz (Angew. Chem. **53** [1940] 258/61, 259). – [4] K. Borer, C. S. G. Phillips (Proc. Chem. Soc. **1959** 189/90). – [5] L. E. Gusel'nikov, N. S. Nametkin (Chem. Rev. **79** [1979] 529/77).

[6] A. Stock (Ber. Deut. Chem. Ges. **49** [1916] 108/11). – [7] F. S. Kipping (J. Chem. Soc. **91** [1907] 209/40, 209). – [8] F. S. Kipping (J. Chem. Soc. **101** [1912] 2106/7). – [9] G. Schott, W. Herrmann (Z. Anorg. Allgem. Chem. **288** [1956] 1/8, 2). – [10] A. W. Hofmann (Aus Justus Liebig's und Friedrich Wöhler's Briefwechsel in den Jahren 1829–1873, Vol. 2, Vieweg, Braunschweig 1888, pp. 1/362, 39).

[11] F. Wöhler (Liebigs Ann. Chem. **127** [1863] 257/741, 263/8). – [12] R. West, M. J. Fink, J. Michel (Science **214** [1981] 1343/4). – [13] F. Fehér, D. Schinkitz, J. Schaaf (Z. Anorg. Allgem. Chem. **383** [1971] 303/13, 311).

2.1.1 First Conjectures on a Solid Hydride

In his first preparation of amorphous silicon, see p. 32, J. J. Berzelius [1] observed in his elementary silicon, which had been treated with hydrofluoric acid ["Flußspatsäure"] and then fully washed, an inflammability which, as he writes, "does not arise from retention of potassium. . ., but can possibly come from a portion of hydrogen with which the silicon is combined. . . the silicon, which is obtained when the brown mass reduced by potassium is thrown into water, is therefore a *hydride* or *hydrogen silicon*" ["rührt nicht von einem Rückhalt von Kalium her. . ., aber sie kann möglicherweise von einer Portion Wasserstoff herrühren, mit der das Silicium verbunden ist. . . Das Silicium, welches erhalten wird, wenn man die durch Kalium reducierte braune Masse in Wasser wirft, ist daher ein *Hydrür* oder *Wasserstoffsilicium*"]. This observation was surprising after A. M. Ampère [2] had clearly stated his expectation (and also that of his contemporaries) in his major attempt to classify the chemical elements in 1816, that silicon, in spite of its similarity to carbon, would probably be unlikely to enter into compound formation with hydrogen. It is therefore not surprising that Berzelius wrote on 12th April, 1824, to P. L. Dulong in Paris [3, 4] and to his former student C. G. Gmelin in Tübingen [5] about his discovery. To the latter he wrote: "It [silicon] may be prepared (as hydride) in such an inflammable state that it burns in air. . ." ["Es [das Silicium] läßt sich jedoch (als Hydrür) in einem solchen Grade von Entzündbarkeit darstellen, daß es an der Luft brennt. . . "]. He also expressed himself quite unambiguously to the former: "The washed substance is a lower hydride of silicon that burns vigorously in oxygen at a red heat, although the silicon is not completely oxidized in the process" ["La substance laveé est un hydrure de silicium qui à une chaleur rouge brûle avec vivacité dans le gaz oxygène, quoique le silicium n'en soit pas entièrement oxydé"].

Somewhat later, on 12th July, 1824, he spoke, in writing to C. G. de la Rive in Geneva only of a "sous-hydrure de silicium" [6]: "Meanwhile, the silicon thus produced retains a small quantity of combined hydrogen and forms a subhydride of silicon" ["Cependant, le silicium ainsi préparé retient en combinaison une petite quantité d'hydrogène, et forme un sous-hydrure de silicium"]. After closing this correspondence Berzelius was not quite sure of the state in which hydrogen was bound in the freshly prepared silicon. This is apparent from his remark [7] that "silicon (from the double salts which hydrofluoric acid forms with silica and potash and soda) contains hydrogen but only in small amounts and perhaps in the same way as Davy [9] considers our ordinary wood charcoal as hydrogen carbon." Then, nine years later in 1833 he writes in the third edition of his textbook [7] that "kiesel [he meant the element silicon] contains some hydrogen" ["der Kiesel [gemeint ist das elementare Silicium] etwas Wasserstoff enthält"]. An interpretation of the occurrence of the gas was ventured a decade later by H. Sainte-Claire Deville [8]. He attributed the hydrogen evolution to decomposition of the excess potassium by acids or water or to attack of the elementary silicon by potassium hydroxide resulting from washing. But, from the results of the most recent investigations, all such ideas can no longer be regarded as correct. It has been demonstrated in experiments on the preparation of semiconductor layers of so-called amorphous silicon that these always contain dissolved hydrogen and that "best silicon" prepared for the purpose mentioned is nothing but a silicon-hydrogen alloy [10]. It may also be noted here that decades later solid silicon-hydrogen compounds have been observed, though in quite other reactions, see p. 81.

References:

[1] J. J. Berzelius (Kgl. Svenska Vetenskaps Akad. Handl. **1824** 46/98 from Ann. Physik Chem. [2] **1** [1824] 169/230, 211/2, 223). – [2] A. M. Ampère (Ann. Chim. Phys. [2] **2** [1816] 5/32, 9). – [3] J. J. Berzelius (Ann. Chim. Phys. [2] **1** [1816] 39/43, 43). – [4] H. G. Söderbaum (Jac. Berzelius, Lettres IV: Correspondence entre Berzelius et P. L. Dulong 1819–1831, Uppsala 1915, pp. 51/6, 55). – [5] J. J. Berzelius (Arch. Gesammte Naturlehre **2** [1824] 115/7).

[6] J. J. Berzelius (Bibliotheque Universelle Sci. Arts Genève [1] **26** [1824] 273/9, 275). – [7] J. J. Berzelius (Lehrbuch der Chemie [in German by F. Wöhler], 3rd Ed., Vol. 1, Dresden-Leipzig 1833, p. 329). – [8] H. Sainte-Claire Deville (Ann. Chim. Phys. [3] **49** [1857] 62/78, 64). – [9] H. Davy (Phil. Trans. Roy. Soc. [London] **99** [1809] 39/104), J. Davy (The Collected Works of Sir Humphry Davy, Vol. 5, London 1840, pp. 140/204, 175). – [10] A. L. Robinson (Science **197** [1977] 651/3).

2.1.2 First Observation and Preparation of Gaseous Silanes

In 1857, F. Wöhler, H. Buff [1], in studying the electrochemical behavior of aluminum as the negative and positive terminal of a galvanic chain in a common salt solution, observed in the latter case contrary to expectations an evolution of gas. Their "attention was still further excited when some of the gas bubbles as they rose and burst in the air ignited spontaneously and burnt with a white flame, producing white smoke" [Ihre "Aufmerksamkeit wurde noch mehr gespannt, als einzelne der aufsteigenden Gasblasen bei dem Zerplatzen in der Luft sich von selbst entzündeten und mit weißer Flamme unter Erzeugung eines weißen Rauches verbrannten"]. Closer examination of the phenomenon showed "that the best yield of spontaneously inflammable gas was obtained particularly from such pieces [of aluminum] as contained rather much silicon" ["daß die reichlichste Ausbeute an selbstentzündlichem Gase besonders von solchen Stücken [Aluminium] erhalten wurde, die ziemlich viel Silicium enthielten"]. Their expectation that "this spontaneously inflammable gas is a compound of silicon and hydrogen" ["daß dieses selbstentzündliche Gas eine Verbindung von Silicium und Wasserstoff ist"] was confirmed when they were able to show that the "white smoke" ["weiße Rauch"] behaved like silica. They were also successful in demonstrating that amorphous brown silicon was formed in the thermal decomposition of the gas mixture by various methods. They attempted "to bring about silicon hydride formation by purely chemical means, without, however, reaching this goal so far" ["die Bildung des Siliciumwasserstoffs auf rein chemischem Wege in die Gewalt zu bekommen, ohne aber bisher den Zweck zu erreichen"]. Shortly after this F. Wöhler [2] observed that small quantities of the spontaneously inflammable gas also result if Mn-Si alloys are treated with dilute nitric acid. As F. Wöhler reported [3], it was then that, by chance, the way opened up for preparing the gas in larger amounts: "It was Mr. [C. A.] Martius*), the son of my famous friend in Munich (Dr. E. W. Martius) who first observed in this laboratory that a slag which had been obtained in preparing magnesium by Deville's method [6] had the property of giving off a spontaneously inflammable gas with hydrochloric acid. In this case the gas could only have been a silicon hydride, produced without doubt from magnesium containing silicon" ["Es war Herr [C. A.] Martius, Sohn meines berühmten Freundes in München (Dr. E. W. Martius), welcher im hiesigen Laboratorium zuerst die Beobachtung machte, daß eine Schlacke, die bei der Darstellung von Magnesium nach Deville's Verfahren [6] erhalten worden war, die Eigenschaft hatte, mit Salzsäure ein selbstentzündliches Gas zu entwickeln. Dieses Gas konnte in diesem Falle nur Siliciumwasserstoffgas sein, entstanden ohne Zweifel aus siliciumhaltigem Magnesium"]. Working from this observation he developed a method of obtaining a suitable starting material for obtaining the "siliconhydrogen gas" ["Siliciumwasserstoffgas"], though he attempted in vain to separate the spontaneously inflammable gas from the accompanying hydrogen and an analysis was therefore impossible. He also observed that the separation of brown elementary silicon on thermal decomposition was accompanied by an increase in the gaseous volume.

Formation of this inflammable gas in connection with the preparation of magnesium silicides was noticed subsequently on a number of occasions [5 to 11] without a better understanding of the gas being reached. C. Winkler [9] also noticed that the concentration of the hydrochloric acid used was significant in determining the degree of inflammability. R. Vogel [11] also arrived at similar results. The first observation on the yield of spontaneously inflammable gas when a so-called "magnesium silicide", which was not exactly stoichiometrical in composition, was dissolved in dilute acid was made in 1902 by H. Moissan, S. Smiles [12]. In their experiments they cooled the crude gas with liquid air and were able in this way to separate hydrogen from the gas, which appeared as a white solid substance. On fractional distillation they believed they had been able to show that the white substance consisted of two compounds, one of which was a gas at room temperature and the other a liquid. The boiling point of the latter was determined as 52°C and the formula Si_2H_6 was assigned to it in the belief that they had found an analog of ethane C_2H_6 [in fact they had obtained trisilane, Si_3H_8], see the following section. That Moissan's liquid was not homogeneous was recognized in 1909 by P. Lebeau [13]. A threefold higher yield of silanes than that of Moissan was obtained 30 years later by W. C. Johnson, T. R. Hogness [14] and W. C. Johnson, S. Isenberg [15] when they decomposed (pure) magnesium silicide at low temperatures in a solution of ammonium bromide in liquid ammonia. In the main, they obtained mono- and disilane as well as small quantities of trisilane.

---

*) The name Martius is altered to Martins in the French translation [4] of Wöhler's original publication [3]. As a result, the originator of this preparative method appears occasionally [5] under this changed name.

The silane mixture prepared in the usual way from magnesium silicide was separated in a chromatographic gas-liquid column by K. Borer, C. S. G. Phillips [27] and the three lower members of the series were isolated in sufficient quantities for identification. The existence of *i*- and *n*-Si_4H_{10} as well as of *i*-and *n*-Si_5H_{12} was established with certainty spectroscopically and it was also possible to recognize that members of this series with $n = 6$, 7, and 8 were present as several isomers.

In the early sixties new reactions were discovered which also gave rise to silane mixtures. In 1963 P. L. Timms, C. S. G. Phillips [28] announced that dissolution of silicon monoxide (obtained by reaction of silicic acid with elementary silicon and condensation from the gas phase) in 10% hydrofluoric acid yielded a mixture of silanes in which they were also able to identify the isomers of the higher members (up to $n = 6$). Two years later Timms and his co-workers [29], reported that decomposition of the rubbery and highly polymeric $(SiF_2)_x$ in 20% hydrofluoric acid gave silanes in 40% yield.

The question of why acid decomposition of silicides, and especially of so-called "magnesium silicide", resulted in several silanes remained unexplained for a long time. In 1923, A. Stock, P. Stiebeler, F. Zeidler [16] thought it likely that, just as the acid decomposition of pure iron carbide, which had been investigated just before by R. Schenk, J. Griesen, F. Walter [17] resulted in a range of hydrocarbons, various silanes could be formed with pure magnesium silicide as the starting material. In 1925, however, R. Schwarz, T. Hoefer [18] had started to clarify the course of the reaction. They had noticed that investigators, especially F. Wöhler [3], A. Geuther [6] and C. Winkler [9] who had dealt with the residues from the decomposition of "magnesium silicide", had arrived at very different results. By using alcoholic hydrochloric acid (instead of pure aqueous), the otherwise vigorous hydrolytic decomposition of a steel-blue magnesium silicide Mg_2Si could be repressed so that short lived intermediate products could be obtained. This led to the assumption of an intermediate unsaturated compound SiH_2, which they named *silen* (see the following sections). They assumed that, as a radical, silen was not stable but immediately condensed, going over gradually to a highly polymerized state though, before this is reached, some underwent hydrolysis to form prosiloxane H_2SiO and various silanes [19].

The reasons for the spontaneous inflammability of the silanes was puzzling for a long time. In 1916, A. Stock, C. Somieski [20] maintained that the phenomenon depended only on chance. C. Friedel, A. Ladenburg [21], however, had already indicated the true way of clarifying the question when, in 1867, they demonstrated that the gas investigated by them did not burn at first when exposed to air, "but that a very small increase in temperature sufficed to bring about ignition. When the hot tip of a knife was brought near to the gas bubbles rising from mercury they burnt immediately with a loud crack if already mixed with air. The resulting warming of the mercury ignited subsequent bubbles" ["doch reicht eine sehr geringe Temperaturerhöhung hin, eine Entzündung zu bewirken. Nähert man den über Quecksilber aufsteigenden Blasen des Gases eine heiße Messerspitze, so verbrennt dasselbe augenblicklich, unter lebhaftem Knall, wenn es schon mit Luft gemengt war. Die so bewirkte Erwärmung des Quecksilbers reicht hin, die folgenden Blasen zu entzünden"]. Reduction of the pressure of a silane-air mixture also leads to self-ignition. K. Adwentowski, E. Drozdowski [22] felt unable to confirm either these results or the views of H. Moissan, S. Smiles [12] and P. Lebeau [13], who thought the self-inflammation to be attributable to contamination by higher silanes or unsaturated silicon hydrides, since the Polish workers had used what in their view was very pure monosilane and explosive oxidation was always observed on contact of the gas with air. Decomposition of the silane under the influence of light was also thought to result in self-ignition. In 1913 D. Berthelot, H. Gaudechon [23] asserted that the silanes decomposed on irradiation with light into hydrogen and elementary silicon, which precipitated as an iridescent yellowish layer on the container walls. Reexamination of this point in 1935 by R. Schwarz, F. Heinrich [24] both with the separate members of the series and with mixtures showed this to be incorrect. Twelve years before this systematic study A. Stock and his co-workers [16] had already shown that silane mixtures were scarcely altered after year-long standing in light, that the pure compounds trisilane Si_3H_8 and tetrasilane Si_4H_{10} were quite stable and that liquid mixtures which contain higher members of the series decomposed only slowly with yellowing. That in fact only a small increase in energy in a silane mixture is required to bring about a spontaneous explosive combustion in contact with air, Hume-Rothery [25] was able to show since the noticeably small increase in the heat of reaction, when he used a concentrated copper(II) chloride solution in metallographic etching of silicon-containing aluminum alloys, sufficed to bring about explosion in the air of all the resulting gas bubbles. In accordance with the above, the oxidation mechanism for the silanes may be readily investigated, as A. Stock, C. Somieski [26] have shown, if there is sufficient cooling and further precautions are taken.

References:

[1] F. Wöhler, H. Buff (Liebigs Ann. Chem. **103** [1857] 218/29). – [2] F. Wöhler (Liebigs Ann. Chem. **106** [1858] 54/9, 56). – [3] F. Wöhler (Liebigs Ann. Chem. **107** [1858] 112/9). – [4] F. Wöhler (Ann. Chim. Phys. [3] **54** [1858] 218/25). – [5] E. Vigouroux (Ann. Chim. Phys. [7] **12** [1887] 153/96, 171/3).

[6] A. Geuther (J. Prakt. Chem. **95** [1865] 424/40, 427). – [7] H. N. Warren (Chem. News **58** [1888] 215/6). – [8] L. Gattermann (Ber. Deut. Chem. Ges. **22** [1889] 186/97, 188). – [9] C. Winkler (Ber. Deut. Chem. Ges. **23** [1890] 2642/68, 2654). – [10] K. Seubert, A. Schmidt (Liebigs Ann. Chem. **267** [1892] 218/48, 235).

[11] R. Vogel (Z. Anorg. Allgem. Chem. **61** [1909] 46/53, 51). – [12] H. Moissan, S. Smiles (Compt. Rend. **134** [1902] 569/75). – [13] P. Lebeau (Compt. Rend. **148** [1909] 43/5). – [14] W. C. Johnson, T. R. Hogness (J. Am. Chem. Soc. **56** [1934] 1252). – [15] W. C. Johnson, S. Isenberg (J. Am. Chem. Soc. **57** [1935] 1349/53).

[16] A. Stock, P. Stiebeler, F. Zeidler (Ber. Deut. Chem. Ges. **56** [1923] 1695/705). – [17] R. Schenk, J. Griesen, F. Walter (Z. Anorg. Allgem. Chem. **127** [1923] 101/22). – [18] R. Schwarz, T. Hoefer (Z. Anorg. Allgem. Chem. **143** [1925] 321/35, 326). – [19] R. Schwarz (Z. Angew. Chem. **53** [1940] 6/11, 7). – [20] A. Stock, C. Somieski (Ber. Deut. Chem. Ges. **49** [1916] 111/57, 112).

[21] C. Friedel, A. Ladenburg (Liebigs Ann. Chem. **143** [1867] 118/28, 124/6). – [22] K. Adwentowski, E. Drozdowski (Anz. Akad. Wiss. Krakau Math. Naturw. Kl. A **1911** 330/44). – [23] D. Berthelot, H. Gaudechon (Compt. Rend. **156** [1913] 1243/5). – [24] R. Schwarz, F. Heinrich (Z. Anorg. Allgem. Chem. **221** [1935] 277/86, 279). – [25] W. Hume-Rothery (Chem. News **143** [1931] 328/30).

[26] A. Stock, C. Somieski (Ber. Deut. Chem. Ges. **55** [1922] 3961/9). – [27] K. Borer, C. S. G. Phillips (Proc. Chem. Soc. **1959** 189/90). – [28] P. L. Timms, C. S. G. Phillips (Inorg. Chem. **3** [1964] 606/7). – [29] P. L. Timms, R. A. Kent, T. C. Ehlert, J. T. Margrave (J. Am. Chem. Soc. **87** [1965] 2824/8).

2.1.3 Individual Silanes

The members of the group of silanes were described as colorless liquids with a disagreeably musty smell, which ignited in air with a loud crack and were barely soluble in water though they dissolved fairly readily in alcohol, benzene and carbon disulfide [1]. Liquid densities are less than one and also about 30 to 50% less than those of the corresponding hydrocarbons. Melting points are low but higher than those of analogous carbon compounds, the same being true of the boiling points. Stock's prediction on the position of the melting points of silanes relative to those of hydrocarbons did not hold for monosilane: he believed the obviously special position of monosilane to result form an incorrect melting point in the case of methane but K. Clusius [36] showed the latter's melting point to be correct.

2.1.3.1 Monosilane SiH_4

The first preparation of the pure compound, which previously had been known only mixed with other silanes, was successfully carried out in 1867 by C. Friedel, A. Ladenburg [2] when they decomposed "silico-ortho-formic acid ethyl ester" (triethoxy-monosilane according to Stock's nomenclature) at higher temperatures with metallic sodium in what A. Stock, C. Somieski [3] called a "remarkable reaction" which takes place according to the equation $4\,SiH(OC_2H_5)_3 \rightarrow SiH_4 + 3\,Si(OC_2H_5)_4$ without change in the sodium. Friedel and Ladenburg obtained the pure compound as a colorless gas which was not spontaneously inflammable at ordinary temperature and pressure. H. v. Wartenberg [4] reported in 1912 that Friedel and Ladenburg's preparative method may be operated in a lower temperature range by using metallic potassium. In 1910, A. Vournassos [5] observed that monosilane resulted on treating silicon-containing aluminum with sodium formate. The chemical behavior was investigated quite fully even by the first to prepare monosilane and, shortly afterwards, R. Mahn [6] studied the action on halogenated compounds, iodine and bromine, and observed the formation of new compounds. The first determination of the boiling point was made by A. Dufour [7] in 1904, but as K. Adwentowski, E. Drozdowski [8] noticed in their preparations for measuring the vapor pressure and critical data, the Friedel-Ladenburg product was not quite pure, and this led A. Stock, C. Somieski [3] in 1916 to use extraordinarily careful separation methods for the silane mixtures and to redetermine physical constants on the best fractions.

Synthesis from the Elements. Synthesis of monosilane from the elements was first attempted in 1871 by C. Friedel [9], without however any marked success. For this purpose he allowed an electric spark to operate in a hydrogen stream between silicon electrodes and thought that he could observe volatilization of the silicon similar to that which Troost and Hautefeuille [37] were able to demonstrate in a stream of chlorine under the same conditions with formation of silicon chlorides, see p. 127. Since, however, he was prevented from studying the transport phenomenon more closely in 1871 "through the painful events which we have just had to take part in" ["par les douloureux événements que nous venons de traverser"] he could speak of the formation of hydrides under the given conditions only as a conjecture. A. Dufour [7] in 1904 was the first to undertake such studies again, finding the compound to be formed, even if in small quantities, on heating silicon in a stream of hydrogen above its melting point. That its formation sets in, however, even at 1600 to 1700°C was first noted by O. Ruff, M. Konschak [10] in measuring the boiling point of the element. In 1933 C. J. Warnke [11] filed a patent according to which, for the preparation of SiH_4, a stream of hydrogen should be blown between a silicon electrode and a mercury or metallic gallium electrode. According to R. Schwarz [12], for silicon, as opposed to carbon, a total synthesis of lower members of the hydrides from the elements at higher temperatures is not possible, as is apparent from the heats of formation (+18 kcal for CH_4 and −6.7 kcal for SiH_4).

Further Methods of Preparation. Monosilane was first prepared with good results from the tetrachloride by hydrogenation with a mixed metal hydride $LiAlH_4$ by A. E. Finholt, A. C. Bond, K. E. Wilzbach, H. I. Schlesinger [13] in 1947. Six years before H. J. Emeléus, A. G. Maddock, C. Reid [14] had shown that monoiodosilane SiH_3I can be reduced to monosilane with metallic magnesium in di-isoamyl-ether, a reaction which E. R. van Artsdalen, J. Gavis [15] later carried out with SiH_3Br, although with quite a small yield.

In 1935, W. C. Johnson, S. Isenberg [44] reverted to the decomposition of magnesium silicides mentioned above for preparing silanes, going over, however, from an acid as the reaction partner to ammonium bromide dissolved in liquid ammonia. In this way they obtained a monosilane with only a very little disilane, its quantity being dependent on the decomposition temperature. Yields could be improved, as F. Fehér, W. Tromm [45] showed in 1955, if one carried out the reaction with hydrazinium chloride in hydrazine, when contamination by disilane was also reduced to about 1%.

Thermal decomposition of monosilane into the elements at temperatures up to a maximum of 400°C, which J. Ogier [25] had already observed in 1880, was first investigated more precisely almost 60 years later by T. R. Hogness, T. L. Wilson, W. C. Johnson [33]. That disilane Si_2H_6 is formed at temperatures which lie about 100°C higher was found by G. Fritz [34] in 1952.

2.1.3.2 Disilane, Silicoethane Si_2H_6

H. Moissan, S. Smiles [16] thought they had observed this compound for the first time in the spring of 1902 when they succeeded in condensing the silane mixture from the acid decomposition of so-called magnesium silicide by cooling with liquid air. Some months later the two discoverers [17] published some physical data for their product and some results on its chemical behavior, the most striking of these being the explosive reaction with tetrachloromethane CCl_4, in which hydrogen chloride results together with elementary carbon and silicon. According to A. Stock, P. Stiebeler [18], however, who were the first to prepare the compound pure and publish its properties, the French investigators definitely did not have disilane in their hands, but rather a mixture of various higher silanes, though chiefly trisilane. Shortly after the first publications on disilane H. Moissan [19] published a new method of preparation in which it could be obtained by acid decomposition of a lithium silicide formulated as Si_2Li_6. A. Stock, C. Somieski [20] were unsuccessful in 1921 in their attempts to obtain disilane by a method analogous to the Wurtz synthesis, involving the action of monochlorosilane SiH_3Cl on alkali metals or their amalgams. They could, however, show that the compound was produced as an intermediate but was immediately decomposed by the alkali metal. In 1947 A. E. Finholt, A. C. Bond, K. E. Wilzbach, H. I. Schlesinger [13] found that disilane can be obtained in very high yield if hexachlorodisilane Si_2Cl_6 is reduced in ether solution with lithium aluminum hydride.

The thermal decomposition, in which monosilane, silicon, and hydrogen are produced, was studied more fully by H. J. Emeléus, C. Reid [35]. Such experiments were first taken up again nine years later in 1948 by K. Stokland [46].

2.1.3.3 Trisilane, Silicopropane Si_3H_8. Tetrasilane Si_4H_{10}. Higher Silanes

Trisilane was first obtained in 1902 by H. Moissan, see above; it was first prepared pure in 1916 by A. Stock, C. Somieski [3] by the very careful fractionation of the crude silane mixture. They were also successful in obtaining *tetrasilane,* some properties of which were described. More accurate physical data were published by A. Stock, P. Stiebeler, F. Zeidler [21] in 1923. They found that in the case of tetrasilane it was a single substance and not a mixture of isomers. The difficultly volatile part of the condensed crude silane, which from its molecular weight should have been a mixture of *penta-* and *hexasilane,* they could not separate further. More recent investigations of the crude silane, which yielded larger quantities of tetrasilane of high purity, confirmed not only the correctness of the physical data but also provided some information on the compound's chemical behavior [22].

The thermal decomposition of trisilane, which leads to monosilane, silicon, and hydrogen, was studied in detail in 1939 by H. J. Emeléus, C. Reid [35], see also [46].

2.1.3.4 Unsaturated (solid) Silicon Hydrides

A *silico acetylene, disilen* Si_2H_2 was thought in 1909 by P. Lebeau [23] to have been found in the mixture of gases obtained in the decomposition of "magnesium silicide" by acids. As the analog of acetylene he assigned the formula given above. Its boiling point was thought to be above 60°C and its chemical character was held responsible for the spontaneous inflammability of the silanes. However, neither A. Stock, C. Somieski [3] in 1916 nor A. Stock in collaboration with W. Siecke [21] seven years later in the course of their extensive and careful work-up of large quantities of the crude silane mixture were able to detect the compound. The statement by L. Wöhler, F. Müller [24] that disilen results from calcium silicide Ca_2Si_2 on acid decomposition is also only a conjecture: volatile unsaturated silicon hydrides appear not to be obtained, see, however, below.

In 1880, J. Ogier [25] had already observed the production of a yellow solid as well as hydrogen on passing a silent electrical discharge through [impure] monosilane. He took the new substance to which he ascribed the formula $(Si_2H_3)_n$ on the basis of the hydrogen formed when it was treated with alkalis, to be an unsaturated silicon hydride. A. Stock, C. Somieski [3], however, showed that at the time no unexceptionable analytical data had been given and furthermore that the homogeneity of the material, which had already been questioned by A. Besson [26] in 1912, had not been established in any way. In Stock's view [3, 21] the same was true for another yellow, solid, unsaturated silicon-hydrogen compound which C. B. Jacobs from New York discovered and studied in 1900 and was described in a publication by C. S. Bradley [27]. It was said to be obtained from alkaline earth silicides by treatment with acids and was formulated as Si_2H_2 by analogy with the corresponding reaction with calcium carbide. In 1933, G. Führ [28] first confirmed Ogier's [25] old findings but two years later R. Schwarz, F. Heinrich [29] showed that when silanes are decomposed in a silent electrodeless ring discharge various highly polymeric silicon hydrides of an unsaturated character resulted, the composition of which fluctuated between $SiH_{1.2}$ and $SiH_{1.7}$. In addition, they were successful in preparing a compound with the formula $(SiH_2)_x$, which could be described as a homogeneous substance, by decomposition of calcium silicide with glacial acetic acid or absolute alcoholic hydrochloric acid. They described this substance, which they called *polysilene,* as a light brown solid which was spontaneously inflammable in air and was decomposed by mineral acids or alkalis without formation of saturated silanes, though it yielded a range of saturated silanes on cracking. The last phenomenon was noticed by A. Stock and his coworkers [3, 21] for all the yellow and apparently almost crystalline substances which they had obtained in the spontaneous decomposition of silane mixtures or by the action of sodium amalgam on chlorosilanes, and for whose composition they found values between $(SiH)_x$ and $(SiH_{1.6})_x$.

In more recent times methods have been published by which well defined unsaturated silicon hydrides, which are now termed *polysilanes (n)* (where n is the hydrogen index in the old formulation $(SiH_n)_x$) may be obtained readily in larger amounts. Thus, *polysilane (2)* was made by M. Schmeisser, M. Schwarzmann [30] by hydrogenating the analogous bromine compound $(SiBr_2)_x$ with lithium aluminum hydride $LiAlH_4$, and *polysilane (1)* was made by decomposing silicobromoform $SiHBr_3$ in ether with magnesium by G. Schott, W. Herrmann [31]. They also studied its properties more closely [32].

It has been shown that the compound first prepared in 1969 by R. C. Chittick, J. H. Alexander, H. F. Sterling [38] by a silent electric discharge in highly diluted monosilane and thought to consist of layers of amorphous silicon and which a few years later has acquired a highly promising significance in semiconductor technology contains a network of Si-Si and Si-H bonds of indefinite composition, as was established by A. G. Revesz [39] on the basis of work in progress. An investigation to determine the hydrogen content more accurately was undertaken by M. H. Brodsky, M. A. Frisch, J. F. Ziegler [40] by bombarding the layers with accelerated $^{15}N^{++}$ ions and measuring the γ-radiation following the reaction $^{15}N+H \rightarrow {}^{12}C+{}^{4}He+\gamma$. The hydrogen content was found to depend markedly on the method of preparing the layers.

The question whether the silicon hydride SiH which can only be detected spectroscopically, see "Silicon" Suppl. Vol. B 1, 1982, pp. 7/59; "Silicium" B, 1959, pp. 227/8, or other unsaturated volatile silicon hydrides exist and can be handled arose in 1958 when R. Schäfer, W. Klemm [41], in investigating the behavior of silicon in a hydrogen stream at higher temperatures, observed a brownish black deposit of silicon in the colder parts of their apparatus. They thought it likely that the "transport" of the element occurred in the form of endothermic Si-H compounds (SiH, SiH_2) which form at temperatures of 1100 to 1200°C from silicon and molecular hydrogen and decompose in the colder parts of the apparatus. They considered that SiH_4 formation could be excluded. W. Mannchen, K. Bornkessel [42] believed silicon "transport" phenomena were also involved in the degassing of metallic magnesium with a small silicon content: they believed that they could deduce Si:H values from the loss ratio and that silicon was transported as SiH_2. These and other questions that emerged K. Bornkessel, J. Pilot [43] attempted to resolve mass-spectroscopically when they conducted the degassing in a mass-spectrometer but it was not possible to decide on the subhydride responsible for the transport.

References:

[1] A. Stock (Z. Elektrochem. **32** [1926] 341/9). – [2] C. Friedel, A. Ladenburg (Liebigs Ann. Chem. **143** [1867] 118/28, 124/5). – [3] A. Stock, C. Somieski (Ber. Deut. Chem. Ges. **49** [1916] 111/57, 112). – [4] H. v. Wartenberg (Z. Anorg. Allgem. Chem. **79** [1912/13] 71/87, 73). – [5] A. Vournasssos (Ber. Deut. Chem. Ges. **43** [1910] 2272/4).

[6] R. Mahn (Jenaische Z. Med. Naturw. **5** [1869] 163 from Jahresber. Fortschr. Chem. **1869** 248). – [7] A. Dufour (Compt. Rend. **138** [1904] 1040/2). – [8] K. Adwentowski, E. Drozdowski (Anz. Akad. Wiss. Krakau Math. Naturw. Kl. A **1911** 330/44). – [9] C. Friedel (Compt. Rend. **73** [1871] 497/9). – [10] O. Ruff, M. Konschak (Z. Elektrochem. **32** [1926] 515/25, 518, footnote 3).

[11] C. J. Warnke, Adam Westlake Co. (U.S. 1967952 [1933/34] from C. **1935** I 286). – [12] R. Schwarz (Chemie [Berlin] **56** [1943] 258/61, 259). – [13] A. E. Finholt, A. C. Bond, K. E. Wilzbach, H. I. Schlesinger (J. Am. Chem. Soc. **69** [1947] 2692/6). – [14] H. J. Emeléus, A. G. Maddock, C. Reid (J. Chem. Soc. **1941** 353/8, 357). – [5] E. R. van Artsdalen, J. Gavis (J. Am. Chem. Soc. **74** [1952] 3196/7).

[16] H. Moissan, S. Smiles (Compt. Rend. **134** [1902] 569/75). – [17] H. Moissan, S. Smiles (Compt. Rend. **134** [1902] 1549/52). – [18] A. Stock, P. Stiebeler (Ber. Deut. Chem. Ges. **56** [1923] 1087/91). – [19] H. Moissan (Compt. Rend. **135** [1902] 1284). – [20] A. Stock, C. Somieski (Ber. Deut. Chem. Ges. **54** [1921] 524/31).

[21] A. Stock, P. Stiebeler, F. Zeidler (Ber. Deut. Chem. Ges. **56** [1923] 1695/705). – [22] H. J. Emeléus, A. G. Maddock (J. Chem. Soc. **1946** 1131/4). – [23] P. Lebeau (Compt. Rend. **148** [1909] 43/5). – [24] L. Wöhler, F. Müller (Z. Anorg. Allgem. Chem. **120** [1922] 49/70, 69). – [25] J. Ogier (Ann. Chim. Phys. [5] **20** [1880] 5/66, 33/4).

[26] A. Besson (Compt. Rend. **154** [1912] 1603/6). – [27] C. S. Bradley (Chem. News **143** [1931] 328/30). – [28] G. Führ (Diss. Univ. Frankfurt a.M. 1933, pp. 1/44, 39/43). – [29] R. Schwarz, F. Heinrich (Z. Anorg. Allgem. Chem. **221** [1935] 277/86). – [30] M. Schmeisser, M. Schwarzmann (Z. Naturforsch. **11b** [1956] 278/82, 280).

[31] G. Schott, W. Herrmann (Z. Anorg. Allgem. Chem. **288** [1956] 1/8). – [32] G. Schott, W. Hirschmann (Z. Anorg. Allgem. Chem. **288** [1956] 9/14). – [33] T. R. Hogness, T. L. Wilson,

W. C. Johnson (J. Am. Chem. Soc. **58** [1936] 108/12). – [34] G. Fritz (Z. Naturforsch. **7b** [1952] 507/8). – [35] H. J. Emeléus, C. Reid (J. Chem. Soc. **1939** 1021/30).

[36] K. Clusius (Z. Physik. Chem. B **23** [1933] 213/25, 216/7). – [37] P. Troost, L. Hautefeuille (Compt. Rend. **73** [1871] 443/7). – [38] R. C. Chittick, J. H. Alexander, H. F. Sterling (J. Electrochem. Soc. **116** [1969] 77/81). – [39] A. G. Revesz (Thin Solid Films **50** [1978] L29/L33). – [40] M. H. Brodsky, M. A. Frisch, J. F. Ziegler (Appl. Phys. Letters **30** [1977] 561/3).

[41] R. Schäfer, W. Klemm (J. Prakt. Chem. **277** [1958] 233/41, 239). – [42] W. Mannchen, K. Bornkessel (Z. Metallk. **51** [1960] 482/5). – [43] K. Bornkessel, J. Pilot (Z. Naturforsch. **16a** [1961] 432/4). – [44] W. C. Johnson, S. Isenberg (J. Am. Chem. Soc. **57** [1936] 1349/53). – [45] F. Fehér, W. Tromm (Z. Anorg. Allgem. Chem. **282** [1955] 29/40).

[46] K. Stokland (Trans. Faraday Soc. **44** [1948] 545/55).

2.2 Silicon-Oxygen Compounds

2.2.1 Silicon Monoxide

In 1954, B. C. Weber, P. S. Hessinger [1] stated that Clemens Winkler had postulated the existence of a compound SiO as early as 1890, but in fact, the opposite is the case. At the time, the discoverer of germanium wrote the following on the basis of his studies on the reduction of silica by metallic magnesium: "From this behavior one may conclude that a *silicon monoxide* SiO does not exist as such, it is also not obtained by strongly heating a mixture of 60 parts by weight (1 mol) of silica with 25 parts by weight (1 at.) of amorphous or crystalline silicon" ["Man darf aus diesem Verhalten schließen, daß ein *Siliciummonoxid* SiO nicht existiert, wie sich dann ein solches auch nicht durch heftiges erhitzen eines Gemenges von 60 Gewichtsteilen (1 Mol) Kieselsäure und 25 Gewichtsteilen (1 At.) amorphem oder krystallisirtem Silicium erhalten ließ"] [2]. This statement that the compound did not exist was, however, modified by H. N. Potter [3] in 1907: "As this experiment was performed in a combustion furnace, under conditions where the maximum temperature probably did not exceed 1000°C, the conclusions based thereon can only be accepted within the temperature range involved. If the Winkler experiment be extended to temperatures attainable with ease in the electric furnace, a brisk reaction takes place, with liberation of a brown smoke which burns to silicon dioxide. If the reaction is effected in an inert atmosphere, in a furnace of the resistance type the smoke can be collected as a soft, brown, very fine and voluminous deposit upon all exposed surfaces within the furnace." Analysis of the brown substance corresponded almost exactly with the values to be expected for SiO: "The proof, therefore, seems conclusive, that silicon can and does reduce its own dioxide, with the production of a lower oxide, which analysis indicates to be the monoxide." He established that carbon and silicon carbide may also be employed as reducing agents. After a detailed study of the reaction product, however, he concluded: "We have never succeeded in finding any chemical test to distinguish silicon monoxide from a mixture of silicon and silica."

H. N. Potter named the new product, for the production of which he designed a new furnace [4], "Monox", and described it as an extremely fine light brown powder which was silky to the touch and for which he suggested all sorts of possible uses. On account of its pleasing color tone and its great fineness it could be used as a pigment either alone or with other pigments. The combination of its fineness and hardness would make it a particularly valuable material for polishing metals. In the pressed and sintered form it could serve for preparing whetstones while, as a simple powder it is probably the worst heat conductor of all the mineral substances [5].

H. N. Potter was, however, not the first to observe silicon monoxide, but he is notable because of his reference to C. Winkler and because he saw possible uses for his "Monox" and also took out patents for the preparation and applications in many countries [6]. C. A. Zapffe, C. E. Sims [20] mention seven patents, the last of which was granted in 1914. The first to publish something on the brown vapor and the solid condensation product in his trap was F. J. Tone [7], who was concerned with the preparation of carborundum. On 27th April, 1905, during a lecture to the general meeting of the American Electrochemical Society in Boston, he reported: "It may be interesting . . . to further mention the product which is formed under certain conditions of operation and which contains a considerable proportion of amorphous silicon. When portions of the furnace charge are too highly heated, volatilized silicon is expelled from the furnace. If this vapor comes freely into contact with the air it oxidizes to SiO_2 in dense white fumes. In cases, however, probably due to deficiency of oxygen, there is found on exposed portions of the charge a brown vitreous substance consisting principally of a homogeneous mixture of silica and amorphous silicon. Considerable analytical work has been done on this product by Mr. E. D. Thebaud, and it was first assumed, on account of the homogeneous character of the substance and the relative proportions of silicon and oxygen, that it was monoxide of silicon. This view has since been found to be erroneous, and the fact is mentioned here simply as being further ground for the belief that the compound silicon monoxide does not exist. Certain experimenters have described the occurrence of this suboxide, but it may be noted that, in the operation of the silicon furnace, neither the incomplete reduction of silicon dioxide nor the partial oxidation of silicon gives a product which can be identified as silicon monoxide. This product containing amorphous silicon analyzes as follows: Si (amorphous) 27.05, SiO_2 65.97, C 4.90, Fe_2O_3 0.42, Al_2O_3 1.26, CaO 0.40. It is very brittle, is apparently without crystalline structure and has a specific gravity of 2.22 at 24°C. It is a non-conductor of electricity." It may be mentioned at this point that Tone's view was confirmed 36 years later by X-ray studies in his old laboratory [8].

Opinions on the existence of a monoxide of silicon, the analog of carbon monoxide, remain as conflicting today as they were right at the beginning after its discovery. In 1965, W. E. Dasent in his book "Nonexistent Compounds" [9], in discussing SiO, and answering the question of the existence of this compound, concluded that it undoubtedly exists as a monomer in the gas phase at high temperatures and may be identified spectroscopically and mass-spectroscopically. The thermodynamic data derived from such studies also lead to the conclusion that silicon monoxide is formed quantitatively at high temperatures from the dioxide and elementary silicon in the gas phase but, on the other hand, there is an extensive and contradictory literature on the solid phase which results on cooling this gas.

Spectroscopic proof of the existence of the gaseous molecule was first furnished in 1927, when K. F. Bonhoeffer [10] was able to photograph its absorption spectrum. An emission spectrum, which, from the conditions of its occurrence, had to be associated with a compound of silicon and oxygen, was attributed to SiO_2 as early as 1908 by A. de Gramont, C. de Watteville [11] and later by W. Jevons [12]. In 1927, however, W. H. B. Cameron [13] suggested, on the basis of similarities to CO, that it was associated with the molecule SiO. Five years previously some arc emission bands had been assigned to this molecule on chemical grounds [14], though this had been ignored at the time [10]. The mass-spectroscopic detection was first carried out by R. F. Porter, W. A. Chupka, M. G. Ingram [15] in 1955.

Whether the condensation product, which varies in appearance according to the conditions (which was first studied in detail in 1949 by G. Grube, H. Speidel [17]), but always conforms analytically to the formula SiO, is a mixture or a chemical individual was considered by E. Zintl and his co-workers [16] in 1940 to be a question which had "not yet been settled". The fine brown powder gave no X-ray interferences and the vitreous solid form usually showed only very weak indications of the interferences of elementary silicon. Electron diffraction studies of the solid substance by H. König [18] in 1948 showed it to be a homogeneous substance, and this was confirmed a year later by the X-ray investigations of G. Grube, H. Speidel [17], who predicted that the solid black form, which chemical analysis showed to have the composition $SiO_{1.1}$, consists chiefly of amorphous silicon monoxide on the basis of the results of X-ray photographs. The brown material, however, clearly consisted of a mixture of silicon and silicon dioxide. In 1953, M. Hoch, H. L. Johnston [19] explored another approach to the detection of a solid silicon monoxide when they studied a mixture of elementary silicon and silicon dioxide roentgenographically in a high-temperature camera during heating. When the mixture had reacted completely at 1300°C they observed only the occurrence of a new cubic lattice, which, they considered, could belong only to silicon monoxide. This interpretation of the research was contradicted two years later by S. Gellen, C. D. Thurmond [21], who showed that the data could be interpreted as due to a mixture of β-cristobalite and β-silicon carbide. This interpretation was also in agreement with the results of a thermodynamic treatment by L. Brewer, R. K. Edwards [22] based on information on SiO in the literature, which was also partly checked. According to this, solid silicon monoxide is unstable below 1450 K and the metastable forms obtained by quenching begin to disproportionate appreciably between 400 and 700 K. A melting point would be expected at temperatures over 1975 K.

From the earliest observation of the „brown vapor" it was to be expected that (gaseous) silicon monoxide would attain great significance in technical processes, particularly in the production of silicon and ferrosilicon as well as of carborundum. In the case of the latter, H. Schäfer and R. Hörnle were able to show that silicon carbide is produced quite readily when gaseous silicon monoxide reacts with degassed sugar charcoal at 1300°C [23]. According to C. A. Zapffe, C. E. Sims [20, 24], the silicon content of steels is closely related to the formation of a monoxide, by the occurrence of which it is possible to explain many of the phenomena described in the relevant literature since 1850 as "anomalies of foundry operation", such as the occurrence of accumulations of silica in a variety of forms in the blast furnace. The first to suggest silicon monoxide formation in other technique processes were E. Zintl and his co-workers [16], who showed that it may be used for the extraction of silicon from silicates. Thus, for example, in the reaction between kaolin and elementary silicon an alumina with only 0.1% SiO_2 is obtained. Formation of SiO also appeared to them as a suitable way of reducing difficultly volatile metal oxides with silicon. In this way they obtained 99% metallic niobium (with 0.3% Si, the remainder oxygen) from, for example, NbO_2.

Finally, special mention must be made of the suitability of silicon monoxide for preparing evaporated protective films on metallic mirrors, a process which, according to H. Schäfer [23] has been operated on a technical scale since 1940 and has been described fully by G. Haß, N. W. Scott [25]. The evaporated SiO film oxidizes superficially to SiO_2. The film can be wiped with a sponge and is stable to heat, protecting the mirror surface from corrosion. Films for electron microscope investigations have also been prepared by a similar process [18].

References:

[1] B. C. Weber, P. S. Hessinger (J. Am. Ceram. Soc. **37** [1954] 267/72, 269). – [2] C. Winkler (Ber. Deut. Chem. Ges. **23** [1890] 2642/68, 2657). – [3] H. N. Potter (Trans. Am. Electrochem. Soc. **12** [1907] 191/214). – [4] H. N. Potter (Trans. Am. Electrochem. Soc. **12** [1907] 223/8). – [5] H. N. Potter (Trans. Am. Electrochem. Soc. **12** [1907] 215/22).

[6] H. N. Potter (Ger. 182082 [1905/07] from C. **1907** I 1764 [for instance]). – [7] F. J. Tone (Trans. Am. Electrochem. Soc. **7** [1905] 243/9, 247). – [8] H. N. Baumann (Trans. Electrochem. Soc. **80** [1941] 95/8). – [9] W. E. Dasent (Nonexistent Compounds. Compounds of

Low Stability, New York 1965, pp. 63/4). – [10] K. F. Bonhoeffer (Z. Physik. Chem. **131** [1928] 363/5).

[11] A. de Gramont, C. de Watteville (Compt. Rend. **147** [1908] 239/42). – [12] W. Jevons (Proc. Roy. Soc. [London] A **106** [1924] 174/94). – [13] W. H. B. Cameron (Phil. Mag. [7] **3** [1927] 110/5). – [14] A. del Campo, J. Estatella (Anales Soc. Espan. Fis. Quim. [Madrid] **20** [1922] 586/8). – [15] R. F. Porter, W. A. Chupka, M. G. Ingram (J. Chem. Phys. **23** [1955] 216/7).

[16] E. Zintl, W. Bräuning, H. L. Grube, W. Krings, W. Morawietz (Z. Anorg. Allgem. Chem. **245** [1940] 1/7). – [17] G. Grube, H. Speidel (Z. Elektrochem. **53** [1949] 339/40). – [18] H. König (Optik **3** [1948] 19/28). – [19] M. Hoch, H. L. Johnston (J. Am. Chem. Soc. **75** [1953] 5224/5). – [20] C. A. Zapffe, C. E. Sims (Iron Age **149** No. 5 [1942] 34/9).

[21] S. Gellen, C. D. Thurmond (J. Am. Chem. Soc. **77** [1955] 5285/7). – [22] L. Brewer, R. K. Edwards (J. Phys. Chem. **58** [1954] 351/8). – [23] H. Schäfer (Chemiker-Ztg. **75** [1951] 48/51). – [24] C. A. Zapffe, C. E. Sims (Iron Age **149** No. 4 [1942] 29/31). – [25] G. Haß, N. W. Scott (J. Opt. Soc. Am. **39** [1949] 179/84).

2.2.2 Other Suboxides and Oxides

O. Hönigschmid [1] in 1909 believed he had discovered two other suboxides, to which he assigned the formulae Si_3O_2 and Si_3O_4. In a renewed preparation and investigation of the substance consisting of yellow flakes, which F. Wöhler [2] had discovered in the decomposition of calcium silicide and had called "silicon", O. Hönigschmid became convinced that the gray-black flakes remaining on decomposing "silicon" in vacuum, which F. Wöhler had held to be a mixture of silica and amorphous silicon, were actually a homogeneous suboxide with the formula Si_3O_2. This would be the analog of carbon suboxide C_3O_2, though he accepted that irrefutable proof of this had not so far been produced. O. Hönigschmid also believed he had found a suboxide Si_3O_4, formed on decomposing leukon, the white substance, produced on irradiation of the "silicon" mentioned above, in vacuum. It was a yellow-brown crystalline material, the formula of which was confirmed by analysis. The existence of the compounds was not confirmed subsequently and, in 1921, H. Kautsky [3] was able to show that "silicon" is not a homogeneous material.

The sesquioxide Si_2O_3 was prepared in 1952 by G. H. Wagner, A. N. Pines [4] by heating silicon oxyhydride $SiHO_{3/2}$ in an inert atmosphere to 900°C. The authors believed the substance to be a true oxide and, on X-ray examination, the amorphous substance gave only a weak SiO_2 line and therefore could not be a mixture of silicon and silica. The compound could be melted (under argon) at 1650°C and then showed disproportionation roentgenographically.

The occurrence of a peroxide $Si(O_2)_2$ was suggested in 1949 by N. Serdaroglu [5], when he decomposed silicon tetrachloride with hydrogen peroxide at low temperatures.

References:

[1] O. Hönigschmid (Monatsh. Chem. **30** [1909] 509/25). – [2] F. Wöhler (Liebigs Ann. Chem. **127** [1863] 255/74). – [3] H. Kautsky (Z. Anorg. Allgem. Chem. **117** [1921] 209/42). – [4] G. H. Wagner, A. N. Pines (Ind. Eng. Chem. **44** [1952] 321/6). – [5] N. Serdaroglu (Istanbul Tek. Univ. Bul. **2** [1949] 87/94 from C.A. **1952** 11000).

2.3 Silicon-Nitrogen Compounds

Introduction. Among the problems which most concerned inorganic chemists at the beginning of this century, the one occupying the first place was the utilization of atmospheric nitrogen for agricultural and industrial purposes. In this connection, the bonding of nitrogen to silicon has had a certain importance for a certain time ever since it became known that silicon is able to take up nitrogen at high temperatures, and to give it up again as ammonia. This property, however, lost its significance when simpler and cheaper processes for the synthesis of ammonia were found, and interest in the compound abated. So it comes about that, in 1926, L. Wöhler [1] could say that "the question of the direct bonding of nitrogen by silicon has been solved neither practically nor at all clearly scientifically, where analogy with carbon would have led us to expect the synthesis of silicocyanogen SiN or even the still simpler elementary synthesis of silicohydrocyanic acid SiNH." According to R. Schwarz [2] a characteristic of silicon-nitrogen compounds is that in most cases they are highly polymeric and involatile. There is a true analogy to the carbon compound trimethylamine in $(SiH_3)_3N$, a readily volatile liquid which is not polymerized, which was first prepared by A. Stock, C. Somieski [3] in 1921. It is especially noteworthy that acid amides are known only for carbon, while, for silicon, the simultaneous bonding of nitrogen and oxygen to the element cannot be realized, so that no silico urea exists. The polypeptide linkage, which is so important in organic chemistry, is so far also unknown for silicon [2].

References:

[1] L. Wöhler (Z. Elektrochem. **32** [1926] 420/3). – [2] R. Schwarz (Z. Angew. Chem. **56** [1943] 258/61). – [3] A. Stock, C. Somieski (Ber. Deut. Chem. Ges. **54** [1921] 740/58).

2.3.1 Silicon Nitrides

F. Wöhler [1] appears to have been successful in preparing a silicon nitride in 1857, when he was able to write to J. C. Poggendorff, the editor of *Annalen der Physik und Chemie* on 19th October of that year: "The new silicon chloride, $Si^2Cl^3 + 2HCl$ [i.e., silicochloroform, see p. 133, discovered a short time before by H. Buff, F. Wöhler [2]] has increased in interest since, with it, one can produce a compound of nitrogen with silicon. I saturated it with ammonia gas, which took place with very strong heating, and obtained a white powdery substance which is a mixture of sal ammoniac with a compound of silicon chloride and ammonia. If this material is heated, a lot of sal ammoniac sublimes. Ammonia escapes and a white heat-stable substance remains. This is silicon nitride ("Stickstoffsilicium"). In order to obtain it quite free from sal ammoniac I put the material in a porcelain crucible standing in a carbon crucible on a strong coke fire. The silicon nitride remained behind as a white, very light and loose mass. It withstood a temperature at which nickel melts without melting or decomposing. It also appears to be unchanged on calcining in the air, or at least does not burn. It is unchanged even by boiling caustic potash, but evolves a large amount of ammonia when fused with caustic potash. Heated with red lead, it reduces the lead with incandescence and formation of nitrous acid. Fused with sodium carbonate, it behaves in the same way as I have shown for boron nitride. With the carbon of the carbonic acid it forms cyanogen. The fused mass [with caustic potash] is a mixture of potassium silicate and cyanate. With it I have prepared well characterized urea. If an excess of silicon nitride is taken, potassium cyanide is also formed, with which I was able to make Berlin blue. Deville has also produced silicon nitride with the ammonia compound of the ordinary chloride, $SiCl_3$ [i.e., $SiCl_4$]. Consequently we continue the closer investigation jointly" ["Das neue [kurz zuvor von H. Buff, F. Wöhler [2] entdeckte] Siliciumchlorür, $Si^2Cl^3 + 2HCl$, [das ist Siliciumchloroform, s. S. 133] hat noch dadurch an Interesse gewonnen, daß man damit eine

Verbindung von Stickstoff mit Silicium hervorbringen kann. Ich sättigte es mit Ammoniakgas, was unter sehr starker Erhitzung stattfand, und erhielt einen weißen, pulverigen Körper, der ein Gemenge von Salmiak mit einer Verbindung von Siliciumchlorür und Ammoniak ist. Wird diese Masse erhitzt, so sublimirt sich viel Salmiak; es entweicht Ammoniak und es bleibt eine weiße, feuerbeständige Substanz. Diese ist Stickstoffsilicium. Um es sicher frei von Salmiak zu erhalten, setzte ich die Masse in einen, in einem Kohlentiegel stehenden Porcellantiegel einem sehr starken Coaksfeuer aus. Das Stickstoffsilicium blieb als eine weiße, sehr leichte und lockere Masse zurück. Es verträgt eine Temperatur, bei der Nickel schmilzt, ohne zu schmelzen oder sich zu zersetzen. Auch scheint es beim Glühen an der Luft unveränderlich zu seyn, wenigstens verbrennt es nicht. Es wird selbst von kochender Kalilauge nicht verändert, aber mit Kalihydrat geschmolzen, entwickelt es eine große Menge Ammoniak. Mit Mennige erhitzt, reducirt es unter Feuerscheinung und Bildung von salpetriger Säure das Blei. Mit kohlensaurem Natron geschmolzen, verhält es sich auf dieselbe Weise, wie ich es vom Stickstoffbor gezeigt habe; es bildet mit dem Kohlenstoff der Kohlensäure Cyan. Die geschmolzene Masse [mit Kalihydrat] ist ein Gemenge von kieselsaurem und cyansaurem Kali. Ich habe damit wohl charakterisirten Harnstoff dargestellt. Nimmt man einen Ueberschuß von Stickstoffsilicium, so bildet sich zugleich Cyankalium, mit dem ich Berlinerblau machen konnte. Deville hat das Stickstoffsilicium auch mit der Ammoniak-Verbindung des gewöhnlichen Chlorids, $SiCl^3$ [das ist $SiCl_4$], hervorgebracht. Wir werden daher die nähere Untersuchung gemeinschaftlich fortsetzen"].

There first followed a short report [3] on this joint work which only confirmed what Wöhler had already published. A full paper first appeared two years later [4]. It dealt chiefly with boron and titanium and their behavior towards nitrogen. In the case of silicon, which was known in general terms not to react with oxygen at high temperatures, to their surprise they established that it was able to combine directly with nitrogen, on the contrary to the classification scheme which was developed in 1816 by A. M. Ampère [5] and generally accepted at the time. When they put crystalline silicon, which was contained in a crucible embedded in carbon powder "for more than an hour over a bright coke fire" they made the following observations: "After cooling the silicon had been largely transformed into a loose bluish mass, covered with a readily detached, coherent and feathery substance like the mineral mountain cork. The latter was white and dark colored only at the surface. Under the microscope it could be seen that this color was due to countless dark tombac-colored crystals situated on excrescences of a white finely crystallized substance. It could not be determined what these crystals with a metallic glance consisted of. When fused with caustic potash both the corky and the bluish material evolved a quantity of ammonia. The latter still contained unchanged silicon and was therefore heated in dry chlorine gas to remove this, as experience had shown that silicon nitride prepared from the chloride was as little decomposed by chlorine at a bright red heat as boron nitride. Silicon nitride so purified now evolved on melting with caustic potash a still greater amount of ammonia, with which sal ammoniac can be prepared and sublimed, this being converted to ammonium chloroplatinate ("Platinsalmiak"). When silicon nitride in a porcelain boat is heated in a Bohemian glass tube to a bright red heat in water vapor, introduced by carbon dioxide gas, which passes through hot water at about 90°C, a large amount of crystalline ammonium carbonate formed in the cold end of the tube. The decomposition proceeded only slowly, though to completion. The resulting silica was quite amorphous. Silicon nitride prepared from the chloride gradually decomposes water even at an ambient temperature. In a moist state it soon starts to smell of ammonia. Its equivalent composition is not yet established" ["Nach dem Erkalten fand sich das Silicium großentheils in eine lockere, bläuliche Masse verwandelt, bedeckt mit einer leicht ablösbaren, zusammenhängenden, faserigen, dem Bergkork ähnlichen Substanz. Letztere war weiß, nur an der Oberfläche dunkel gefärbt. Unter dem Mikroscop erkannte man, daß diese Farbe von zahllosen, dunkel-tomback-farbenen Krystallen herrührte, aufsitzend auf Warzen einer weißen, fein krystallisirten Substanz. Woraus diese metallglänzenden Krystalle bestanden, war nicht zu ermitteln. Mit Kalihydrat geschmolzen entwickelte sowohl die korkartige als auch die bläuliche Masse eine Menge Ammoniak. Da letztere noch unverändertes Silicium enthielt, so wurde sie zur Entfernung des letzteren in trockenem Chlorgas erhitzt, denn die Erfahrung hatte gezeigt, daß selbst das aus dem Chlorid bereitete Stickstoffsilicium so wenig wie das Stickstoffbor bei starker Glühhitze durch Chlor zersetzt wird. Das so gereinigte Stickstoffsilicium entwickelte nun beim Schmelzen mit Kalihydrat eine noch größere Menge Ammoniak, mit welchem Salmiak dargestellt und dieser sublimirt und in Platinsalmiak verwandelt werden konnte. Als Stickstoffsilicium auf einem Porcellanschiffchen in einem böhmischen Glasrohr bis zum starken Glühen in Wasserdampf erhitzt wurde, zugeführt durch Kohlensäuregas, das durch ungefähr 90° heißes Wasser strich, bildete sich im kalten Ende des Rohrs eine große Menge krystallisirtes kohlensaures Ammoniak. Die Zersetzung ging nur langsam, aber vollständig vor sich; die entstandene Kieselsäure war ganz amorph. Das aus dem Chlorid dargestellte Stickstoffsilicium zersetzt schon bei gewöhnlicher Temperatur allmälig das Wasser. Im feuchten Zustande fängt es bald an, stark nach Ammoniak zu riechen. Seine Aequivalentzusammensetzung steht noch nicht fest"].

This unexpected observation led the two chemists to some interesting reflections: "One can probably assume that silicon, next to oxygen, is the element present in the greatest amount by weight and the most widespread. Since it is not oxidizable in the dense crystalline state at the highest white heat, even in oxygen gas, it is indeed striking that it does not occur native in the free state. Water is also not decomposed by it at an ordinary red heat. It does, though, as the above experiment showed, combine directly with nitrogen. If one wished to give oneself up to geological phantasies, one could imagine that, at the time when our planet was formed when the elements combined together to the compounds which now make up its crust and mountain masses, silicon entered into combination with nitrogen and the still red-hot silicon nitride, coming into contact with water, decomposed to silica and ammonia. It was in this way that the ammonia was originally produced through which, on the first occurrence of living nature, nitrogen was introduced into the organic compounds formed" ["Man kann wohl annehmen, daß das Silicium, nächst dem Sauerstoff, das in der größten Gewichtsmenge vorhandene und verbreitetste der Elemente ist. Da es im dichten krystallinischen Zustand bei der höchsten Weißglühhitze selbst in Sauerstoffgas nicht oxydirbar ist, so ist es in der That auffallend, daß es nicht im freien Zustande, nicht gediegen vorkommt. Auch zersetzt es nicht bei gewöhnlicher Glühhitze das Wasser. Es vereinigt sich aber, wie der obige Versuch zeigt, direct mit dem Stickstoff. Wollte man sich geologischen Phantasieen hingeben, so könnte man sich vorstellen, das Silicium sei in der Bildungsperiode unseres Planeten, wo sich die Elemente zu den Verbindungen vereinigten, die jetzt seine Rinde und Gebirgsmassen ausmachen, mit Stickstoff in Verbindung getreten, und das noch glühende Stickstoffsilicium habe sich, mit Wasser in Berührung kommend, in Kieselsäure und Ammoniak zersetzt. So sei ursprünglich das Ammoniak entstanden, durch welches bei dem ersten Auftreten der lebenden Natur der Stickstoff in die entstehenden organischen Verbindungen eingeführt worden ist"] [4].

The two investigators did not comment on the composition of their product obtained in two different ways, apart from calling it "Stickstoffsilicium". It was not until twenty years later that P. Schützenberger [6], in repeating Deville and Wöhler's experiments, tried to analyze the substances obtained in such different ways, finding them not to be identical. Even the products of direct nitriding appeared not to be homogeneous. He believed in this case that he had obtained a nitride which was green and insoluble in hydrofluoric acid and boiling caustic potash solution, which, from its Si content, had the formula SiN. This could be distinguished from a white substance, soluble in hydrofluoric acid without gas evolution and for which he assumed the formula Si_3N_4. Further investigations in collaboration with A. Colson led P. Schützenberger [7] to the conclusion that the green material was not a nitride, but that he was dealing with a compound containing carbon with the composition Si_2C_2N, which they called "azotocarbure". This always forms if crystalline silicon at a white heat comes into contact simultaneously with carbon or carbon-containing gases and nitrogen. L. Weiss, T. Engelhardt [8] confirmed this in 1910, though they found the correct formula to be Si_3C_3N. In 1920, however, W. Moldenhauer [9], in describing the reactions of free nitrogen, doubted if this material was a uniform substance.

The joint work of F. Wöhler and H. Sainte-Claire Deville offered two routes for obtaining "Stickstoffsilicium", the first by treating a silicon-halogen compound with ammonia, which will be discussed later, and the second, and more interesting, by direct synthesis from the elements. L. Weiss, T. Engelhardt [8] showed, in their major investigation, that this second route is in fact possible, but requires temperatures between 1300 and 1400°C and very long reaction times. Then, however, when saturated with the gas, crystalline silicon yields a homogeneous product with the composition Si_3N_4. In this work they made the surprising observation that, depending on the method of purification of samples not saturated with nitrogen, different compounds were obtained, though it is not possible to go into this here.

In 1897, E. Vigouroux [12] was able to confirm, as H. Sainte-Claire Deville had already indicated, that very high temperatures had to be used to initiate reaction when so-called amorphous silicon was employed. Thirty years earlier A. Genther [11] had undertaken experiments for nitriding silicon with various silicides, though without any success. His experiments with pure silicon, on the other hand, had yielded "Stickstoffsilicium", which he did not analyze, though only at temperatures at which the iron tube used to protect the porcelain reaction tube started to melt. He described his product as a greenish substance, the chief component of which was white, though mixed with microscopic grains of silicon. After repeating the experiment in 1879, P. Schützenberger [6] also undertook the analysis of this white substance and gave it the formula, as Si_3N_4, a result which, in collaboration with A. Colson [7], he later believed had to be corrected to Si_2N_3. Subsequently many investigators studied the nitriding of silicon, stimulated by these variable analytical results and the observation by L. Weiss, T. Engelhardt [8] of an apparent lack of consistency in the nitriding. Like H. Funk [28] in 1924, however, they all came to the conclusion that only the compound Si_3N_4 had been obtained. In 1959, M. Billy [29] published a careful investigation of the course of the reaction and proved that the other silicon-nitrogen compounds, postulated earlier, do not exist.

A. Genther's experiments [11] to simplify the nitriding of silicon by using silicides of various metals, which were mentioned briefly above, have often been repeated and modified. They may be passed over here, especially as, for the most part, they were undertaken with the object of obtaining a compound analogous to calcium cyanamide, $CaCN_2$. In 1909, however, a patent was granted to the Norwegians A. S. Larsen, O. J. Storm [13], according to which they sought to obtain silicon nitride by treating liquid ferrosilicon with nitrogen. The nitride obtained by this method was in such a finely divided state that it could be used directly as a fertilizer or for the preparation of nitrogen compounds. The process was also said to be suitable for freeing other metal melts from (unwanted) silicon.

The first to propose, in nitriding silicon, to go over from the then expensive elementary form of silicon to a suitable mixture of silica and carbon was H. Mehnert [10] in Charlottenburg in 1895. He used an electric furnace and took out a patent on the process. In 1909 the Badische Anilin- und Soda-Fabrik (BASF), a firm taking a leading part in ammonia synthesis, applied for a patent [14] for a similar process, according to which the uptake of nitrogen in the process could be greatly accelerated by addition of various substances such as soda, iron, other metals, and metal oxides. Two years later, a further patent by the Norwegians mentioned above [15] claimed that the same result could be obtained by additions of chlorine or sulfur. When, however, L. Weiss, T. Engelhardt [8, p. 56] checked the process in a Moissan furnace in 1910, they obtained only colored sublimates which were free of nitrogen. As the authors correctly surmised, this unsatisfactory result was brought about only by unsuitable conditions in the furnace used [17]. The first scientific evaluation of this nitriding reaction appears to have been undertaken in 1913 by C. Matignon [16], who reported the reaction temperature as 1500°C and also calculated the heat of formation of the compound indirectly. More recent work by A. G. Nasini, A. Cavallini [18] showed that the reaction temperature specified was set too high. In 1925, E. Friederich, L. Sittig [17] studied the reaction somewhat more fully, and, inter alia, showed that addition to the reaction mixture of, for example, 10% of iron oxide, lowers the reaction temperature considerably and increases the yield.

The second route for the preparation of "Stickstoffsilicium" which the collaborative work of F. Wöhler and H. Sainte-Claire Deville opened up was based on the treatment of silicon halides with ammonia, see p. 87. In 1958 it was asserted [20] that the method had been "investigated as a new technique" in 1937 by S. Satoh [19] when he allowed silicon tetrachloride to react with ammonia in the cold. After removal of the ammonium chloride formed, a silicon amide $Si(NH_2)_4$ was obtained, which he decomposed in several stages at increasing temperatures until, finally, he had the pure silicon nitride Si_3N_4 in his hands. However, S. Satoh [19], by his literature citations, which cover all of the separate stages of the work, showed clearly that his method was not in any way new. It was also not quite new in the case of Wöhler and Sainte-Claire Deville, for 27 years earlier, in 1830, J. F. Persoz [21] had studied the saturation of silicon tetrachloride with ammonia and obtained a white and, as he stresses, heat stable substance which decomposed in water forming ammonium chloride and a "sort" of ammonium silicate. He found for the composition: "1 atom silicon chloride and 6 atoms of ammonia" ["1 Atom Chlorsilicium und 6 Atome Ammoniak"], that is $SiCl_4 \cdot 6\,NH_3$, a formula which, sixty years later, A. Besson [22] held to be correct for this material which is often called "Persoz's compound" ["Persozscher Körper"]. Meanwhile, in 1879, P. Schützenberger [23] turned his attention to the preparation of Persoz's compound, without, however, analyzing it. He was the first to observe the splitting off of ammonium chloride on heating in a stream of hydrogen and attributed to the substance remaining a complicated formula involving silicon, chlorine, nitrogen, and hydrogen. When he removed the chlorine completely by very long heating in a stream of ammonia, a white compound insoluble in water remained to which he assigned the formula Si_2N_3H. Ten years later L. Gattermann [24] formulated Persoz's compound as the diimide $Si(NH)_2$ or silicocyanamide $NSiNH_2$ on the basis of his own analyses and expressed the hope (which was not fulfilled) that he would be able to obtain the compound SiNH, silicohydrocyanic acid, by treating hexachlorodisilane Si_2Cl_6 with ammonia. When, however, F. Lengfeld [25] in 1899 prepared Persoz's compound in benzene solution at low temperatures so as to exclude the action of water completely, he obtained a solid white material which, after quantitative control of its formation, he considered to be silicon tetraamide $Si(NH_2)_4$. He observed that this material changed into Gattermann's diimide even at room temperature with loss of ammonia. The same result was also obtained in 1903 by E. Vigouroux, C. Hugot [26] by a somewhat simpler experimental method. A. Besson's claim [22], made in 1890, that in the reaction between ammonia and a tetrahalogenosilane exactly 7 moles NH_3 participate and not 6 or 8, is, however, directly opposed to these observations. Reexamination of the problem by A. Stock, F. Zeidler [27] in 1923 confirmed Besson's claim and established that the silicon compound richest in nitrogen to result is not a tetraamide but an analog of guanidine, di-amino-imido-monosilane, $Si(NH_2)_2NH$, which then goes over to the diimide. The reason for the mistake made by Lengfeld, who believed 8 moles of NH_3 participated in the reaction, was seen by A. Stock, F. Zeidler [27] to be the fact that benzene solutions of ammonia in contact with air very rapidly give off some gas and incorrect figures are consequently obtained. They also blamed the French workers for the absence of convincing analytical data. A. Stock, F. Zeidler [27] also believed that silicon tetraamide, if it existed, would not be associated and might even be volatile. A detailed study of reactions in the system $SiHal_4$-NH_3 was given in 1959 by M. Billy [29], followed by a careful analysis of the step-wise thermal decomposition of the reaction product, the final stage of which is Si_3N_4.

E. Vigouroux [30] maintained in 1897 that nitriding of silicon will occur even at a red heat. When this was checked by M. Billy [29], however, it was found that, although ammonia is decomposed on elementary silicon, nitride formation is first detectable at temperatures around 1250°C.

The first X-ray investigation of what the optical microscope showed to be very small thin lamellae of Si_3N_4, which were not birefringent and whose form could not be determined more exactly, was reported in 1934 by A. G. Nasini, A. Cavallini [18], though they were unable to determine the crystal structure. W. C. Leslie, K. G. Carroll, R. M. Fisher [31] in 1952 believed they had established its isomorphism with germanium nitride Ge_3N_4, to which an orthorhombic structure was attributed. The occurrence of two different forms of silicon nitride was established in 1957 by B. Vasiliou, F. G. Wilde [32] and, in the same year, these were more fully characterized by O. Glemser, K. Beltz, P. Naumann [33], who found that one form, which they called α-Si_3N_4, is obtained in a pure state by decomposing $Si(NH)_2$ at 1350°C. Mixed with a little β-Si_3N_4, this was also obtained on nitriding with ammonia at 1350°C, while synthesis from the elements (at the same temperature) yields about equal parts of the two forms. They designated α-Si_3N_4 as the low temperature form and believed it to be an superstructure of the high temperature stable β-Si_3N_4. In the following year it was shown that the α-form, which has a hexagonal structure, is transformed by prolonged heating into the β-form (which also has a hexagonal structure, but a unit cell which is half as big) [20].

L. Weiss, T. Engelhardt [8] were rightly very sceptical in expressing themselves on the possibility of a technical application for silicon nitride, especially, at that time, as a possible means of synthesizing ammonia: "Since, however, the silicon-nitrogen compounds are extremely stable, we doubt if these substances will become significant for technical purposes." However, as was shown fifty years later [34] this property, which was undesirable at the time, has "made silicon nitride to a new industrial material which has proved to have properties far superior to those of the usual materials." In fact, in the course of investigations it has been shown that the compound, which F. Wöhler showed in 1857 to be characterized by its heat resistance, possesses a remarkable resistance to molten non-ferrous metals, and will also resist molten steel and cast iron for a short time. This has made it suitable for the preparation of crucibles and boats for zone melting, casting channels and protective tubes for thermoelements and also for the lining of combustion chambers in gas turbines, high performance motors and nozzles for rockets [34]. In preparing such objects parts of the desired shape are prepared from commercial silicon powder with at the most 2% of impurities by slip casting, pressing, or sintering and then nitrided directly in suitable furnaces, usually with a high pressure of the gas [34, 35]. The parts produced in this way are characterized by permanency of dimension and high breaking modulus [35]. Silicon nitride proves to be specially suitable as a ceramic binder for carborundum objects [34], which are far superior in every respect to those made of pure silicon carbide [36].

References:

[1] F. Wöhler (Ann. Physik Chem. [2] **102** [1857] 317/8). – [2] H. Buff, F. Wöhler (Liebigs Ann. Chem. **104** [1857] 94/104, 94/8). – [3] H. Sainte-Claire Deville, F. Wöhler (Liebigs Ann. Chem. **104** [1857] 256). – [4] H. Sainte-Claire Deville, F. Wöhler (Liebigs Ann. Chem. **110** [1859] 248/50). – [5] A. M. Ampère (Ann. Chem. Phys. [2] **2** [1816] 5/32, 9).

[6] P. Schützenberger (Compt. Rend. **89** [1879] 644/6). – [7] P. Schützenberger, A. Colson (Compt. Rend. **92** [1881] 1508/11). – [8] L. Weiss, T. Engelhardt (Z. Anorg. Allgem. Chem. **65**

[1910] 38/104, 98). – [9] W. Moldenhauer (Die Reaktionen des freien Stickstoffs, Berlin 1920, p. 58). – [10] H. Mehnert (Ger. 88995 [1895] from C. **1897** I 142).

[11] A. Genther (Arch. Pharm. **173** [1865] 24/35; Jenaische Z. Med. Naturw. **2** [1866] 203/19). – [12] E. Vigouroux (Ann. Chim. Phys. [7] **12** [1897] 5/74, 43). – [13] A. S. Larsen, O. J. Storm (Ger. 217037 [1909] from C. **1910** I 306/7). – [14] Badische Anilin & Soda-Fabrik (Ger. 234129 [1909/11] from C. **1911** I 1467). – [15] A. S. Larson, O. J. Storm (Ger. 249246 [1911/12] from C. **1912** II 397/8).

[16] C. Matignon (Bull. Soc. Chim. France [4] **13** [1913] 791/3). – [17] E. Friederich, L. Sittig (Z. Anorg. Allgem. Chem. **143** [1925] 293/320, 313/4). – [18] A. G. Nasini, A. Cavallini (Proc. 9th Congr. Intern. Quim. Pura Aplicata, Madrid 1934, Vol. 3, pp. 280/93). – [19] S. Satoh (Sci. Papers Inst. Phys. Chem. Res. [Tokyo] **34** [1937/38] 144/54). – [20] T. Turkdogan, P. M. Bills, V. A. Tippett (J. Appl. Chem. **8** [1958] 296/302).

[21] J. F. Persoz (Ann. Chim. Phys. [2] **44** [1830] 315/25, 318/9). – [22] A. Besson (Compt. Rend. **110** [1890] 240/1). – [23] P. Schützenberger (Compt. Rend. **89** [1879] 644/6). – [24] L. Gattermann (Ber. Deut. Chem. Ges. **22** [1889] 186/97, 194). – [25] F. Lengfeld (Am. Chem. J. **21** [1899] 531/7, 534).

[26] E. Vigouroux, C. Hugot (Compt. Rend. **136** [1903] 1670/2). – [27] A. Stock, F. Zeidler (Ber. Deut. Chem. Ges. **56** [1923] 986/97, 992/7). – [28] H. Funk (Z. Anorg. Allgem. Chem. **133** [1924] 67/72). – [29] M. Billy (Ann. Chim. [Paris] [13] **4** [1959] 795/851). – [30] E. Vigouroux (Ann. Chim. Phys. [7] **12** [1897] 5/73, 44/5).

[31] W. C. Leslie, K. G. Carroll, R. M. Fisher (J. Metals **5** [1952] 204/6). – [32] B. Vasiliou, F. G. Wilde (Nature **179** [1957] 435/6). – [33] O. Glemser, K. Beltz, P. Naumann (Z. Anorg. Allgem. Chem. **291** [1957] 435/6). – [34] W. Schreiter (Seltene Metalle, Vol. 2, Leipzig 1961, p. 412). – [35] J. F. Collins, R. W. Gerby (J. Metals **7** [1955] 612/5).

[36] H. L. Read, F. C. Roe, H. S. Schroeder, W. L. Wroten (Ind. Eng. Chem. **47** [1955] 2513/6).

2.4 Silicon–Fluorine Compounds

2.4.1 Early Observations. The Etching of Glass

The earliest observations on silicon-fluorine compounds relate to the property of hydrofluoric acid of attacking glass und are to be found in old reports on the supposed softening and actual etching of glass by "corrosive liquor", which was often of complicated composition. According to M. Merrifield [1], the first report of this sort is in a manuscript from the first half of the 15th century in italianized latin entitled *Segreti per Colori* [Secrets about Colors] (MS No. 165 in the library of the San Salvatore cloister in Bologna) where it is No. 217 in a comprehensive collection of recipes: "To make a water for cutting glass. – Take vitriol, which comes from walls [probably wall saltpeter, calcium nitrate] and make a distilled water from it and keep it in a well closed vessel. Then take Roman vitriol [Roman alum] and pound it well, distil it and keep the water also in a closed vessel; then take sal-ammoniac and distil it, and keep this also. When you want to use the liquor, take equal quantities of each of these waters, mix them together, and draw with the mixed liquor upon the glass, and it will cut exactly as you like wherever it is wetted with this water" ["A fare aqua da tagliare el vetrio. – Tolli vitriolo che nasce per le mura et fanne aqua a lambico et serbala bene turata poi tolli vitriolo romano et pistalo bene et fanne aqua a lambico et serbala bene turata poi tolli sale armoniaco et fanne aqua alo lambico et serbala bene et quando la voraj operare tolli de le ditte aque de omne una tanto et mistale insiemj et disegna lo vetrio cum dita aqua et taliarasse dove sera bagnato cum dita aqua a tuo piacere"].

It is striking that "tagliare = to cut" is used in this recipe and not "intagliare = to etch", as is the case in another manuscript on the working of copper and iron, which is only a little more recent [1], especially as, in the same recipe No. 217, and immediately following the text quoted above, glass cutting with a diamond is described in detail (for the first time?). It is probable, therefore, that this recipe represents only an improvement on a method of cutting glass which was already known to Theophilus Presbyter in the 11th century in *Schedula Diversarum Artium* [2], in which a red hot cleaving iron was placed on the desired dividing line, which has previously been moistened. The "corrosive liquor" prepared by the new formulas ensures a more accurate separation of the panes by better wetting of the glass and by its corrosive action.

J. R. Partington [3] has drawn attention to a report on the attack of glass by chemical agents which is more than 200 years more recent. According to this, T. Birch [4] reported in his Supplements to the *Philosophical Transactions of the Royal Society of London* that in 1671 John Ray (1628–1705), the english clergyman, botanist, and discoverer of formic acid has informed the Society that a certain "Samuel Fisher had newly found out a menstruum [liquid solvent] that dissolved glass, and reduced it into a white calx; and that after the glass is well moistened with the menstruum, it may be shaved with a sharp knife almost like horn, though it is much more brittle than horn". Since this short report by J. R. Partington [3] has been included in the introduction to his discussion of Scheele's discovery of silicon tetrafluoride and hydrofluoric acid, it seems that the author considered it likely that Fisher's ingredients contained hydrofluoric acid, especially as it was known that the mineral fluorspar occurs in various locations in England (see "Fluor", 1926, p. 4), though any indication of the composition of the menstruum is completely lacking in the text cited.

The position is similar for another mention of a preparation with glass etching properties, which is only about eight years later. According to a study by Johann Beckmann (1739–1811) [5], professor of economic sciences in Göttingen, this is to be found in the *Teutsche Academie* of Joachim v. Sandrart for 1679 [6]: "Heinrich Schwanhardt, with his subtle understanding, has probed what up till now was considered to be impossible, and has invented such a "corrosive" as causes glass, which is otherwise so hard, and equally other metals and stones, to be etched at will both inwardly [that is, deeply] and in relief. These have until now otherwise been the best containers for all strong spirits [spirits means: acids and other substances obtained by distillation]. He has already portrayed complete pictures of people, in part naked and partly clothed, and some other kinds of animals, flowers, and plants which are quite natural, and has brought them into very high relief" ["Heinrich Schwanhardt hat mit seinem subtilen Verstande, dasjenige, was man bisher für unmöglich geschätzt, ergründet, und ein solches Corrosiv erfunden, dem das sonst so harte Crystalline Glas gehorsamen und gleich anderem Metall und Stein einwärts [das heißt: tief] und erhoben [erhaben] sich ätzen lassen muß, da es doch sonst aller starken Spirituum [Geister, das bedeutet: durch Destillation gewonnene Säuren und anderes mehr] beste Behältniß bisher gewesen ... Er hat bereits vollkommene Menschen-Bilder, theils nackend, theils bekleidet, auch allerhand Thiere, Blumen und Kräuter ganz natürlich [ab-]gebildet und es im Erheben sehr hoch gebracht [das heißt: die Bilder stark erhaben ausgeführt"]. H. Schwanhardt (sometimes written as Schwankhart [7]) had, as the Nürnberg historiographer J. G. Doppelmayr [8] reports, discovered the "corrosive" by chance: "After A[nno] 1670, fortunately and quite by chance he [H. Schwanhardt] discovered how to etch on glass panes, the ground on which appeared matt while all the writing put on was clear (the occasion for this was provided by his spectacles, which became quite frosted after aqua fortis [a distillate containing nitric acid] got onto them unintentionally, since they were of soft glass) ... He also developed the art greatly as the first one so as to cut the figures in relief on the panes ... as he was actually able to demonstrate this after many trials. Died October 2nd, 1693" ["Nach A[nno]. 1670 fande er [H. Schwanhardt] unvermutet (die Occasion hierzu gabe seine Brille, die, nachdem ein Scheidewasser [ein Salpetersäure haltiges Destillat] wider seine

Intension darauf kame, als ein weiches Glas, ganz matt wurde, worauf er dann seine Kunst noch weiter perfectioniret) glücklich auf gläserne Scheiben zu ätzen, auf welchen sich der Grund matt, dabey aber jede angebrachte Schrifft gantz hell ergabe . . . triebe auch die Kunst, um die Figuren erhoben auf die Gläser zu schneiden am ersten sehr weit . . . gleich wie er dieses alles durch viele Proben in der That sehr rühmlich erwiesen. Starb den 2. Oktober A. 1693"]. J. R. Partington [9] gave the following further biographical information on this pioneer of glass etching: Heinrich Schwanhardt, a son of Georg Schwanhardt (1601–1667), a pupil of the "Court-Gem Cutter" Caspar Lehmann – the title was bestowed on him by Kaiser Rudolf II in 1609 –, moved to Nürnberg before 1670, having previously lived in Prag. C. Lehmann is believed to be the one who discovered how to cut precious stones with the wheel. References to Schwanhardt's discovery by other contemporary historians concerned with the history of the free city of Nürnberg are also met with [10, 11], though they provide no additional information. An engraved glass plate made in Nürnberg with the etched inscription "Auxilium Iesu Christi adveniat" also bears the date 1686 and is in the Germanic Museum in Nürnberg. According to R. Schmidt [12], this can in all probability be attributed to H. Schwanhardt himself. In R. Schmidt's opinion the depth of the etching excludes the use of a fluorine-free "corrosive". The most recent interpretation of the round and slightly bowed plate appeared in 1963, together with the results of an investigation by the State Conservator at the museum, Dr. E. Meyer-Heisig, according to whom the plate is really etched, and all other frosting processes can be excluded [30].

T. J. Seebeck [13] reported as follows on three further glass panels with etched inscriptions when he handed them over to the President of the Verein zur Beförderung des Gewerbefleisses in Preussen for the "Exhibition of Glass Products in the possession of the Royal Crafts Institute" ["Ausstellung in der Sammlung von Glasfabrikaten, welche das Königl. Gewerbe-institut besitzt"] in 1828: "All three panels are round and made of green glass, two being bowed like clock glasses and the third plane. The ground on all three is etched and the written characters are raised" ["Sämmtliche 3 Scheiben sind rund, von Grünglas, zwei wie Uhrgläser gebogen, die dritte eben; auf allen dreien ist der Grund geätzt, die Schriftzüge sind erhaben"]. Like the first, two of the panels bear the date 1686 and the third, which is later, names another craftsman as having made it: "Anno 1703 den 23 Appriel ist bey den Herrn Conradt Rüssen dieses Fenster gemacht worden von Johann Helmhackh". According to T. J. Seebeck [13] this J. Helmhackh was born in 1679 in Nürnberg, became "Glasmeister" there in 1704 and also died there in 1760. T. J. Seebeck probably obtained the panels during his sojourn in Nürnberg [14]. He made known his view on the nature of the "corrosive" employed in relating the following experience: "Fluorspar must already have been found in commerce then, for in several old pharmacies in Nürnberg I found among old trade goods which had been put aside not only zircons, garnets, and chrysolites but also fluorspar, the latter, if I am not mistaken, with the lable "Bohemian Emerald". I still have a small quantity of this fluorspar and it is greenish throughout and consists of small clear hand-picked fragments, which luminesce fairly brightly when warmed" ["Flußspath muß damals schon im Handel vorgekommen sein, denn in ein paar alten Apotheken in Nürnberg fand ich unter den älteren zurückgestellten Handelsartikeln außer Zirkonen, Granaten, Chrysolithen, auch Flußspath, letzteren, wenn ich nicht irre, mit der Ueberschrift "Böhmischer Smaragd". Von diesem Flußspath besitze ich noch eine kleine Quantität, er ist durchgehends grünlich und besteht aus lauter kleinen ausgesuchten Bruchstücken, welche durch Wärme ziemlich lebhaft leuchtend werden"].

H. Cassebaum [15] refers specifically to the fact that the chemical composition of none of these panels is known and that they might also be made of a soft glass, like that used in the spectacles mentioned above, which can be attacked by fluorine-free aqua fortis. He was unable to decide therefore if the "perfected corrosive" had been a mixture containing hydrofluoric acid or not. J. Beckmann [5] had already regretted that he "had nowhere been able to discover a report on perfecting" ["die Perfectionirung nirgendwo habe gemeldet habe finden können"] and felt himself obliged to draw attention to the sensitivity of soft glasses to concentrated acids by referring to A. Baumé's *Experimentalchemie* [16]. H. Cassebaum [15] was able to show that familiarity with this property of glasses went much further back, when he found that recipes for etching glass were given in an anonymous book, which appeared in Hamburg in 1686 [20] under the title *"Schatzkammer rarer und neuer Curiositäten in den allerwunderbahresten Würckungen der Natur und Kunst..."* One such recipe concluded with the words: "On hard glass it is necessary to have stronger and sharper water" ["Auf härterem Glas muß man auch stärker und schärffer Wasser haben"]. From the identity of the dates on the panels and year in which the book appeared, H. Cassebaum [15] concluded that Schwanhardt's "corrosive" had been fluorine-free. J. R. Partington [9] was also very surprised when, in the English translation of the fourth edition of Gmelin *Handbook* the view was expressed – as he believed, for the first time [17] – that the "perfected corrosive" of the Nürnberg glass-cutter had contained fluorine. L. Gmelin [18] had, however, believed this to be correct almost thirty years earlier and, in so doing, was only following his father, Johann Friedrich Gmelin (1746–1804) in his *Geschichte der Chemie* [19], which appeared in 1798, and also the view of M. H. Klaproth in his *Chemisches Wörterbuch* of 1807 [29]. There was also a relevant observation in 1791 in *Hildt's Handlungs-Zeitung* [31]: "Aqua fortis does not attack glass. However, it clarifies it so much that Schwanhardt must already have been familiar with the art of etching glass in 1660 [!], although the statement about aqua fortis is in error" ["Scheidewasser greift Glas nicht an. Soviel aber erhellet daraus, daß Schwanhardt bereits 1660 [!] die Kunst, das Glas zu ätzen gewußt habe, obgleich die Angabe mit Scheidewasser irrig ist"]. J. R. Partington [9] was also convinced that "there is not the slightest foundation for this statement", although he did not refer to a corresponding study of the etching of old glasses.

According to J. H. M. Poppe [33], Schwanhardt and the few others in the know kept the art very secret, so that it was soon lost. It reappeared in 1725 in the form of a set of instructions which certainly involved hydrofluoric acid. In 1790, K. Mönch (1744–1805), a professor of chemistry at Marburg drew attention to a report [21] which had been taken over by J. Beckmann [5] and later by Friedrich Christian Accum (1769–1838), though the latter did not acknowledge its origin [22]; he was probably still professor for chemistry and mineralogy at the Survey Institute in London. Mönch himself wrote: "Schwanhardt and his pupils probably kept the preparation of this etching water secret, as the recipe for it was first disclosed in 1725. It may well be, however, that an older one is hidden away in some so-called book of crafts. During the year in question Dr. Joh. Georg Weygand from Goldingen in Curland [a town on the Abave in the Lettish Socialist Soviet Republic] sent a set of instructions [23] to the editor of the *Breslauer Sammlung zur Natur- und Medicingeschichte* [the physician Johann Kanold] which were said to have come originally from Dr. Matth. Pauli from Dresden, who was then deceased, and who, so it was said, had etched all kinds of figures, armorial bearings and landscapes on glass by this means" ["Vermutlich haben Schwanhardt und seine Schüler die Zurichtung dieses Ätzwassers geheim gehalten, denn erst im Jahre 1725 ist die Vorschrift dazu bekanntgemacht worden; doch kann es wohl seyn, daß schon eine ältere in irgend einem sogenannten Kunstbuche versteckt ist. In dem genannten Jahre schickte Doct. Joh. Georg Weygand aus Goldingen in Curland [Stadt an der Abave in der Lettischen Sozialistischen Sowjetrepublik] an den Herausgeber der *Breslauer Sammlung zur Natur- und Medicingeschichte* [den Arzt Johann Kanold] eine Vorschrift [23], die sich von dem damals verstorbenen Doct. Matth. Pauli aus Dresden herschreiben sollte, welcher, wie gemeldet wurde, auf diese Weise, manigfaltige Figuren, Wap[p]en und Landschaften in Glas geätzt hat"]. Mönch followed with an extract from these instructions, "as the *Breslauer Sammlungen* are not in the hands of every expert" ["da die *Breslauer Sammlungen* nicht in jedes Kenners Händen sind"]. The recipe itself reads: *"Invention of a sharp etching water by means of which all kinds of figures may be etched and corroded on glass at will* . . . Namely, when the spiritus nitri per destillationem [nitric acid obtained by distillation] has already passed down into the receiver, it is finally subjected to strong heat and poured when well dephlegmated [i.e., well dehydrated] into a Waldenburg (clay) flask (as it attacks ordinary glass). After this, some powdered Bohemian green emerald (otherwise known as hesphorus which, when powdered, gives a green glow when heated), is shaken in and is placed on the sand again for 24 hours. In the meantime a piece of glass which has been made free of all fat and clean with lye is taken and protected or covered properly round the glass border with wax, so that the border or rim is about one finger high. After this, the above sharp etching liquor is poured onto it, so that it is well and uniformly covered everywhere. It is allowed to remain on it, the longer the better. In this way it attacks the glass and the part covered with sulfur and vernis [linseed oil varnish] is left standing out and is anaglyptic [i.e., in relief]" [*"Invention von einem scharfen Ätzwasser, womit man ins Glas allerhand beliebige Figuren radiren und corrodiren kann* . . . Nemlich, wenn der Spiritus nitri per destillationem [Salpetersäure, durch Destillation erhalten] bereits in den Rezipienten hinuntergegangen, so treibt man ihn zuletzt mit starkem Feuer, und gießt ihn, wohl dephlegmirt [das heißt: gut entwässert] (weil er das ordinaire Glas angreift), in eine Waldenburgische [Ton-]Flasche; nachgehends schüttet man einen pulverisierten Böhmischen grünen Smaragd (sonst Hesphorus genannt, welcher pulverisirt in der Wärme grün leuchtet) darein, setzt es wieder 24 Stunden in den Sand; inzwischen nimmt man ein mit einer Lauge von allem Fett sauber und rein gemachtes Glas, und verwahret, oder fasset dasselbe rings um des Glases Rand mit Wachs sauber ein, daß die Zarge oder Bort ungefähr eines Fingers hoch sey; nachgehends gießet das obige scharfe Ätzwasser also darauf, daß dasselbige fein gleich allenthalben bedeckt sey, lässet sich darauf je länger je besser stehen; so greifet es dann das Glas an, und bleibt das mit Schwefel und Vernis [Leinölfirnis] Bedeckte erhaben und anaglyphisch [das heißt: reliefartig] stehen"] [23]. This passage was used by J. G. Krünitz [24] in 1777 under the heading "Etching Liquor" ["Etzwasser"] in his *Oeconomische Encyclopädie* without giving its origin, and this certainly did a great deal towards making the process known and spreading it.

It is quite certain that the "green Bohemian emerald" ["Böhmischer grüner Smaragd"] used in preparing the "sharp etching water" was a green variety of the known mineral fluorspar, the color of which "can sometimes be very beautiful, approaching emerald green" [26], as there is a specific mention of its characteristic thermoluminescence, which was discovered in Berlin in 1677 by J. S. Elsholz [27], who was Physician in Ordinary to the Elector, Friedrich Wilhelm von Brandenburg, sometimes referred to as the Great-Elector. Furthermore, A. S. Marggraf [28] in his investigations on fluorspar goes into the question of its other names, saying that the mineral is called pseudoemerald, among other things, by apothecaries. He also includes the name hesperus [greek: evening star] in the title, without going into it more closely. The name is said to go back to G. C. Kirchmaier (1635–1700), "professor of rhetoric and an alchemist" in Wittenberg [48, 49], who described the four "phosphors" which were then known, and also introduced the latin name for the star, Vesperugo. He did not give a reason for the name. In spite of his belief, which has been quoted above, that H. Schwanhardt could have carried out glass etching only with distillates containing strong nitric acid, H. Cassebaum [15] did not totally exclude the possibility that solutions containing hydrofluoric acid were already known to him. In fact, G.C. Kirchmaier, who came from Uffenheim (a town in Franken, Germany) undertook a journey to Regensburg in 1678 to visit relatives and also went to Nürnberg on a commission for the Brandenburg Elector. There, according to his own account, he learned from two craftsmen [this is more likely to be the term for analytical chemists (Scheide-Künstler) rather than that for hand workers with glass] about the "use of hesperus" [which, from the title of his publication *De Phosphoris et Natura Lucis* refers to the possibility of making the mineral luminescent, rather than to its use in glass etching]. This led A. Cassebaum [15] to suppose that, after he had been instructed by the Nürnberg glass-cutters, G. C. Kirchmaier had told Dr. M. Pauli about etching glass with hydrofluoric acid, either directly or through Johannes Kunckel. The relationship between these two savants appeared to H. Cassebaum to be confirmed by the fact that, in Dr. M. Pauli's library, which comprised more than 3000 volumes, the writings of the savants proved to have been present.

The report of the Marburg professor K. Mönch [21], which is discussed above, may have been included by Lorenz v. Crell (1744–1816) in his *Annalen der Chemie* in 1790 only for the sake of historical accuracy, as he had already accepted a short communication on "a process for etching figures on glass" ["Verfahren, Zeichnungen in Glas zu ätzen"] four years earlier [34]. It spoke quite clearly of hydrofluoric acid: "Herr Graf von G., an expert and lover of the natural sciences, has used the peculiar property of hydrofluoric acid of dissolving glass to etch letters, designs etc. on glass in a similar way to that in which copper is etched with aqua fortis" ["Herr Graf von G., ein Kenner und eifriger Liebhaber der Naturwissenschaften, hat die der Flußspatsäure eigene Fähigkeit, Glas aufzulösen, dazu angewandt, Buchstaben, Zeichnungen u.a.m. damit auf ähnliche Weise in Glas zu ätzen, als man mit Scheidewasser in Kupfer ätzt . . ."]. The editor comments on this: "I thank Mr. Assessor [M. H.] Klaproth for his kindness in communicating this useful method" ["Die Mitteilung dieser sinnreichen Methode verdanke ich der Gefälligkeit des Herrn Assessors [M. H.] Klaproth"]. At the time M. H. Klaproth was still an apothecary in Berlin und was busy with experiments on glass etching, during which he found the process which subsequently was to prove to be the most suitable for all applications of this art to be put forward in the course of the years, namely etching with gaseous hydrofluoric acid. In September 1787, when he had already become professor of chemistry at the Artillery School and at the Royal Mining and Foundry Institute, he communicated a paper to the annual assembly of the Berlin Academy of Arts entitled *"Über die Kunst, in Glas und Porzellän zu ätzen" ["On the art of etching in glass and porcelain"]*, which was printed a year later [35]. In this he first described in detail the known method of dipping or covering the surface to be treated with a fluorspar-acid mixture. He then added: "A second process, which is preferable, involves exposing it [the glass] only to attack by the vapor arising from such a mixture instead of bringing the glass or porcelain plate to be etched into direct contact with the stirred mass. Etching produced only by the vapor of the spar acid has the advantage over that [the first method] that, among other things, the graduations and fine hatchings appear to be more regular and clearer" ["Die zweite Verfahrensart, welche jener vorzuziehen ist, besteht darin, daß man, anstatt die zu ätzende Glas- oder Porzellanplatte mit der angerührten Masse in unmittelbare Berührung zu bringen, jene nur dem Angriff des von solcher Mischung aufsteigenden Dunstes bloßstellt. Es empfiehlt sich, diese durch den bloßen Dunst der Spatsäure bewirkte Einätzung vor jener [ersten] u.a. dadurch, daß die Striche und feinen Schraffierungen regelmäßiger und reiner ausfallen"]. Klaproth then gave practical instructions and recommended the use of this method strongly: "Making a glass micrometer for the astronomer's use is one of the most ingenious and difficult tasks in the art of glass cutting, because of the extraordinary degree of accuracy and sharpness with which the lines have to be engraved. Even with care, it is still not possible to prevent glass splinters from breaking away at the angles of the small squares which make up the network of intersecting lines. This detrimental happening is not possible if the lines are etched on the glass with spar acid vapor instead of being engraved with a diamond" ["Die Verfertigung der Glas-Micrometer zum Gebrauch der Astronomen ist, wegen der ungemeinen Genauigkeit und Feinheit, womit die Linien eingeschnitten werden müssen, eine der künstlichsten und mißlichsten Arbeiten der Glasschneidekunst; denn bei aller Vorsicht kann doch nicht verhütet werden, daß nicht in den Winkeln der kleinen Quadrate, welche von den netzförmig sich durchschneidenden Linien gebildet werden, Glassplitterchen ausspringen sollten. Dieser nachteilige Zufall ist aber nicht möglich, wenn die Linien, statt der Eingrabung mit dem Demant [Diamant], mit Spatsäuredunst aufs Glas geätzt werden"]. Later he stated: "For etching very fine lines, for example the lines on a glass micrometer, a few minutes [exposure] already suffice" ["Zur Einätzung sehr feiner Züge,

z. B. der Linien eines Glasmikrometers, wenige Minuten [der Einwirkung] schon hinreichen werden"]. The process caused a sensation and, in June of the same year, an extract had already appeared in the *Hildt's Handlungs-Zeitung* [36]. The Berlin professor of history, Johann Samuel Halle (1727–1810), also adopted the process, which he described as an "English discovery that had been kept very secret", in full detail in his collection *Fortgesetzte Magie* [25]. According to O. Vogel [32], Halle was the first to draw attention to the danger associated with glass etching: "Inhalation of the vapor is damaging to the health, as with all strong acids, and consequently a hearth, an open window is chosen for the experiment or, in winter time, a place near an open furnace door, so that the flame is able to draw the vapor to it" ["Das Einathmen der Dämpfe ist wie bei allen heftigen Säuren der Gesundheit nachtheilig, daher wählt man zum Versuche einen Feuerherd, ein offenes Fenster, und zur Winterszeit die Nähe einer geöffneten Ofenthüre, damit die Flamme die Dämpfe an sich ziehen möge"].

At the same time as Klaproth's article appeared it was claimed by L. v. Crell in a note in his *Chemische Annalen* [37] that the discovery of glass etching had also been made in France [in reality it was no new discovery for, as W. H. S. Bucholtz [51] reported, P. J. Macquer in 1778 had already given an account of the etching of glass with hydrofluoric acid-containing mixtures in his *Dictionnaire de Chymie*]: "It has been publicly announced by M. le Compte de Puymorin [more exactly: Jean Pierre Casimir de Marcassus, Baron de Puymaurin, 1757–1841] from Toulouse that he has made the same discovery of etching on glass with fluorspar acid (as our Herr Graf von G. in the *Annalen*), probably without in any way knowing about the latter's process. We let this last assurance rest on Mr. de Puymaurin's reputation, except that it is then just as certain that our Mr. Graf v. G. can have had no knowledge of the former's process. Furthermore the latter method also differs from the first in that he chose the acid which had already been set free, while Herr von G. applied a brew of fluorspar and vitriolic acid" ["Vom Herrn Grafen von Puymorin [genau: Jean Pierre Casimir de Marcassus, Baron de Puymaurin, 1757 bis 1841] aus Toulouse wird öffentlich bekannt gemacht, er habe dieselbe Entdeckung gemacht, auf Glas durch Flußspatsäure zu ätzen (wie unser Herr Graf von G. in den *Annalen*): ohne wahrscheinlicher Weise vom Verfahren des letzteren Kenntnis gehabt zu haben. Wir lassen diese letzte Versicherung auf des Herrn de Puymaurin Ansehen beruhen: nur daß dann noch so viel gewiß ist, daß unser Hr. Graf von G. keine Kenntnis von jenes Verfahren gehabt haben kann. Übrigens unterscheidet sich auch des letzteren Methode von der des ersteren, daß er die schon entbundene Säure wählt, Herr von G. den Brei aus Flußspat und Vitriolsäure auftrug"]. O. Vogel [32] commented on this somewhat peculiar report to the effect that the commission which had to scrutinize Baron de Puymaurin's communication before its acceptance for the *Observations sur la Physique* [38], and which included C. L. Bertholet as well as other members of the Paris Academy, had drawn attention to Count von G's publication, but that de Puymaurin had replied that he was not a master of the German language and that the report in *Crell's Annalen* had therefore remained unknown to him. De Puymaurin described how he came to discover the process as follows [38]: "In the laboratory of M. de Fourcroy I saw a window pane which had become frosted and eaten away by the gas which came from a retort in which there was a residue from the hydrofluoric acid distillation. Astonished by this rapid and peculiar effect, I wanted to see if I could produce a similar one with dilute hydrofluoric acid. In this I was successful, and it convinced me that hydrofluoric acid exerts almost the same action on glass as does nitric acid on copper and other metals. Only one step remained for me in order to make that property of hydrofluoric acid useful in the arts" ["Ich sah im Laboratorium des Herrn von Fourcroy eine Fensterscheibe, die matt geworden und angefressen war von dem Gas, das aus einer Retorte kam, in der sich ein Rückstand von der Flußsäure-Destillation befand. Erstaunt über diese rasche und eigenartige Wirkung, wollte ich versuchen, ob ich eine ähnliche mit verdünnter Flußsäure zu Wege bringen könnte. Dies gelang mir in der Tat und überzeugte mich, daß Flußsäure auf Glas fast die gleiche Wirkung ausübt, wie Salpetersäure auf Kupfer und sonstige Metalle. Es blieb mir nur noch ein Schritt zu tun, um von jener Eigenschaft der Flußsäure eine für die Künste nützliche Anwendung zu machen"]. Baron de Puymaurin exhibited various designs etched on glass to the Paris Academy of Sciences and to that at Toulouse, and in every case they were surprised by the clearness and perfection of the etched lines. One of these glass etchings was a symbolical portrayal of "Chemistry" and "Genius" weeping at C. W. Scheele's grave [C. W. Scheele died on 21st or 26th May, 1786] [32]. "This immortal chemist", de Puymaurin wrote [38] "succeeded in discovering hydrofluoric acid, one of whose properties, which he was the first to recognize, is now applied in the arts" ["Diesem unsterblichen Chemiker ist die Entdeckung der Flußsäure gelungen, deren eine Eigenschaft, die er als Erster erkannt hat, jetzt in den Künsten Anwendung findet"]. He continued: "My first attempt to etch glass with hydrofluoric acid was made in Toulouse on 17th May, 1787. Mr. [A. F.] de Fourcroy had stated in [the second edition of] his *Éléments d'Histoire Naturelle et de Chimie* [39]: The spar acid has found no sort of use until now, but its property of dissolving silica, will make it very useful later on. I have made a start in fulfilling part of this learned chemist's prediction in that I used hydrofluoric acid for etching glass. It can very readily be made useful in physics ... for matting glasses ... as well as for graduating ... Perhaps one day we can also be able to use it with thick glass for engraving plates for printing geographical maps etc." ["Mein erster Versuch, mit Hilfe von Flußsäure Glas zu ätzen, wurde am 17. Mai 1787 in Toulouse durchgeführt. Herr [A. F.] de Fourcroy hat in seinen *Éléments d'Histoire Naturelle et de Chimie* [39] noch [in der zweiten Auflage] angegeben: L'acide spathique n'a été jusqu'actuellement employé à aucun usage; mais sa propriété de dissoudre la terre silicieuse, le rendra vraisemblablement très-utile par la suite ... Ich habe den Anfang gemacht, einen Teil der Vorhersage dieses erfahrenen Chemikers zu erfüllen, indem ich die Flußsäure zum Glasätzen gebrauchte. Man kann sie sehr leicht für die Physik nutzbar machen ... zum Mattieren der Gläser ... sowie zum Graduieren ... Vielleicht wird man sie eines Tages auch verwenden können, um dicke Gläser zum Gravieren von Druckplatten für geographische Karten u.a.m. herzurichten"]. De Puymaurin had himself attempted printing, as he relates in a postscript, but the glass had broken during the second impression. O. Vogel [32] comments in a footnote that this had been remedied later by glueing the glass plate, which had already been etched, to an iron plate, when printing could be carried out with a copperplate press. This technique, which was also later often called hyalography [from the greek hyalos = glass] was subsequently used repeatedly in the course of the 19th century, but it is not possible to go into it here, and reference must be made to specialist handbooks [40]. The importance of the etching ground for the quality of the etching was appreciated quite early on, for example by Tuhten in Wolfenbüttel [41] in 1790. Here too, reference must be made to handbooks for an account of the development [40]. The sort of interest shown in the process of glass etching towards the end of the 18th century is shown by the fact that in the *"Sammlung chemischer Präparate zu unterhaltenden und nützlichen Experimenten für Liebhaber der physischen Scheidekunst und vorzüglich für Jugendlehrer beym Unterricht"*, which was sold in 1791 at a price of 5 ducats [a gold coin with a fine gold content of $\sim$3.1 g] by the Jena professor of chemistry, pharmacy, and technology Johann Friedrich Göttling (1755–1809) whose patron J.W. v. Goethe was, among the "tools" of this supplement to his first experimental chest was under No. 10 "Various small glass plates coated with varnish on which small figures are drawn in order to etch these figures in the glass with hydrofluoric acid vapor" ["Verschiedene kleine Glasscheiben mit Firniß überzogen, und worauf kleine Figuren gezeichnet sind, um mit der Flußspathluft diese Figuren in das Glas zu ätzen"] [50]. The significance which was attached to the process is clear from an anonymous report [42] in 1803: "At the dedication of the monument to generals [J.-B.] Kléber [murdered 1800] and [L. C. A.] Desaix [fallen at the battle of Marengo 1800] there was placed, among other things which testify to the states of the arts etc. etc. in France, a glass plaque etched in this manner. Thus no record is lost to posterity, for the resolve to erect this monument in honor of both heroes has been entrusted to such a lasting and uncorrodable material as glass. The first Consul himself [Napoleon] placed this plaque under a stone" ["Bey der Einweihung des den Generalen [J.-B.] Kléber [ermordet 1800] und [L. C. A.] Desaix [gefallen in der Schlacht bei Marengo 1800] errichteten Monuments hat man unter anderen Dingen, die den gegenwärtigen Zustand der Künste etc. etc. in Frankreich bezeugen sollen, auch eine auf diese Art geätzte Glastafel hineingelegt. So ging für die späteste Nachwelt keine Nachricht verlohren, da man einem so festen und unzerfreßbaren Körper, als Glas ist, den Beschluß der Regierung, dieses Monument zu Ehren beyder Helden zu errichten, anvertraute. Der Erste Consul selbst [Napoleon] legte diese Tafel unter einem Stein hinein"]. As can be learned from an entry in *the Dictionnaire de Biographie Française* [43], the monument seems to have disappeared in the meantime.

The first publication on glass etching to appear in the United States was in 1823 and had B. Silliman as author [44] and the interesting title: *Fluoric Acid of Gay-Lussac, and its application to the Etching of Glass.* It begins with the words: "As we have not seen any notice that this powerful acid has been obtained in this country, we will briefly mention that we procured it in full strength during the late winter . . .". It then describes the preparation of the acid, the lead vessels recommended in the literature being replaced by some made of pure silver: "The cost of the entire apparatus, alembic-head [for a distillation flask with capacity of 16 fluid ounces], tube, cap, and receiver was about sixty dollars". The glass etching which is then described was carried out with dilute liquid acid. The author reports that several years earlier he had tried to etch glass with gaseous hydrofluoric acid with more or less success, and was able to recommend the dilute liquid acid unreservedly, though its poisonous character must be borne in mind. C. Kampmann [45], in the historical section of his book on *Die Dekorirung des Flachglases,* believes this to have been the source for two publications which appeared almost simultaneously in 1832 and 1833, the first by the Cisterian monk Leopold Schmid [46] and the second by Heinrich Anton Poller [47], which were "as similar to one another as one egg is to another, so that one does not know who had copied from the other". The first part of Kampmann's assertion is certainly not correct and the second cannot be checked. The wealth of relevant literature, and the fact that it never goes into the interesting silicon compounds which result, precludes further discussion of the history of glass etching at this point.

References:

[1] M. Merrifield (Original Treatises Dating from the XIIth to the XVIIIth Centuries on the Art of Painting, Vol. 2, London 1849, pp. 332, 494/5, 666). – [2] Theophilus Presbyter (in: A. Ilg, Schedula Diversarum Artium, Book 2, Chapter 18, Vienna 1874, pp. 122/3). – [3] J. R. Partington (A History of Chemistry, Vol. 3, London 1962, p. 213). – [4] T. Birch (A History of the Royal Society for Improving of Natural Knowledge, A Supplement to the Philosphical Transactions, Vol. 2, London 1756 [Reprint Brussels 1967], p. 495). – [5] J. Beckmann (Beyträge zur Geschichte der Erfindungen, Vol. 3, Leipzig 1792, pp. 536/58, 546/51).

[6] J. v. Sandrart (L'Academia Todesca della Architectura, Scultura & Pittura or: Teutsche Academie der edlen Bau- Bild- und Malerey-Künste, Vol. 3, Pt. 2, Chapter 240, Nürnberg 1679, p. 346). – [7] L. Darmstaedter (Handbuch zur Geschichte der Naturwissenschaften und der Technik, 2nd Ed., Berlin 1908, p. 139). – [8] J. G. Doppelmayr (Historische Nachrichten von den Nürnbergischen Mathematicis und Künstlern, Pt. 1, Nürnberg 1730, pp. 249/50). – [9] J. R. Partington (Mem. Proc. Manchester Lit. Phil. Soc. **67** No. 6 [1922/23] 73/87). – [10] J. C. Wagenseil (De Sacri Romani Imperii Libera Civitate Noribergensis Commentatio, Altdorf b. Nürnberg 1697, p. 154).

[11] P. J. Zahn (Oculus Artificialis sive Telediophricus sive Telescopium, 2nd Ed., Nürnberg 1702, p. 516 from [9]). – [12] R. Schmidt (Das Glas, 1st Ed., Berlin 1912, p. 237, 2nd Ed., Berlin–Leipzig 1922, pp. 248, 405). – [13] T. J. Seebeck (Verhandl. Vereins Beförderung Gewerbefleißes [Berlin] **7** [1828] 246/7). – [14] J. C. Poggendorff (Biographisch-Literarisches Handwörterbuch zur Geschichte der exacten Wissenschaften, Vol. 2, Leipzig 1863, Column 889). – [15] H. Cassebaum (Silikattechnik **34** [1983] 213/4).

[16] A. Baumé (Chymie Expérimentale et Raisonnée, Vol. 3, Paris 1773, p. 392). – [17] L. Gmelin (Handbook of Chemistry [translated by H. Watts], Vol. 2, London 1849, p. 268). – [18] L. Gmelin (Handbuch der theoretischen Chemie, 2nd Ed., Vol. 1, Frankfurt a. M. 1821, p. 268). – [19] J. F. Gmelin (Geschichte der Chemie, Vol. 2, Göttingen 1798, p. 590). – [20] Anonymous (Schatzkammer rarer und neuer Curiositäten in den allerwunderbahresten Würckungen der Natur und Kunst, Hamburg 1686, pp. 467/8 from [15]).

[21] K. Mönch (Chem. Ann. Crell **1790** II 133/5). – [22] F. Accum (J. Nat. Phil. Chem. Arts Nicholson **4** [1800/01] 1/4). – [23] J. G. Weygand (Sammlung von Natur- und Medicin- wie auch hierzugehörigen Kunst- und Litteratur-Geschichten, Classis V, von physikalischen und medicinischen Erfindungen, so Mense Jan. 1725, entdeckt oder bekannt worden [Breslauer Sammlung], Breslau 1725, pp. 107/8). – [24] J. G. Krünitz (Oekonomische Encyclopädie oder Allgemeines System der Land- und Staatswirtschaft, Vol. II, Berlin 1777, p. 678). – [25] J. S. Halle (Fortgesetzte Magie oder die Zauberkräfte der Natur, Vol. 1, Vienna 1788, p. 516).

[26] M. Bauer (Edelsteinkunde, 2nd Ed., Leipzig 1909, p. 651). – [27] J. S. Elsholz (Miscellanea Curiosa Med. Phys. Acad. Naturae Curiosorum **8** [1677] 30). – [28] A. S. Marggraf (Hist. Acad. Roy. Sci. Belles Lettres Berlin **1768** 3/11, 3/4). – [29] M. H. Klaproth, F. Wolff (Chemisches Wörterbuch, Vol. 2, Berlin 1807, p. 324). – [30] J. P. Altmann (Das neue Lehrbuch der Glasätzerei, Stuttgart 1963, pp. 25/6).

[31] Anonymous (Hildt's Handlungs-Ztg. [Wöchentliche Nachrichten von Handel Manufakturwesen und Oekonomie Gotha] **8** June 11 [1791] 188 from [32]). – [32] O. Vogel (Glashütte **72** [1942] 37/42). – [33] J. H. M. Poppe (Geschichte der Technologie, Vol. 3, Göttingen 1811, p. 117). – [34] M. H. Klaproth (Chem. Ann. Crell **1786** II 494). – [35] M. H. Klaproth (Monatsschr. Akad. Künste Mech. Wiss. Berlin **1788** 85/91 from [32]).

[36] Anonymous (Hildt's Handlungs-Ztg. [Wöchentliche Nachrichten von Handel Manufakturwesen und Oekonomie Gotha] **5** No. 24 [1788] pp. 190/2 from [32]). – [37] L. v. Crell (Chem. Ann. Crell **1788** I 567). – [38] Baron de Puymaurin (Observations sur la Physique sur l' Histoire Naturelle et sur les Arts **1788** 419/90 from [32]). – [39] A. F. de Fourcroy (Éléments d' Histoire Naturelle et de Chimie, 2nd Ed., Vol. 2, Paris 1786, p. 27). – [40] E. Tscheuchner (Handbuch der Glasfabrikation nach allen ihren Haupt- und Neben-zweigen, 5th Ed., Weimar 1885, pp. 498/507), F. E. Fischer (Das Gesamtgebiet der Glasätzerei, Braunschweig 1892, pp. 1/77), J. B. Miller (Die Glasätzerei, 5th Ed., Vienna – Leipzig 1928, pp. 1/132).

[41] Tuhten (Chem. Ann. Crell **1790** II 241/3). – [42] Anonymous (Das Neueste und Nützlichste der Chemie Fabrik-wissenschaft Apotheker-Kunst und Oekonomie [Nürnberg] **6** [1803] 33/8). – [43] E. Francheschini (in: R. d'Amat, R. Limouzin-Lamothe, Dictionnaire de Biographie Française, Vol. 10, Paris 1965, Column 1173/5). – [44] B. Silliman (Am. J. Sci. Arts **6** [1823] 354/6). – [45] C. Kampmann (Die Dekorirung des Flachglases durch Aetzen und Anwendung chemigraphischer Reproductionsarten, Halle a. d. Saale 1889, pp. 125/41, 128).

[46] L. Schmid (Practische Anleitung auf Glas zu ätzen, Wien 1832, pp. 1/32). – [47] H. A. Poller (Die Kunst aller Arten von Zeichnungen als Blumen Thiere Landschaften Porträts usw. in Glas zu ätzen: Eine Erfindung der neuesten Zeit für Zeichnen Silhouttirer etc., Quedlinburg 1833 from [45]). – [48] G. C. Kirchmaier (De Phosphoris et Natura Lucis nec non de Igne Commentatio Epistolica, Chapter 1, § 3, Wittenberg 1680, pp. 6/7). – [49] F. Ferchl-Mittenwald

(Chemisch-Pharmazeutisches Bio- und Bibliographicon, Mittenwald 1938 [Reprint Wiesbaden 1971], p. 273). – [50] J. F. Göttling (Mag. Neueste Phys. Naturgesch. **7** No. 3 [1792] 190/9, 198).

[51] W. H. S. Bucholtz (Neueste Entdeckungen Chem. [Crell] **3** [1781] 50/64).

2.4.2 The Discovery of Fluorosilicic Acid and of Silicon Tetrafluoride

On 6th September, 1768, Andreas Sigismund Marggraf (1709–1782) presented the results of an investigation on fluorspar, which had started four years earlier, to the Berlin Academy of Sciences under the title: "Observation concernant une volatilisation remarquable d'une partie de l'espèce de pierre, à laquelle on donne les noms de flosse, flüsse, flus-spaht et aussi celui d'hesperos; laquelle volatilisation a été effectuée au moyen des acides" ["Observation on a remarkable volatilization of part of a type of rock which is called flosse, flüsse, flus-spath and also hesperos, the volatilization being brought about by means of acids"] [1]. According to H. Cassebaum [8] this investigation arose from a dispute between J. H. Pott, the professor for theoretical and practical chemistry at the Collegium Medicum in Berlin, and J. H. G. v. Justi, the Prussian Chief of Mines and Superintendent of the Prussian Glass and Steel works, on the fusibility of "fluorspar"-marble mixtures, in which the two opponents must have started out with different minerals going under the name "fluorspar". This probably accounts for the involved way in which the title is spelt out. Marggraf describes his interesting and decisive experiment as follows: "This experiment carried out on a small scale induced me to repeat it with a larger amount of this rock. I accordingly took some of the green variety, which was very finely powdered but had not been calcined for too long. I put 8 ounces of this green fluorspar or pseudo-emerald into a clean retort, poured 8 ounces of clear English oil of vitriol over it and rinsed down what had remained adhering to the neck [of the retort] with about three ounces of distilled water, when I brought everything to the bottom [of the retort]. I then shook it up well to mix every part and a sulfurous vapor at once resulted and after that a white sublimate, which settled as a dust on cooling, and was deposited in the neck of the retort. Over this I first put a receiver which I had held ready, sealed it well and placed the retort in a crucible with sand. Initially I applied gentle heat and, after excess moisture had gone over, still more of a similar sublimate came up, ever more of which became attached to the receiver in proportion to the increase in the heating. And, as in the experiments on a small scale, it took on the consistency of butter of antimony [$SbCl_3$] and could be melted with a glowing coal. After the furnace had been raised to incandescence, a sublimate which a glowing coal did not melt appeared, again as in the first [small] trial. But after later experiments this sublimate was found to be different from that which had been produced in the preceding experiments. When the cooled retort had been broken up I found there 12 ounces of residue, so that 4 ounces of the oil of vitriol had penetrated into the rock. The bottom of the retort had holes here and there, as if it had been fired at with shot, which is sufficient evidence of the fluxing property with which this rock is endowed. The receiver contained the liquid which had passed over, with a good quantity of sublimate, a very considerable quantity of which had, however, gone up into the neck of the retort towards the end [of the distillation]. This liquid, as well as the sublimate, had a strong smell of sulfur. I detached all the sublimate with a glass implement and combined it with the liquid in the receiver. I poured a little hot water over it, as well as over the sublimate which was also in the receiver. I rinsed and detached the sublimate and tried to dissolve it by triturating it for a very long time with hot distilled water, which succeeded fairly well. I filtered it all and, as a good quantity of insoluble material remained in the filter, I washed it as well as possible with hot distilled water. After drying, a light friable powder remained in the filter. Thereupon I oversaturated the liquid mentioned above which had passed through the filter with a solution of fixed alkali obtained from tartar [K_2CO_3]. I obtained a precipitate which settled to the bottom with much difficulty. I washed it carefully with hot water and, having dried in the usual way I had a very beautiful white powder which weighed two ounces and two drachms, which melted to a porcelain-like mass both in a crucible on charcoal and also with the blowpipe on a candle flame, which did not happen in any way with the white powder which had remained in the bottom of the filter. What is peculiar here is that the vitriolic acid, which otherwise 'fixes' other bodies, for example mercury or the volatile arsenic, which are volatile on heating, here brings about a completely opposite effect on this mineral, which otherwise resists heat strongly, and causes part of it to volatilize. Yet another fact deserves to be noticed, and that is that in this work with vitriolic acid, and also in the following with other acids, the glass of the receiver and that of the retort have been so strongly attacked that both are eaten away to a considerable extent".

["Cet essai fait en petit m'anima à le réitérer sur une plus grande quantité de cette pierre. J'en pris donc de celle de l' espèce verte qui avoit été pilée fort fine, mais qui n' avoit pas été si longtems calcinée. Je mis de ce Flus-Spaht verd, ou Pseudo-Smaragdus, 8 onces dans une retorte de verre net; je versai dessus 8 onces d'une huile de vitriol d'Angleterre claire; je rincai ce qui étoit demeuré attaché au col avec environ trois onces d'eau distillée, en le faisant tomber tout à fait au fond; je secouai bien le tout pour en mêler les parties ensemble, & aussitôt il s'éleva une vapeur de souffre suffoquante, & d'abord après un sublimé blanc, qui en se refroidissant tomba en poussiere & s'attacha au col de la retorte; surquoi j'adaptai d'abord un récipient que je tenois tout prêt, je le lutai bien, je mis la retorte dans une coupelle de sable; je donnai d'abord un feu doux, & après que l'humidité superflue eut passé, il s'éleva encore davantage d'un semblable sublimé, dont il s'attacha toujours plus au récipient à proportion de l'augmentation de la chaleur; &, comme dans l'essai en petit, il prit la consistance d'un beurre d'antimoine, & se laissa fondre au moyen d'un charbon embrasé. Quand le feu eut été augmenté jusqu'à l'incandescence, il parut, encore comme dans le premier essai, un sublimé que le charbon ardent ne fondoit pas; mais, après des recherches ultérieures, ce sublimé se trouva différent de celui qui avoit été produit dans l'expérience précédente. Quand la retorte refroidie eut été brisée, il s'y trouva un résidu de 12 onces, de facon que 4 onces de l'huile de vitriol s'étoient insinuées dans la pierre; le fond de la retorte avoit des trous cà & là, comme si l'on y avoit tiré avec de la dragée; ce qui témoigne assez la propriété fondante dont cette pierre est douée. Le récipient contenoit le liquide qui avoit passé, avec une bonne quantité de sublimé, dont pourtant il s' étoit élevé à la fin une portion assez considérable dans le col de la retorte; ce liquide avoit, aussi bien que le sublimé, une forte odeur de souffre. Je détachai tout le sublimé avec un outil de verre, & je le joignis au liquide dans le récipient; je versai dessus, aussi bien que sur le sublimé qui se trouvoit pareillement dans le récipient, un peu d'eau chaude; je rincai & détachai le sublimé; je cherchai à le dissoudre en le pilant fort longtems dans un mortier de verre avec de l'eau chaude distillée; ce qui réussit assez bien. Je filtrai le tout; & comme il restoit sûrement encore dans le filtre une bonne quantité de matiere insoluble, je l'édulcorai au mieux avec de l'eau chaude distillée; & après le desséchement il me resta dans le filtre une poudre legere & friable. Là-dessus je faoulai le liquide susdit qui avoit passé par le filtre avec une solution de sel alcali fixe tiré du tartre; j'obtins un précipité qui eut beaucoup de peine à se poser au fond; je l'édulcorai soigneusement avec de l'eau chaude distillée, & l'ayant fait convenablement sécher, j'eus une fort belle poudre blanche, pesant deux onces & deux dragmes, qui tant au creuset qu'au charbon, ou avec le chalumeau à la flamme de la chandelle, se fondoit en une masse semblable à de la porcelaine; ce qui n'arrivoit en aucune façon à la poudre legere qui étoit demeurée au fond du filtre. Ce qu'il y a de particulier ici, c'est que l'acide du vitriol, qui autrement fixe assez au feu les autres corps volatils, par exemple le Mercure ou l'Arsenic volatil, produit ici un effet tout contraire sur cette pierre qui résiste d'ailleurs fort au feu, & en volatilise ainsi une partie. Encore une chose qui mérite d'être remarquée, s'est que, dans ce travail avec l'acide du vitriol, aussi bien que dans les suivans avec d'autres acides, le verre du recipient & celui de la retorte soient si fortement attaqués, & l'un & l'autre considérablement rongés"].

Thus A. S. Marggraf in this work, which is often noticed only on account of his interpretation of the reactions described as involving the occurrence of a "new earth" [2], was the first to have had silicon tetrafluoride in his hands and to have observed its decomposition with water vapor. He was also the first to recognize the liquid, which is formed during the distillation in the receiver filled with water, as an acid and to prepare from it by neutralization with potash that very "flußspathsaure Kieselkali" which sixty years later J. J. Berzelius was to recommend as the starting material for the ready preparation of amorphous silicon, see p. 32. The existence of this acid and its difficultly soluble alkali salt was soon confirmed. This was done by C. W. Scheele [3], P. C. Abilgaard [4], T. Bergman [5], and J. C. F. Meyer [6], see pp. 111, 122.

"Marggraf's [observed] phenomenon on distilling fluorspar with vitriolic acid (1768)", so Johann Christian Wiegleb (1732–1800) believed in discussing [7] the most important publications in the chemical literature of the year 1771, "has probably stimulated C. W. Scheele to initiate a more exact investigation of the supposedly especially volatile earth, since Marggraf had not gone into it at all. Scheele [3], however, had a quite different picture of his investigation. He formed the opinion, namely, that the fluorspar contained a special acid, which was driven out by the vitriolic acid (and here he had judged correctly), and that this had the special property that, when it combined with water vapor, it formed the earth which appears in the process (this was quite wrong)" ["Marggrafs [beobachtete] Erscheinung bey Destillation des Flußspats mit Vitriolsäure (1768) hat wahrscheinlich C. W. Scheele angereizt, mit der vorgegebenen [angeblichen] besonderen flüchtigen Erde eine genauere Untersuchung anzustellen, weil Marggraf darauf gar nicht eingegangen war. Scheele [3] machte sich aber von seiner Beobachtung einen ganz anderen Begriff. Er urtheilte nemlich, daß der Flußspat eine besondere Säure enthielte, welche durch die Vitriolsäure ausgetrieben würde (und darin hat er sehr richtig geurtheilet), welche aber die besondere Eigenschaft besäße, daß sie mit Verbindung der Wasserdünste die dabei erscheinende Erde bilde (dies war ganz falsch)"] [7].

Whether it was really correct, as J. C. Wiegleb thought, that it was Marggraf's publication and the volatile earth in it or, as H. Cassebaum [8] supposed, the etching ot the glass vessel in the experiments (since C. W. Scheele appears to have already observed something of the sort in the spring of 1770 at the start of his investigations on phosphoric acid from bone ash) or a more general interest in the "phosphors", which stimulated him to study phosphorus smaragdinus, fluorspar, can hardly be decided now, since C. W. Scheele mentioned Marggraf's work in none of his three publications on fluorspar, so doubt may arise as to whether he knew about it at all in his earliest investigations. The year in question and the publication by Carl Wilhelm Scheele (1742–1786) on his *"Untersuchungen des Flußspats und seiner Säure"* ["Investigations on Fluorspar and its acids"] [3] are generally associated in the literature both with the discovery of hydrofluoric acid and with that of silicon tetrafluoride [9], while contemporaries, as the quotation given shows, only recognized the former. C. W. Scheele was in advance of A. S. Marggraf only in so far as he realized [3, § 11] that "the white material settling in the receiver in working with fluorspar" ["bey der Bearbeitung des Flußspat in der Vorlage sich absetzenden weißen Materie"] was not a new "volatile earth" but "Kieselerde" (SiO_2) for, he continued: "§ 12. That this must first have been produced during the operation [the distillation of fluorspar with oil of vitriol or in the rectification of the distillation product from the receiver] is proved to me by several experiments" ["§ 12. Daß diese Kieselerde erst unter der Arbeit [der Destillation von Flußspat mit Vitriolöl oder bei der 'Rectifikation' des Destillationsprodukts aus dem Rezipienten] erzeugt sein müsse, beweisen mir mehrere Versuche"] (the description of which may be omitted here). C. W. Scheele's "completely false" picture of the production of silica, according to J. C. Wiegleb [7], is then surprising: "[3, § 12]. From this phenomenon it therefore follows that all the fluorspar acid combines gradually with water vapor and, in so doing, sets free the silica powder. If it is objected that the fluorspar acid is perhaps already combined from its nature with fine silica powder, which volatilizes with it, and finally separates on its combination with water, as for example hydrochloric acid separates out the 'Spießglanzkönig' [antimony] if butter of antimony [antimony trichloride] is mixed with water, fluorspar acid must deposit all of the silica combined with it during the first distillation, and can therefore retain no more for the subsequent distillations" ["Aus diesen Erscheinungen folgt also, daß sich alle Flußspatsäure nach und nach mit Wasserdünsten vereinigt, und damit das Kieselpulver ausgemacht [freigesetzt] hat; denn wollte man einwenden, daß die Flußspatsäure vielleicht schon von Natur, mit einem feinen Kieselmehle vereinigt sey, daß sie sich mit verflüchtige, und endlich bei ihrer Vereinigung mit dem Wasser absetze, wie z. B. die Salzsäure den Spießglanzkönig [Antimon] absetzt, wenn Spießglanzbutter [Antimontrichlorid] mit Wasser gemischt wird, so müßte die Flußspatsäure schon bei der ersten Destillation, alle mit ihr verbundene Kieselerde absetzen, und konnte also für die nachfolgenden Destillationen, nichts mehr übrig behalten"]. C. W. Scheele specifically rejected the glass of the apparatus as the source of the silica: "I put a mixture of powdered fluorspar and oil of vitriol in a brass cylinder and, after I had hung an iron nail and some wood charcoal over this mixture, the cylinder was closed with a well fitting lid. After two hours the wood charcoal was unchanged on opening the vessel. The charcoal was now moistened, hung up again in the cylinder and it was then opened after it had stood for two hours when covered. The nail and carbon were now covered with a white powder, which behaved exactly like silica. Since I used no glass vessel in this experiment, the silica obtained in the preceding work [experiments] has not been dissolved from the glass by the fluorspar acid and deposited subsequently, as would otherwise have been suspected from the strong attack that the inner surface of the glass had undergone" ["In einen messingnen Cylinder that ich eine Mischung von gepulverten Flußspat und Vitriolöl, nachdem ich über dieser Vermischung einen eisernen Nagel und eine Holzkohle aufgehangen hatte, wurde der Cylinder mit einem gutpassenden Deckel verschlossen. Nach zwey Stunden fand sich aber die Holzkohle bey Eröffnung des Gefäßes unverändert. Die Kohle wurde jetzt angefeuchtet, abermals in den Cylinder aufgehangen, und derselbe, nachdem er zwey Stunden bedeckt gestanden hatte, geöfnet: jetzt waren Nagel und Kohle mit einem weißen Pulver bezogen, welches sich ganz wie Kieselerde verhielt. Da ich zu diesem Versuche keine gläserne Gefäße gebrauchte; so konnte also die in den vorhergehenden Arbeiten [Experimenten] erhaltene Kieselerde, nicht durch die Flußspatsäure aus dem Glase aufgelößt, und nachher abgesetzt worden seyn, wie man sonst wohl aus dem starken Angriff, den die innern Flächen des Glases erlitten hatten, wohl hätte vermuten können"] [3, § 12]. C. W. Scheele's thinking on the "origin" of the silica in his experiment is perhaps somewhat better explained by a remark in his *Brown Book,* a note book which has been traced again, where he cites the discoveries that have just been described as an example of the fact that the [element] water makes the mineral acids "more fixed than they are in the pure state" ["mehr figiret [fix, d. h. fest macht] als sie in reinem Zustand sind"] [10].

C. W. Scheele's statements about the origin of the silica brought back to his contemporaries many reports by alchemists and earlier chemists on the possibility of transforming various "elementary earths" into one another (though these will not be discussed here). J. C. Wiegleb's contrary view, which has been quoted above, is also expressed more explicitly in another publication [11]: "It appeared to me right from the start when I came to read Mr. Scheele's communication, that his explanation was somewhat risky and improbable, for as yet no example is known where water can be converted by acids into an earth in such a way, still less that it can be changed to a siliceous earth" ["Es schien mir gleich von Anfang an, da ich Herrn Scheelens Abhandlung zu lesen bekam, dessen Erklärung etwas gewagt, und unwahrscheinlich zu seyn. Denn es ist noch kein Beyspiel bekannt, daß Wasser auf eine solche Art durch Säuren zur Erde verwandelt werden könnte; noch weniger, daß es sich zu einer kieseligten Erde sollte verändern lassen"]. C. W. Scheele, however, found support in influential quarters. Torbern Bergman (1735–1784), the professor of chemistry and pharmacy at Uppsala, wrote in his great *Dissertatio de Terra Silicea,* which was published in 1779 [5]: "In order to learn about the production of kiesel earth, at least as far as one has been able to understand this mystery till now, we will observe the distillation of fluorspar acid closely. The phenomena in this operation are unusual, and require the most careful study so that they can be properly explained" ["Ut genesin terrae siliceae cognoscamus, quantum scilicet hactenus illud mysterium introspicere licuit, destillationem acidi fluoris attenti contemplemur. Insolita operationis phaenomena sunt, &, ut rite enodentur, sollertissima requirant inquisitionem"]. In the summary he adhered to C. W. Scheele's point of view: "One must ask where the kiesel earth which occurs in this operation has its origin. Is it drawn out of the mineral fluorspar and carried up on the wings of the fluorspar acid? Or has it merely dissolved from the glass vessel? There are several reasons, which will soon be explained, for favoring neither of these views, and all the experiments so far set up indicate rather a material produced. 1. It is not difficult to see that it is wrong to associate the kiesel earth with the fluorspar mineral itself. Sometimes the fluorspar mineral may contain kiesel earth, but that is not always so . . . The green Garpenberg mineral [used] is mostly completely free of kiesel earth, so that it can be dissolved perfectly in both nitric and hydrochloric acids. 2. Since the glass is corroded by the vapor of fluorspar acid, the not improbable surmise arises at first sight that the origin of the kiesel earth is to be sought in this. But it must be observed that kiesel earth can also be obtained with small [experimental] quantities and gentle heat, without any visible corrosion. Furthermore the [amount] corroded does not agree with the kiesel earth. 3. That the kiesel earth dissolves from the surface of the retort, or volatilizes over into the receiver, and is deposited in the water, is contradicted by various experiments. If, namely, wine alcohol is introduced into the receiver in place of water the acid is collected in it, but no silica powder appears, which, however, must happen according to the hypothesis of volatilization. For this "coagulation", therefore, in addition to the hot acid vapor, water is needed, and indeed in the vapor form, for in its condensed state coagulation occurs only with difficulty, and most readily at its surface, at which it is constantly vaporizing as long as liquid is there" ["Quaeritur igitur unde siliceum in operatione obvium sit derivandum? Num e fluore minerali eductum & alis acidi fluoris allevatum? Num ex ipso vase vitreo extricatum? Plura mox adferenda momenta neutri favent opinione, & hanc materiem potius productam innuunt omnia in hanc rem huc usque instituta tentamina. 1:o Quod siliceum hocce immerito adscribatur ipsi fluori minerali haud difficulter elucet, nam licet hic fluor interdum foveat siliceum, hoc tamen non semper accidit, . . . Garpenbergensis viridis e.g. omni siliceo plerumque est orbatus, ideoque totus quantus solvi potest acido tam nitroso, quam muriatico. 2:o Quum vitrum acidi fluoris vaporibus corrodatur, primo intuitu non inprobabili conjectura in illo quaeritur siliceum prodiens. Sed observetur oportet, quod parca dosi & leni igne siliceum etiam adquiri possit sine ulla corrosione. Praeterea corrosio quantitati silicei non congruit. 3:o Quod siliceum, e vase retorto in recipiens, acido solutum vel volatilisatum transeat, & in aqua deponatur, variis experimentis contrariatur. Si loco aquae alcohol vini recipienti immittitur, acidum in eo colligitur, sed nullus comparet pulvis siliceus, quod tamen fieri deberet ex hypothesi volatilisationis . . . Requiritur itaque ad hanc concretionem, praeter vapores acidos calidos, aqua, & quidem in vapores resoluta, nam intra massam difficilius, facillime vero et ejus superficiem, quae numquam non vaporat, quamdiu liquida est, coagulatio peragitur"]. Th. Bergman finally summed up: "After carefully weighing up these arguments, which have not been collected hastily from only one or the other distillation, but from distillations which were often repeated, modified, and carefully studied, it seems to be necessary to revert to the [idea of a] new product, which should never happen rashly" ["Hisce perpensis momentis, quae non unica alterave destillatione leviter visa sunt collecta, sed saepenumero repetita, variata & adcurate examinata, ad novam productionem heic recurrere necessarium videtur, quod numquam temere fieri debet . . ."].

What C. W. Scheele had to say in his first publication about the gas produced in the distillation is to be found in the summary [3, § 13]: "From all the observations which have been carried out on the behavior of fluorspar acid it therefore follows: a) that it dissolves the kiesel earth; b) that it sets this earth free again if it . . . combines . . . with some other substances . . .; d) that the fluorspar acid is to be obtained pure only with difficulty, but is always contaminated with some kiesel earth . . ." ["Aus jenen sämmtlichen Bemerkungen, welche über die Verhältnisse der Flußspatsäure angeführt worden sind, folgt also: a) daß sie Kieselerde auflöst; b) daß sie diese Erde wieder fahren läßt, wenn sie sich mit einigen anderen Körpern . . . vereiniget . . .; d) daß die Flußspatsäure nur schwer rein zu erhalten seyn wird, sondern allemal mit etwas Kieselerde verunreinigt . . ."]. S. F. Hermbstädt, professor of chemistry and pharmacy in Berlin, who translated and published C. W. Scheele's *Sämmtliche Werke* in German, reproached the scientist for his poor understanding of the processes taking place in his experiments a few pages before the sentences quoted: "From this result [namely the corrosion of the glass vessel used in successive distillations ("rectification") by what he held to be hydrofluoric acid] Herr Scheele should have been readily able to deduce that it is a property of this acid to dissolve kiesel earth, and he would then have made the discovery, which he made several years later, earlier, as another quotation from him on this subject will show in what follows" ["Aus diesem Erfolge [nämlich der Korrosion der benutzten Glasgefäße in aufeinander folgenden Destillationen ("Rectification") dessen, was er für Flußsäure hielt] hätte Herr Scheele leicht schließen können, daß es eine Eigenschaft dieser Säure sey, Kieselerde aufzulösen, und er würde sodann diese Entdeckung früher gemacht haben, die er erst mehrere Jahre nachher machte, wie ein anderer Aufsatz von ihm über diesen Gegenstand in der Folge zeigen wird"] [12]. T. Bergman was quite explicit in speaking of the gaseous character of the distillation product [5]: "This acid vapor, when collected over mercury, is an elastic liquid in the form of air, which cannot be liquefied by cold, but water absorbs it and thereby becomes acid, though water vapor brings about flocculation" ["Ipse vapor acidus in hydrargyro collectus fluidum elasticum aëriforme praebet, quod frigore in liquidum coarctari nequit, sed aqua illum absorbet acescens, aquei vero vapores coagulant"].

C. W. Scheele's next publication [13] on this subject, which appeared in 1780 under the title *"Fortgesetzte Untersuchungen über den Flußspat"* brought out nothing new on the composition of the gaseous reaction product from the treatment of fluorspar with vitriolic acid in glass vessels. In it he only defended himself against attacks by A. G. Monnet (1734–1817), the Paris apothecary and General Inspector of French Mines and also by an otherwise unknown Mr. Boullanger [14] (according to A. Lavoisier [15] this was a cover name for the Duke of Liancourt, de la Rochefoucauld (1747–1827) though, according to A. G. Monnet [16], the paper was written "in the style of M. d'Arcet", a professor at the Collège de France, Director of the Royal Porcelain Factory of Sèvres and General Inspector of the Mint; H. Kopp [17] favored the latter). Both had questioned C. W. Scheele's supposed existence of a "characteristic fluorspar acid". His anxiety to counter Boullanger's attack went back to the first half of that year, as is apparent from a letter of 26th December, 1774, to his friend the Mine Controller J. G. Gahn (1745–1818), who was later the teacher of J. J. Berzelius. In it he also reported on his collaboration with T. Bergman and disclosed, without realizing it, why his experiments, which are mentioned above and which led to his mistake, had had to take the course described [18]: "I eagerly await his [Boullanger's] remaining experiments, for he will show there what this salt acid has changed [a further publication by Boullanger is unknown]. He says that he has found a considerable quantity of kiesel earth in the fluorspar, which I have never observed. Prof. Bergman and I have now once more digested some pulverized fluorspar with nitric acid and it dissolved completely. Consequently, there is certainly no silica in our fluorspar . . . He says further, that acidum nitri concentratum and acidum salis conc. [concentrated nitric acid and concentrated hydrochloric acid] do not decompose the fluorspar, since he has not obtained any fluorspar acid. He is right about this, for spiritus nitri and salis com. fum. do not decompose it on their own. If, however, some water is there, it is decomposed. I was therefore wrong, for spiritus nitri fum. and salis comm. fum. have been mentioned in my paper. At the time I lixiviated the fluorspar in a glass mortar with water and allowed the material poured off to settle. I distilled the fine powder while still moist with spiritus nitri fum. and acido salis comm. fum. [fuming nitric acid and fuming hydrochloric acid] and then obtained decomposition, which could be deduced from the skin of silica which was apparent in the receiver above the water. It appears that a double affinity must be [present] if these volatile acids are to decompose the fluorspar, namely that of the acidi nitri or salis comm. for the lime earth and that of the water for the fluorspar acid. As long as the fluorspar acid has quite other properties than the salt acid it is and remains a distinguishable acid. I hardly want to dabble further with fluorspar. Others can do more, for the way is open to them" ["Ich warte mit Verlangen auf seine [Boullangers] übrigen Versuche, denn da will er zeigen, was eigentlich diese Salzsäure so verändert hat. [Eine weitere Veröffentlichung Boullangers ist nicht bekannt.] Er sagt, daß er eine ansehnliche Quantitet Kieselerde im Flusspathe gefunden hat, welche ich nicht observirt habe. Ich und Herr Prof. Bergman haben nun von Neuem pulverisirten Flusspath mit Salpetersäure digerirt und selbigen gänzlich aufgelöst, folglich ist gewiß kein Kiesel in unserem Flusspath ... Er sagt ferner, daß acidum nitri concentratum et acidum salis conc. [konzentrierte Salpetersäure und konzentrierte Salzsäure] den Flusspath nicht decomponiren, denn er hat keine Flusspathsäure über bekommen. Er hat hierin Recht, denn spiritus nitri et salis com. fum. decomponiren selbigen nicht. Aber ist etwas Wasser dabei, so wird er decomponirt. Ich habe deswegen Unrecht gehabt, denn es sind spiritus nitri fum. et salis comm. fum. in meiner Abhandlung erwähnt worden. Ich pflegte damalen den Flusspath im Glasmörtel mit Wasser auszulaviren und das Abgegossene alsdann sinken zu lassen. Dieses gesunkene feine Pulver habe ich noch feucht mit spiritu nitri fum. et acido salis comm. fum. [rauchende Salpetersäure und rauchende Salzsäure] destillirt und da die Decomposition erhalten, was aus der Kieselhaut, welche sich im Recipienten über dem Wasser zeigte, zu schliessen war. Es scheint, dass eine doppelte Affinitet [vorhanden] sein muß, wenn diese flüchtige acida den Flusspath decomponiren sollen, nämlich die des acidi nitri oder salis comm. zur Kalkerde und die des Wassers zu der Flusspathsäure. So lange die Flusspathsäure ganz andere Eigenschaften wie die Salzsäure hat, so lange ist und bleibt sie auch eine unterschiedene Säure. Ich habe kaum Lust weiter mit dem Flusspathe zu schmieren. Andere können mehr thun, denn der Weg steht ihnen nun offen"].

That his disinclination to "dabble with fluorspar" could not have persisted is shown in a letter of 10th May, 1776, which is likewise to J. G. Gahn: "So far, I know of no method of obtaining acidum fluoricum [fluorspar acid] pure, without having kiesel [-earth, silica] dissolved in it" ["Ich weiß bis dato keine Methode, das acidum fluoricum [Flußpatsäure] rein, ohne Kiesel [-erde] in sich aufgelöst zu haben, zu bekommen"] [18]. It was in his third publication in 1786 [19] entitled *"Neue Beweise der Eigenthümlichkeit der Flußsäure"* that C. W. Scheele succeeded for the first time in distinguishing exactly between hydrofluoric acid and its silica-containing "impurity", preparing both gases in a pure form and showing that the silica of the "impurity" was taken up from the glass. This third publication was induced by a suggestion from Franz Carl Achard (1753–1821), the director of the mathematical-physical class of the Academy of Sciences in Berlin and later the one who discovered beet sugar, that the white earth, as it occurs in the preparation of fluorspar acid, is not silica at all [20], as C. W. Scheele and, after him, the "celebrated court apothecary in Stettin", J. C. F. Meyer (1733–1811) had supposed [6] but a "new earth peculiar to fluorspar", a view which F. C. Achard very soon gave up [21]. C. W. Scheele now substantiated the correctness of his earlier statements in the most varied ways, among others by preparing "pure fluorspar acid" from fluorspar and sulfuric acid in a retort made of tin, with a wax-coated receiver containing water. Then "I repeated the experiment, only with the difference that I added some rock crystal which had been well ground in a brass mortar and also attached the receiver coated on its inner surface with wax. After an hour I took it away again and found that the water contained in it was quite thick with the silica which had gone over and had been rendered volatile" ["darauf wiederholte ich den Versuch nur mit dem Unterschiede, daß ich zu der Mischung des Flußspatpulvers und Vitriolöls, etwas im messingenen Mörser zart zerriebenen Bergkrystall setzte, und legte auch die mit Wachs auf der innern Fläche überzogenen Vorlage vor. Nach einer Stunde nahm ich sie wieder ab, und fand, daß das darin enthaltene Wasser von dem übergegangenen und flüchtig gemachten Kiesel ganz dick war"].

In this third publication on fluorspar C. W. Scheele now also mentioned Priestley's investigation [22], which had appeared ten years previously, without however quoting it exactly. He treated it disparagingly, as he had already done in the draft of the publication in his *Braune Buch* [10], in discussing a single one of the experiments of the English investigator: "... Perhaps here [on pouring oil of vitriol over lead fluoride PbF_2, when "fluorspar acid was driven out with foaming"] the volatile fluorspar earth combines with the oil of vitriol and changes this fixed or almost fixed acid into acid air? That Mr. Priestley appears to arrive at such a conclusion I can readily concede to this eminent man, for he does not claim to be an analytical chemist [Scheidekünstler]" ["... Doch vielleicht verbindet sich hier [beim Übergießen von Bleifluorid PbF_2 mit Vitriolöl, bei dem "flußspatsaure Dämpfe mit Schäumen ausgetrieben werden"] die flüchtige Flußspatherde mit dem Vitriolöl, und verwandelt diese fixe oder beynahe fixe Säure in sauere Luft? Daß Hr. Priestley einen solchen Schluß zu machen scheint, kann ich diesem vortrefflichen Manne leicht zugeben; denn Er giebt sich nicht für einen Scheidekünstler aus"]. Apart from this passage which C. W. Scheele treated with irony, Priestley had also written the following [22] when he reported his experiments with the "Swedish acid", hydrofluoric acid: "This phenomenon [the decomposition by water of the gas evolved in a glass vessel from fluorspar and oil of vitriol] I explain to myself in the following way. The vitriolic acid, when it attacks the spar, becomes partly volatile through the phlogiston contained in the spar and goes over into vitriolic acid air [as he had studied it earlier when obtained from olive oil and oil of vitriol] and base earth of the spar is therefore mixed with it as long as it is being evolved until it reaches the water, with which the acid then combines allowing the base earth to fall out" ["Diese Erscheinung [die Zersetzung des in Glasgeräten aus Flußspat und Vitriolöl entwickelten Gases in Wasser] erkläre ich mir auf folgende Art: Die Vitriolsäure wird, indem sie den Spath angreift, durch das in dem Spath enthaltene Phlogiston, zum Theil flüchtig und geht also in vitriolsaure Luft über [wie er sie früher, erhalten aus Olivenöl und Vitriolöl, untersucht hatte], und die Grunderde des Spathes ist ihr daher, so lange sie sich entwickelt, beygemischt, bis sie das Wasser erreicht, mit dem sich hernach die Säure verbindet und die Grunderde fahren [ausfallen] läßt"]. C. W. Scheele, who was clearly concerned only with a knowledge of hydrofluoric acid and the composition of the mineral, did not consider at all the sentence by Priestley which follows immediately afterwards, in which he characterized the gas as a new substance. J. Priestley in fact continued: "I will give the reasons for this opinion as soon as possible. I must first, though, acquaint my reader with observations I have made on the acid air in its combined state, before the earthy material is set free from it, so that you are able to follow the further decomposition step by step. For through this admixture it almost becomes quite another substance with quite special properties, which would not be observed for the pure acid air if it were no longer combined with the earthy material. It therefore has every right to merit its own name, irrespective of the fact it is a compound which can be broken up into its components. I have named it **fluorspar acid air**" ["Ich werde die Gründe für diese Meynung, so bald es angehen wird, angeben; ich muß nur erst meine Leser mit denjenigen Beobachtungen bekannt machen, die ich über die sauere Luft in ihrem zusammengesetzten Zustande, ehe sich noch die steinige Materie von ihr losgemacht hatte, angestellt habe, damit sie mir bey der weiteren Zergliederung derselben Schritt vor Schritt folgen können. Denn durch diese Beymischung wird sie nahezu zu einem ganz anderen Körper, und erhält ganz besondere Eigenschaften, die man an der reinen, saueren Luft, wenn sie garnicht mehr mit dieser steinigten Materie verbunden ist, nicht im mindesten gewahr wird. Und sie verdient daher, ohngeachtet sie eine zusammengesetzte Substanz ist, und in ihre Bestandtheile zergliedert werden kann, doch mit allem Rechte eine eigene Benennung. Ich habe ihr den Namen **Flußspathsaure Luft** beygelegt"].

Following this he studied the chemical behavior of his "fluorspar acid air" towards a range of compounds, among which was ammonia, "alkali-like air" in his terminology, observing a phenomenon which was to become important, though for another silicon tetrahalide that is more readily prepared [$SiCl_4$], namely the extraordinarily dense cloud formation in the reaction between the two gases: ". . . A white cloud resulted as I mixed these two kinds of air" [". . . es entstand, so wie ich diese zwo Luftgattungen mischte, eine weiße Wolke"]. J. Priestley did not investigate the "fluorspar crust" – this is what he called the silicic acid deposit*) from the decomposition of the gas – more closely. He paid equally little attention to tracing the reason for the destruction of glass vessels in his experiments with "fluorspar acid air", of which he complained repeatedly: "The strongest flasks which I was able to obtain were often completely eaten through in a quarter of an hour if I applied a considerable degree of heat and the [fluorspar acid] air was liberated very rapidly. This eating through of glass is a very remarkable property of this air, but it appears to possess it only when it is heated and then to a substantial degree" ["Es wurden sehr oft die stärksten Flaschen, die ich nur erhalten konnte, wenn ich einen beträchtlichen Grad von Feuer gab, und die [flußspathsaure] Luft sich sehr schleunig entband, in Zeit von einer Viertelstunde gänzlich durchfressen. Dieses Zerbeißen des Glases ist eine sehr merkwürdige Eigenschaft dieser Luft, allein sie scheint sie doch nur, wenn sie erhitzt, und zwar zu einem beträchtlichen Grade erhitzt ist, zu besitzen"].

C. W. Scheele's first publication [3] in 1771 led many to repeat the work, as can be seen from what has already been said. This was in complete contrast to Marggraf's investigation [1], which was three years earlier, and was due, not only to the newly discovered acid, but to the "generation of silica" which had been assumed in it. One of the first to publish his results was the Director of the Veterinary School in Kopenhagen, P. C. Abilgaard (1740–1801), though he only confirmed C. W. Scheele's observations [4]. Among the publications which have become more generally known, there followed in 1779/80 T. Bergman's dissertation [5], which has been quoted above, the results of which he had already made known to interested scientists. Thus J. G. Leonhardi, professor of medicine in Leipzig and the one who translated P. J. Macquer's *Dictionnaire de Chimie* into German, reported [27]: "I received the first report of this discovery through a letter from Mr. Bergman, with which excellent Swedish chemist I exchange letters, and he sent me a small quantity of this earth. I submitted it at once to all the tests which could give me further light on its nature. I found . . . [after carrying out experiments with the earth which are enumerated] and could no longer doubt, that it did not possess the nature of quartz or kiesel-earth" ["Ich erhielt die erste Nachricht von dieser Entdeckung durch einen Brief von dem Herrn Bergman, mit welchem trefflichen schwedischen Chemisten ich in einem Briefwechsel stehe, und der mir auch eine kleine Menge von dieser Erde übersandte. Ich unterwarf sie sogleich allen Prüfungen, die mir von ihrer Natur mehreres Licht geben konnten. Ich fand . . . [nach Durchführen der aufgezählten Experimente mit der "Erde"] und zweifelte nicht mehr daran, daß sie nicht die Natur der Quarz- oder Kieselerde haben sollte"].

Other workers like J. C. F. Meyer [6], W. H. Bucholtz [28], J. C. Wiegleb [11], and J. F. Westrumb [24], who were critical of ideas on the silica produced in the processes of distilling fluorspar with acids, came to the conclusion that the glass of the apparatus was the source of

*) This is not silicic acid. It is a mixture with some Si-N compounds.

the silica. How C. W. Scheele learned of this, and, in so doing, corrected his earlier viewpoint, is shown by J. C. Wiegleb in his book [23], with J. F. Westrumb (1751–1819), an apothecary in Hameln and Great Britain's Commissioner of Mines, as his informant: "So far nothing further about Scheele's picture of the peculiar action of fluorspar acid (1771, [3]) had been written openly. About this time, however, Wiegleb set up an experiment on the nature of fluorspar acid, by which it has been proved that the earth appearing in the distillation of this acid is not freshly formed, in accordance with Scheele's ideas, but is silica which the fluorspar acid carries over from the glass vessel during the distillation as the result of a hitherto totally unknown property" [11]. Soon afterwards this was also confirmed by J. C. F. Meyer's experiments [6]. Westrumb gave the following report on the course of the controversy: "In connection with the history of the dispute about this earth, it is important to know that Priestley and Macquer were familiar with the property of eating through glass which the spar acid possessed (though both failed to realize that the silica which had dissolved out would distil over with it). Wgb. [Wiegleb]: Mr. Ehrhart, the Royal Botanist in Hannover, was the first to explain the true nature of this earth from an observation of the corrosion of the vessel in the same way as Wiegleb has explained it here. As early as 1778 I saw in his home in Hannover a paper which Mr. Ehrhart had sent to Mr. Scheele on this point. Mr. Scheele, who received a letter from the Court Apothecary Meyer in Stettin at the same time, in which he mentioned the experiment which he had done with ammonium fluoride in lead vessels, finally let himself be persuaded to study the formation of this earth anew. And see, he found that when he avoided all siliceous material, be it quartz, sand, or glass, in preparing the spar acid, no more earth appeared. He set up the experiment in metal cylinders and hung wet sponges in these after he had put fluorspar and oil of vitriol in one while in the other this mixture had been shaken with sand. After several hours the sponge in the latter vessel held a great deal of silica powder, while in the first there was none at all. Now he finally admitted to his friend, and this was in March 1781, that he had been wrong and that the fluorspar acid was able to volatilize the silica. Thus it was just at this time that Mr. Scheele, since Mr. Wiegleb had been at pains to make him change his mind, was diverted from his earlier beliefs by Mr. Ehrhart, Mr. Meyer and by his own experiments" ["Ueber Scheelens Begriff von der sonderbaren Wirkung der Flußspatsäure (1771, [3]) war bisher nichts weiter öffentlich geschrieben worden. Um diese Zeit aber stellte Wiegleb Versuche über die Natur der Flußspatsäure an, wodurch bewiesen wurde, daß die bey Destillation dieser Säure erscheinende Erde nicht nach Scheelens Begriff neu entstehe, sondern Kieselerde sey, welche die Flußspatsäure, nach einer bisher ganz unbekannten Eigenschaft, bey der Destillation von dem gläsernen Gefäße mit überführe [11]. Eben dies wurde auch bald darauf durch Herrn J. C. F. Meyers Versuche [6] bestätigt. Ueber den Verlauf dieses Gegenstands ertheilte Westrumb folgende Nachricht: "Zur Geschichte des Streits über diese Erde ist es wichtig zu wissen, daß schon Priestley und Macquer die Eigenschaft der Spatsäure kannten, das Glas zu zerfressen (beyden kam aber noch nicht in Sinn, daß die aufgelöste Kieselerde mit überdestilliren sollte). Wgb. [Wiegleb]: daß aber der Königl. Botanikus in Hannover Herr Ehrhart der erste war, welcher die wahre Natur dieser Erde aus einer von ihm beobachteten Zerfressung der Gefäße so erklärte, wie sie Wiegleb hier erkläret hat. Schon im Jahre 1778 sah ich in Hannover bey ihm einen Aufsatz, den Herr Ehrhart an Herrn Scheelen über diesen Punkt sandte. Herr Scheele, der damals zugleich einen andern Brief vom Herrn Hofapotheker Meyer aus Stettin erhielt, worinn dieser die Versuche erwehnte, die er mit Flußspatsalmiak in bleyernen Gefäßen angestellt hatte, ließ sich endlich bewegen, von neuem über die Entstehung dieser Erde nachzuforschen; und siehe, er fand, daß, wenn er alles kieselartige, es sey Quarzsand oder Glas, bey Verfertigung der Spatsäure vermied, keine Erde zum Vorschein kam. Er stellte die Versuche in metallenen Cylindern an, hieng in diese nasse Schwämme, nachdem er in den einen Flußspat und Vitriolsäure, in den andern aber eben diese Mischung und Sand geschüttet hatte. Nach Verlauf von einigen Stunden hielt der Schwamm des letztern Gefäßes sehr viel Kieselpulver, der im ersten aber gar keines. Nun gestand er endlich, und zwar schon im Merz 1781 seinem Freunde, daß er sich geirret habe, und daß die Flußspatsäure im Stande sey, die Kieselerde zu verflüchtigen. Also war Herr Scheele gerade in der Zeit, da Herr Wiegleb ihn zu widerlegen bemüht war, durch Herrn Ehrhart, Herrn Meyer und eigene Versuche schon von seiner vorigen Meynung abgebracht [24]"].

That this version is essentially correct is confirmed by a letter from C. W. Scheele to T. Bergman on 9th March, 1781, in which he expressed his alarm at the mistake which he and Bergman had made [18]: "... Meyer in Stettin has discovered my mistake about the generation of kiesel earth from fluorspar acid and water. He prepared Flusspatsalmiak [ammonium fluoride] in such a way that he allowed fluorspar acid saturated with alkali volatili [volatile alkali] to evaporate ad siccum [to dryness] in a lead still (as he found silica dissolved in this ammonium salt if he allowed it to evaporate in glass). He poured oleum vitrioli [oil of vitriol] on to this ammonium salt in an iron retort, attached a lead receiver, in which he had poured water, and obtained no trace of silica in the distilled acid in this way. I remembered (as soon as I read Mr. Meyer's letter) that last summer I poured acidum vitrioli [vitriolic acid] on to fluorspar in a tin cylinder and afterwards found no silica on a wet sponge which had been hung up. This puzzled me at the time and I have now repeated this operation. I took two tin cylinders. In both I put fluorspar powder, which had been ground in a metal mortar, but I mixed into one some finely ground sand. After both cylinders, with the acido vitrioli poured in, had stood for a few hours covered, and with the wet sponge hung up inside, I opened both, finding on one sponge no silica at all and on the other, where the sand was, everything was choked with silica dust. What then can be the reason why I obtained silica nine years ago by this method in my first experiment? Has the fluorspar this earth in it, or was it derived from the glass mortar in which I had ground it to a fine powder? It is remarkable that this acid is also able to volatilize the silica in the cold" ["... Meyer in Stettin hat meinen Irrthum von der Generation der Kieselerde aus Flusspathsäure und Wasser entdeckt. Er bereitete sich Flusspathsalmiak auf die Art, dass er die mit alkali volatili [flüchtiges Alkali] saturirte Flusspathsäure in einen Bleikessel ad siccum evaporiren liess (denn er fand Kiesel in diesem Salmiak aufgelöst, wenn er das Salz in Glas evaporiren liess); auf diesen Salmiak goss er oleum vitrioli in einer eisernen Retorte, legte einen bleiernen Recipienten vor, in welchen er Wasser gegossen, und erhielt auf solche Art keine Spur von Kiesel in der destillirten Säure. Ich erinnerte mich (sobald ich Herrn Meyers Brief gelesen), letzten Sommer auf Flusspath in einem zinnernen Cylinder acidum vitrioli [Vitriolsäure] gegossen, und nachhero an dem aufgehangenen nassen Schwamm keine Kieselerde gefunden zu haben. Dieses wunderte mich damals nicht wenig, und nun habe ich diesen Process wiederholt. Ich nahm zwei zinnerne Cylinder; in beide legte ich Flusspathpulver, das in metallenen Mörsern gerieben war, mischte aber in den einen etwas zart geriebenen Sand; nachdem beide Cylinder mit zugegossenem acido vitrioli und eingehängtem nassen Schwamm ein paar Stunden wohl zugedeckt gestanden hatten, öffnete ich beide und fand an dem einen Schwamm gar keine Spur von Kieselerde, an dem andern aber, in welchem der Sand war, sass alles voll von Kieselstaub. Was kann dann wohl die Ursache sein, dass ich nach dieser Methode vor neun Jahren bei meinem ersten Versuch Kiesel erhielt? Hatte der Flusspath diese Erde in sich, oder kam sie von dem gläsernen Mörser her, in welchem ich ihn zum zarten Pulver rieb? Es ist wunderlich, dass diese Säure den Kiesel auch in der Kälte volatilisiren [flüchtig machen] kann"].

Friedrich Ehrhart (1742–1795) was not named here by C. W. Scheele, but, in a collection of the correspondence of the Swedish chemist that has still been preserved [25], there are three examples of what was clearly a confidential exchange of letters between him and the Hannoverian "Botanikus". One, No. 133, which falls in the period in question, was a letter by Ehrhart dated 14th August, 1779, but it appears from the (abbreviated) quotation to contain nothing about experiments with fluorspar acid. Two further letters stem from later (No. 219 from 12th March, 1784; No. 281 from 6th January, 1786). F. Ehrhart, however, was not the first to associate corrosion of the glass with the occurrence of silica on treating fluorspar with sulfuric acid. This was, rather, the English physician John Hill (1716–1775) from London in the second bilingual edition of the book *"On stones" of Theophrastus Eresius* (372/369–288/85 B.C.) in a section entitled *"On the new Swedish acid and of the stone from which it was obtained"*. In this, in dealing with the treatment of the mineral with sulfuric acid, he writes, inter alia: "The corrosion of the glass of the retort seems to be an effect of that peculiar sublimation which rises in the distillation; nay, and begins to rise, even without the operation . . ." The book, which appeared as early as 1774 was clearly unknown to contemporary chemists [26].

It may also be added for the sake of completeness that T. Bergman's unsuccessful attempt to find a quantitative relationship between the glass corroded and the silica obtained, which he reported in a single sentence, see above, was repeated in 1781 both by J. C Wiegleb [11] and by the Weimar Court Physician W. H. Bucholtz [28], who was a scientific councillor to J. W. v. Goethe as well as, two years later, by the famous chemist of the Kurfürstlich Sächsische Schmelzadministration, C. F. Wenzel [29] with carefully executed weighings and apparent success. J. G. Leonhardi [30] was already justificably critical of these results at the time.

References:

[1] A. S. Marggraf (Hist. Acad. Roy. Soc. Belles Lettres Berlin **1786** 3/11). – [2] J. R. Partington (A History of Chemistry, Vol. 2, London 1961, p. 728). – [3] C. W. Scheele (Kgl. Svenska Vetenskaps Akad. Handl. **33** [1771] 120/38 in: Sämmtliche physische und chemische Werke [German by S. F. Hermbstädt], Vol. 2, Berlin 1793, pp. 3/32). – [4] P. C. Abilgaard (Schriften der Societät der Wissenschaft Kopenhagen Jahr 1779 from Neuesten Entdeckungen Chem. [Crell] **2** [1781] 168/70). – [5] T. Bergman (Diss. de Terra Silicea [24.11.1779] in Opuscula Physica et Chemica, Vol. 2, Uppsala 1780, pp. 26/53, 34, 38/47).

[6] J. C. F. Meyer (Schriften Ges. Naturforschender Freunde Berlin **2** [1781] 319/33). – [7] J. C. Wiegleb (Geschichte des Wachsthums und der Erfindungen in der Chemie in der neueren Zeit, Vol. 2, Berlin–Stettin 1791, p. 106). – [8] H. Cassebaum (Carl Wilhelm Scheele, Biographien hervorragender Wissenschaftler Techniker und Mediziner, Vol. 58, Leipzig 1982, p. 27). – [9] J. R. Partington (A. History of Chemistry, Vol. 3, Macmillan, London 1962, pp. 1/854, 214). – [10] U. Boklund (Carl Wilhelm Scheele, Bruna Boken, Pt. II, Manuscript 3.3, about 1771, pp. 28/9, Manuscript 36.5, pp. 296/7).

[11] J. C. Wiegleb (Neuesten Entdeckungen Chem. [Crell] **1** [1781] 3/15). – [12] S. F. Hermbstädt (in: C. W. Scheele: Sämmtliche physische und chemische Werke, Vol. 2, Berlin 1793, p. 23 footnote). – [13] C. W. Scheele (Kgl. Svenska Vetenskaps. Akad. Handl. **1** [1780] 18/26 in: Sämmtliche physische und chemische Werke [German by S. F. Hermbstädt], Vol. 2, Berlin 1793, pp. 235/46). – [14] Boullanger (Expériences et Observations sur le Spath Vitreux ou Fluor Spatique, Paris 1773, pp. 1/32). – [15] A. Lavoisier (Traité Élémentaire de Chimie, Vol. 1, Paris 1789, p. 263).

[16] A. G. Monnet (Observ. Phys. Hist. Nat. Arts [Rozier] **10** [1777] 106/16). – [17] H. Kopp (Geschichte der Chemie, Vol. 3, Braunschweig 1845, p. 369). – [18] A. Nordenskiöld (Carl Wilhelm Scheele: Efterlemnade Bref och Antekningar, Stockholm 1892, pp. 152/4, 176, 324/5). – [19] C. W. Scheele (Chem. Ann. Crell **1786** I 3/17 in: Sämmtliche physische und chemische Werke [German by S. F. Hermbstädt], Vol. 2, Berlin 1793, pp. 407/22). – [20] F. C. Achard (Observ. Phys. Hist. Nat. Arts [Rozier] **23** [1783] 37/40; Bestimmung einiger Bestandtheile einiger Edelgesteine, Berlin 1779, p. 110).

[21] F. C. Achard (Vorlesungen über Experimentalphysik, Vol. 2, § 821, Berlin 1791, pp. 135/6). – [22] J. Priestley (Versuche und Beobachtungen über verschiedene Gattungen der Luft [German translation by C. Ludwig], Pt. 2, Sect. 11, Wien–Leipzig 1778, pp. 186/209 [1st Ed.

in English 1776]). – [23] J. C. Wiegleb (Geschichte des Wachsthums und der Erfindungen in der Chemie in der neueren Zeit, Vol. 2, Berlin–Stettin 1791, pp. 246/7). – [24] J. F. Westrumb (in: L. Crell, Auswahl aller eigenthümlichen Abhandlungen aus den Entdeckungen in der Chemie, Vol. 1, Leipzig 1785, p. 2 from [11]). – [25] O. Zekert (Carl Wilhelm Scheele: Sein Leben und seine Werke, Pt. 3/7, Nemayer, Mittenwald 1933, pp. 1/303, 250, 265).

[26] J. Hill (Theophrastus's History of Stones with an English Version and Notes, 2nd Ed., London 1774, pp. 267/78 from J. F. T. Berliner, Science [2] **66** [1927] 192/3). – [27] J. G. Leonhardi (Peter Joseph Macquer's Chymisches Wörterbuch, 2nd Ed., Vol. 3, Leipzig 1789, p. 152). – [28] W. H. Bucholtz (Neuesten Entdeckungen Chem. [Crell] **3** [1781] 50/64). – [29] C. F. Wenzel (Chymische Untersuchung des Flußspats, Dresden 1783, pp. 32/4). – [30] J. G. Leonhardi (Peter Joseph Macquer's Chymisches Wörterbuch, 2nd Ed., Vol. 6, Leipzig 1790, pp. 198/205).

2.4.3 Silicon Tetrafluoride. Tetrafluorosilane. Siliceous Fluoric Acid Gas

In experiments on the reaction of fluorspar with acids which have been discussed earlier, chemists were interested in the main with preparing the new and interesting hydrofluoric acid in a pure state. In the course of this work they had hit on the fact that, when working in glass apparatus, another gas was produced which contained silica. In preparative experiments they all noticed that this compound was more readily obtained on adding glass than with sand or quartz powder. To illustrate this point some observations by J. C. F. Meyer [1] may be quoted: "§ 8. Not yet satisfied with this experiment, I poured two drachms of oil of vitriol over one drachm [3.75 g] of green, calcined and finely ground fluorspar in a gun barrel. I hung a moistened piece of carbon inside, covered it with a lid, warmed the barrel on a sand bath and, on opening it after a quarter of an hour, found the carbon to be dry with no earth on it. I now threw a scrupel [1.25 g] of finely ground sand in, hung the moistened carbon inside and again found nothing after a quarter of an hour. Some small pieces of green glass were now introduced. It began to foam so vigorously that it overflowed" ["§ 8. Mit diesen Versuchen noch nicht zufrieden, übergoß ich eine Drachme [3.75 g] grünen, geglüeten und zerriebenen Flußspaths in dem Flintenlaufe mit zwo Drachmen Vitrioloel; hieng eine naßgemachte Kohle hinein, bedeckte es mit einem Deckel, erwärmte den Lauf in einer Sandcapelle, und fand nach einer Viertelstunde da ich ihn öfnete, die Kohle trocken und keine Erde daran. Ich warf nun noch einen Scrupel [1.25 g] fein zerriebenen Sandes dazu, hing die naßgemachte Kohle hinein, und fand nach einer Viertelstunde abermals nichts. Nun wurden einige Stückchen grünes Glas in die Mischung geworfen; Sie fing so heftig an zu schäumen, daß es überlief"]. That the gas prepared in this way, or simply by treating fluorspar with sulfuric acid in glass vessels, was not identical with hydrofluoric acid which had been discovered by C. W. Scheele was not recognized by scientists in 1808. They were trying to prepare the base substance of this acid from it and obtained amorphous silicon without realizing it, see p. 22. The gas [i.e., SiF_4] was, however, called "gaz fluorique" by L. J. Gay-Lussac, J. L. Thenard [2], while H. Davy [3] used the name "fluoric acid gas" or, somewhat more precisely "fluoric acid which has been in contact with glass". H. Davy's brother, J. Davy, who edited the *Collected Works of Sir Humphry Davy*, confirms specifically in a footnote [4] that the presence of silicon in the gas was not realized, in spite of the second of the two names: "... at this time [1808] ... it was not then known that silica or its basis, is an essential ingredient of the gas which in the text is named fluoric acid gaz". J. R. Partington [5] believed, that "Davy's mistake was due to his hasty work at this time" after he had pointed out that the two gases had been distinguished from one another as early as 1781, as can be seen from what has been said above. The two French scientists made a further observation on the evolution of the gas in the course of their work in 1809: "It is noticeable that, when fluorspar which contains a lot of silica is taken for this process, decomposition (by the action of silica on the hydrofluoric acid gas) is greatly accelerated and gives siliceous hydrofluoric acid gas in quantity" ["Merkwürdig ist es, daß, wenn man zu diesem Proceß Flußspat nimmt, der viel Kieselerde enthält, die Zersetzung (durch Einwirkung der Kieselerde auf das flußsaure Gas) sehr beschleunigt wird und kieselig flußsaures Gas in Menge hergibt"] [2]. In 1885 W. Hempel [44] published a well worked out set of instructions for preparing the gas in which powdered fluorspar and quartz powder in a 2:3 ratio were mixed together. This is encountered in the literature with many variations (due to the difficulties which arise in the reaction of the mixture with sulfuric acid), see "Silicium" B, 1959, p. 616. As early as 1812, however, John Davy (1790–1868), the brother of Humphry Davy, had shown a preference for using powdered glass instead of quartz [6] and there were many who followed him in this, for example, A. Maus [7] in 1827 and E. Moles, T. Toral [8] in 1937. The latter also stored the reaction mixture over broken glass to remove the last traces of hydrofluoric acid. In 1812, J. Davy [6] claimed that the gas obtained by his method was "perfectly saturated with silex" after no sign of corrosion had been detected in the glass storage vessels, even after some weeks. He did, nevertheless, observe a dark deposit, which was readily removable, the nature of which he was unable to explain.

J. J. Berzelius [9] published another preparative method in 1823 involving the thermal decomposition of "flußspatsaurem Kieselbaryt", $BaSiF_6$. He obtained "siliceous fluorspar acid gas" without any trace of free fluorspar acid, since it had not attacked the glass" ["ohne alle Spur von freier Flußspathsäure, weil es das Glas nicht angegriffen hatte"]. Some estimates of the yield in this operation – 2 l of gas from 30 to 40 g of the barium salt in half an hour – were given in 1884 by C. Truchot [10], and a more detailed study of the thermal dissociation was first undertaken by G. Hantke [11] in 1936. He also included the decomposition of the corresponding alkali compounds in his investigation. Until then, the latter had been used for the preparation of silicon tetrafluoride only through their decomposition by acids, and working instructions for this, leading to a very good yield and a high degree of purity, were published by H. S. Booth, C. F. Swinehart [12] in 1935. In 1869, G. Gore [13], in preparing silicon tetrafluoride, had started with the simple acid potassium fluoride KHF_2, which was mixed with precipitated silicon dioxide and decomposed with concentrated sulfuric acid. He claimed that all of the combined fluorine was obtained as SiF_4. Using sand or quartz powder, however, yields were poorer.

The formation of silicon tetrafluoride by fluorinating halogenosilanes has frequently been observed, though probably first by O. Ruff, C. Albert [14] in 1905. Since, however, this compound is very difficult to separate in a pure state from the mixture of mixed halides resulting from such reactions, this method of preparing it is of little importance. Direct fluorination of silicon dioxide to tetrafluorosilane has also been tried. It was successful in 1886, when V. Olivieri [15] used chromyl fluoride. Similar reactions, using other reagents which split off fluorine, are frequently described in the literature, but none has become important, see "Silicium" B, 1959, p. 615.

H. Moissan [16] synthesized tetrafluorosilane from the elements in 1891 in a fluorspar tube. Reaction occurred with inflammation and melting of the silicon, as he had observed in his original preparation of fluorine five years earlier without, at the time, attaching any value to the reaction product [17]. J. R. Partington comments on this observation [34]: "He [i.e., H. Moissan] tells us that he anticipated that silicon would burn spontaneously in fluorine and he kept a crystal of silicon which he presented to the gases formed in his experiments. The crystal lasted a long time but on 26th June, 1886, on presenting it to the gas from the anode in this experiment, the silicon took fire and Moissan knew that, for the first time in the long series of experiments going back to Davy's researches, free fluorine had been obtained". This method of preparation from the elements, which requires expensive apparatus, appears to have been only once used, when H. v. Wartenberg, R. Schütte [18] measured the heat of reaction directly in 1933, using a platinum vessel lined with fluorspar. H. Moissan [19] also fluorinated elementary silicon successfully with iodine pentafluoride. After initially heating the mixture externally, the reaction again proceeded with inflammation and yielded only iodine vapor as the secondary product. H. Moissan [20] found when he introduced lithium silicide, which he had prepared for the first time, into an atmosphere of fluorine, reaction took place less violently. G. S. Newth [21] had already shown in 1895 that crystalline silicon burned very brilliantly in pure gaseous hydrogen fluoride, when the latter was hot following its generation by the thermal decomposition of acid potassium fluoride. The products are silicon tetrafluoride and hydrogen. Another preparative method, which is also useful technically, owes its origin to the observation by metallurgists [35] that, on adding fluorspar during the melting of siliceous ores, part of the silicon is driven off as silicon fluoride. Based on this, E. Karcher and C. M. Tessier du Motay put forward a process which was patented in England on 14th June, 1864, under the name of R. A. Brooman [36]. In it silica, fluorspar, and carbon were heated to melting, when the gas was evolved. The yield, referred to the fluorine content, was said to amount to 68% [37].

Physical properties of the gas, which has an unpleasant and choking smell (which is probably why early workers such as A. S. Marggraf [33] spoke of its sulfurous smell) were soon studied more closely. As early as 1812, J. Davy [6] found a value of 3.5735 for the density referred to air and this was improved only marginally later by A. Dumas [22] and later still in 1884 by C. Truchot [10]. Determinations of the liter weight in the first quarter of this century gave the value 4.69 [23]. M. Faraday (1794–1867) was successful in liquefying the gas in 1844, and reported [25]: "Fluorsilicon condenses to a liquid, but only with the help of the lowest temperature [i.e., −160°F]. It is clear and colorless and, at the same time, extraordinarily watery and mobile, like hot ether. It then produces a pressure of about nine atmospheres and shows no sign of solidification" ["Fluorkieselsäure [im Original: Fluorsilicon] verdichtet sich zur Flüssigkeit, aber erst mit Hülfe der niedrigsten Temperatur [d.i. −160° Fahrenheit]. Sie ist klar und farblos, dabei äußerst dünnflüssig und beweglich, wie heißer Aether. Sie erzeugt alsdann einen Druck von etwa neun Atmosphären, und giebt keine Anzeige von Erstarrung"]. Solid tetrafluorosilane was first obtained forty years later by K. Olszewski [26] by cooling to −120°C as a white amorphous mass, and he was able to show that the compound vaporizes slowly on raising the temperature without forming a liquid. The triple point was determined in 1930 by W. I. Patnode, J. Papish [27] as −90.2°C at 1318 torr. They also determined the vapor pressure curve. A year later densities of tetrafluorosilane in the liquid phase were also measured [28]. The critical data were determined for the first time by H. Moissan [19] in 1904. J. J. van Laar [29] on the basis of the additivity of the atomic values for the constants in his equation of state, which are named after him, believed the value of −1.5°C for the critical temperature given by H. Moissan to be too high and that the critical pressure was more likely to be 33 atm (compared with Moissan's value of 50 atm). In fact H. S. Booth, C. F. Swinehart [12] in 1935 found the values $t_K = -14.15$°C and $p_K = 36.66$ atm. Both A. F. O. Germann, H. S. Booth [24] in 1917 and W. I. Patnode, J. Papish in 1930 [27] found values for the molecular weight which were significantly greater than would correspond with the formula. According to O. Ruff, E. Asche [30], whose experiments gave the theoretical value, this was only because the extraordinary water sensitivity of the silicon fluoride had led to formation of fluorosilicic acid from moisture in the tap grease. The earliest crystallographic studies with solid silicon tetrafluoride were undertaken in 1930 by G. Natta [31], his data being refined twenty-four years later by M. Atoji, W. N. Lipscomb [32]. The chemical behavior has been investigated so frequently that it will not be possible to give an account of it here, apart from referring to the fact that its striking behavior in forming a thick white fog with water vapor, and the subsequent separation of solid silica led not only to the discovery of the compound itself, but also to that of a range of other substances, see above. The behavior of the compound towards hot metallic potassium resulted in the initial discovery of elementary silicon, see p. 29. Both J. J. Berzelius [9] and H. Sainte-Claire Deville [38] found that white hot iron and aluminum can also be used. The fact that tetrafluorosilane is also able to attack the mercury of a manometer in the cold was reported by E. Moles, T. Toral [8], thus correcting earlier reports to the contrary.

The view of most chemists on the composition of the compound was that, as J. Davy believed [6], it consisted only of "kiesel" and the base substance of fluorspar acid [HF]. As R. Caillat pointed out [39], J. J. Berzelius was an exception and in 1824 still believed that it also contained oxygen [40] on the basis of his experiments with boric acid and silicon fluoride. The formula assumed changed in the course of time according to the valency attributed to silicon. Thus in 1858 C. Marignac [41] believed the formula SiF_2 to be correct, while W. Knop [42] wrote it as SiF_3. As soon as it was recognized that the element was four valent, the composition SiF_4 was accepted. J. J. Berzelius, however, went so far as to express his incorrect view that oxygen was present in numerical terms [43]: "I believe that these experiments [the preparation of fluorosilicic acid] have proved that in gaseous siliceous fluorspar acid [i.e., SiF_4] the acid and the silica contain equal amounts of oxygen, that is 3 atoms [he meant molecules] of the acid [which contains oxygen because of its acid character] combine with 2 atoms of silica . . ." ["Ich sehe es durch diese Versuche [Herstellung der Kieselfluorwasserstoffsäure] als bewiesen an, daß in der gasförmigen kieselhaltigen Flußspatsäure [das ist SiF_4] die Säure und die Kieselerde gleich viel Sauerstoff enthalten, also 3 Atome [gemeint: Moleküle] der [wegen ihres Säure-characters sauerstoffhaltigen] Säure sich mit 2 Atomen Kieselerde verbinden . . ."]. In a footnote it is clear from the symbols that he assigns two oxygens to the "atom" of hydrofluoric acid and three to the silica.

References:

[1] J. C. F. Meyer (Schriften Ges. Naturforschende Freunde Berlin **2** [1781] 319/33, 327). – [2] L. J. Gay-Lussac, J. L. Thenard (Ann. Chim. [Paris] Févr. 1809 from Ann. Physik [1] **32** [1809] 1/15, 8). – [3] H. Davy (Phil. Trans. Roy. Soc. [London] **98** [1808] 333/70 from J. Davy, The Collected Works of Sir Humphry Davy, Vol. 5, London 1840, pp. 102/39). – [4] J. Davy (The Collected Works of Sir Humphry Davy, Vol. 5, London 1840, p. 191 footnote). – [5] J. R. Partington (Mem. Proc. Manchester Lit. Phil. Soc. **67** No. 6 [1923] 73/87, 86).

[6] J. Davy (Phil. Trans. Roy. Soc. [London] **102** [1812] 352/69, 353 footnote). – [7] A. Maus (Ann. Physik Chem. [2] **11** [1827] 83/7, 86). – [8] E. Moles, T. Toral (Z. Anorg. Allgem. Chem. **236** [1938] 225/31, 228). – [9] J. J. Berzelius (Kgl. Svenska Vetenskaps. Akad. Handl. **1823** 284/350, 332 from Ann. Physik Chem. [2] **1** [1824] 169/230, 182/3, 225). – [10] C. Truchot (Compt. Rend. **98** [1884] 821/4).

[11] G. Hantke (Z. Angew. Chem. **39** [1936] 1064/71). – [12] H. S. Booth, C. F. Swinehart (J. Am. Chem. Soc. **57** [1935] 1337/42). – [13] G. Gore (J. Chem. Soc. **22** [1869] 368/406, 373/4). – [14] O. Ruff, C. Albert (Ber. Deut. Chem. Ges. **38** [1905] 53/64, 60). – [15] V. Olivieri (Gazz. Chim. Ital. **16** [1886] 218/22).

[16] H. Moissan (Ann. Chim. Phys. [6] **24** [1891] 224/88, 244). – [17] H. Moissan (Compt. Rend. **102** [1886] 1543/4). – [18] H. v. Wartenberg, R. Schütte (Z. Anorg. Allgem. Chem. **211** [1933] 222/6). – [19] H. Moissan (Compt. Rend. **139** [1904] 711/4). – [20] H. Moissan (Compt. Rend. **134** [1902] 1083/7).

[21] G. S. Newth (Chem. News **72** [1895] 278). – [22] A. Dumas (Ann. Chim. Phys. [2] **33** [1826] 337/91, 369; Ann. Physik Chem. [2] **9** [1827] 416/41, 418). – [23] A. Jacquerod, M. Tourpaian (J. Chim. Phys. **11** [1913] 3/28, 28). – [24] A. F. O. Germann, H. S. Booth (J. Phys. Chem. **21** [1916/17] 81/100). – [25] M. Faraday (Phil. Trans. Roy. Soc. [London] **135** [1845] 155/77 in:

Experimental Researches in Chemistry and Physics, London 1859 [Reprint Brussels 1969], pp. 16/124, 104/5; Ann. Physik Chem. [2] **64** [1845] 467/71, 469).

[26] K. Olszewski (Monatsh. Chem. **5** [1884] 127/8). – [27] W. I. Patnode, J. Papish (J. Phys. Chem. **34** [1930] 1494/6). – [28] L. Le Boucher, W. Fischer, W. Biltz (Z. Anorg. Allgem. Chem. **207** [1932] 61/72, 64/5). – [29] J. J. van Laar (Die Zustandsgleichung von Gasen und Dämpfen, Leipzig 1924, p. 183). – [30] O. Ruff, E. Asche (Z. Anorg. Allgem. Chem. **196** [1931] 413/20, 413/5).

[31] G. Natta (Gazz. Chim. Ital. **40** [1930] 911/22). – [32] M. Atoji, W. N. Lipscomb (Acta Cryst. **7** [1954] 597). – [33] A. S. Marggraf (Hist. Acad. Roy. Sci. Belles Lettres Berlin **1768** 3/11, 9). – [34] J. R. Partington (A History of Chemistry, Vol. 4, Macmillan, London 1964, pp. 913/4). – [35] B. Kerl (Handbuch der metallurgischen Hüttenkunde 2nd Ed., Vol. 1, Freiberg 1861, p. 168).

[36] R. A. Broomann (C. **1865** 942). – [37] J. R. Wagner (Jahresber. Chem. Technol. II [1866] 277). – [38] H. Sainte-Claire Deville (Ann. Chim. Phys. [3] **49** [1857] 62/78, 76). – [39] R. Caillat (Ann. Chim. Phys. [11] **20** [1945] 367/420, 367). – [40] J. J. Berzelius (Ann. Physik Chem. [2] **2** [1824] 113/50, 142).

[41] C. Marignac (Compt. Rend. **46** [1858] 854; J. Prakt. Chem. **76** [1858] 161/4). – [42] W. Knop (C. **1858** 388/99, 404/7). – [43] J. J. Berzelius (Ann. Physik Chem. [2] **2** [1824] 113/50, 142). – [44] W. Hempel (Ber. Deut. Chem. Ges. **18** [1885] 1434/40).

2.4.4 Higher Homologs of Silicon Tetrafluoride

The first attempts to obtain higher fluorinated silicon compounds were made in 1902 by H. Moissan, S. Smiles [1], when they brought sulfur hexafluoride and disilane together. As they were transferring the reaction product for further investigation, a violent explosion occurred, accompanied by a blue flame and leaving behind a deposit of elementary silicon. More than sixty years later P. L. Timms, R. A. Kent, T. C. Ehlert, J. L. Margrave published a simpler and safer method for preparing perfluorsilanes [2]. The lower members of the series were obtained by destructive distillation of highly polymeric silicon difluoride $(SiF_2)_x$ in vacuum at 250 to 300°C. They were described as liquids or white crystals, the compounds being very reactive, decomposing with water and igniting in air. The melting points of these fluorides were considerably lower than those of the corresponding hydrides. According to H. P. Hopkins, J. C. Thompson, J. L. Margrave [3], homologous compounds up to $Si_{16}F_{34}$ may be obtained by cracking the polymer in this way.

Hexafluorodisilane Si_2F_6 was first prepared in 1931 by W. C. Schumb, E. L. Gamble [4] by another method of fluorination. Under ordinary conditions it was a gas, which could be condensed to a snow white solid, the melting point of which was −18.5°C. For its preparation they fluorinated hexachlorodisilane carefully with anhydrous zinc fluoride. These workers studied the chemical behavior of the compound in detail and characterized it by further physical data. For the molecular weight (theoretical: M = 170.1) they found the value M = 173.8 [5], which was somewhat too high. The reason for this was realized more than thirty years later when M. Schmeisser, K.-P. Ehlers [6] were able to show that, with the procedure used, the reaction product was still always contaminated with $SiCl_3F$ and $SiCl_4$, and that these impurities could not be separated by distillation, so that pure Si_2F_6 was obtainable only by preparative gas chromatography. If, however, hexabromodisilane Si_2Br_6 is fluorinated in the same way, the reaction products can then be separated by distillation.

Octafluorotrisilane Si_3F_8 and **Decafluorotetrasilane Si_4F_{10}** were discovered and isolated in the course of the pyrolytic experiments described above [2]. Physical properties were measured (melting point of Si_3F_8: −1.2°C and of Si_4F_{10}: 66 to 68°C) using preparations which massspectroscopical studies showed to be more than 95% pure. The impurity in Si_3F_8 was mainly Si_3OF_8 and, in Si_4F_{10}, the next higher homolog, Si_5F_{12}. The very high degree of supercooling possible for each of the two compounds, which could amount to as much as 100°C, was particularly striking.

The compound $\mathbf{Si_{10}F_{22}}$, an even higher homolog of the series with n = 10, was obtained by M. Schmeisser [7] in 1954 as a solid white involatile compound when he fluorinated the corresponding chloro compound, see p. 131, in benzene solution with ZnF_2. The residual fluorinating agent remains as an impurity which can not be removed, while other fluorinating agents prove to be of no use in this case.

References:

[1] H. Moissan, S. Smiles (Compt. Rend. **134** [1902] 1549/52). – [2] P. L. Timms, R. A. Kent, T. C. Ehlert, J. L. Margrave (J. Am. Chem. Soc. **87** [1965] 2824/8). – [3] H. P. Hopkins, J. C. Thompson, J. L. Margrave (J. Am. Chem. Soc. **90** [1968] 901/2). – [4] W. C. Schumb, E. L. Gamble (J. Am. Chem. Soc. **53** [1931] 3191/2). – [5] W. C. Schumb, E. L. Gamble (J. Am. Chem. Soc. **54** [1931] 583/90).

[6] M. Schmeisser, K.-P. Ehlers (Z. Angew. Chem. **76** [1964] 781/2). – [7] M. Schmeisser (Silicon Sulphur Phosphate Colloq., Münster 1954 [1955], pp. 28/31).

2.4.5 Unsaturated Silicon Fluorides

H. Sainte-Claire Deville [1] was the first to obtain evidence of the existence of an unsaturated silicon fluoride. In 1857 he had attempted to reduce silicon tetrafluoride with aluminum at a red heat. At the beginning of the experiment he obtained a very small quantity of a quite volatile liquid, which wetted the glass of the apparatus very poorly. He was not, however, able to explain how it had been formed or its composition, though he suggested that it might be a "lower protofluoride" for which he assumed the formula SiF_2. Almost two decades later L. Troost, P. Hautefeuille [2] observed transport phenomena with molten silicon under the influence of a stream of gaseous silicon tetrafluoride, and the formation of amorphous silicon outside the melting zone. To explain this phenomenon they assumed the formation of a sub-fluoride in the hot part of the apparatus, which decomposed again in the cooler parts. They believed that they had isolated the compound as a very volatile white powder in a cooled tube and termed it "sesquifluorure" of silicon. The next relevant observation was made in 1913 by A. Jaquerod, M. Tourpaian [3] when they were able to show that the liter weight of silicon tetrafluoride, which they had passed through glass wool at a dull red heat to free it from hydrofluoric acid, deviated greatly from the theoretical value. All these speculations on the occurrence of unsaturated silicon fluorides were checked in 1920 by O. Ruff [4], who was not able, however, to confirm their existence, as W. C. Schumb [5] reaffirmed in 1942.

Not until 1954 was M. Schmeisser [6] successful in preparing unsaturated silicon fluorides with the composition $(SiF)_n$ and $(SiF_2)_n$ as solid, highly polymeric substances by reaction of the mixed silicon bromofluorides $SiFBr_3$ and SiF_2Br_2 with magnesium in etheral solution. The same compounds may also be obtained by fluorinating silicon subchlorides and subbromides with zinc fluoride, ZnF_2, though the use of other common fluorinating agents always leads only to SiF_4. Four years later D. C. Pease [7] confirmed the suggestion made by Troost and Hautefeuille [2], which is mentioned above, on the occurrence of subfluorides when SiF_4 vapor is passed over glowing silicon and patented this process. He passed SiF_4 vapor at a greatly reduced pressure (0.04 mm Hg) over lumps of silicon or metallic silicides at 1200°C at a rate of 250 cm^3/h and obtained the monomer SiF_2 by quenching the reaction product to −196°C. On warming to

−78°C, this went over to a pale yellow plastic polymeric solid $(SiF_2)_n$. The solid ignited spontaneously when air was admitted. In 1964 M. Schmeisser, K.-P. Ehlers [8] were able to show that the polymer $(SiF_2)_n$ may be obtained by the catalytic disproportionation of hexafluorodisilane Si_2F_6 in a reaction similar to that for the corresponding chloro- and bromo-compounds. A year later P. L. Timms, R. A. Kent, T. C. Ehlert, J. L. Margrave [9] studied the polymer prepared by Pease's method more closely, finding it to react with water forming a white material, hydrogen and silanes. It is insoluble in organic and inorganic solvents (apart from perfluorosilanes) and on heating in vacuum at 200 to 300°C it melts, decomposing to perfluorosilanes and a silicon-rich polymer $(SiF)_n$. At 400°C decomposition is explosive, SiF_4 and elementary silicon being the main products. The fact that the polymer becomes red-brown in color at very low temperatures was observed by H. P. Hopkins, J. C. Thompson, J. L. Margrave [10] and indicates the presence of unsaturated valencies.

Once P. L. Timms, R. A. Kent, T. C. Ehlert, J. L. Margrave [9] had shown in 1965 that monomeric SiF_2 prepared by Pease's method [7] is surprisingly long lived – according to V. M. Khanna, R. Hauge, R. F. Curl, J. L. Margrave [11] its half-life at low pressure is about 2 min – its physical and chemical properties were elucidated in great detail subsequently by a number of workers, though this can not be discussed here. It should, however, be mentioned that J. L. Margrave, D. L. Perry [12] raised the question of the intermediate occurrence of the dimer $[F_2Si{=}SiF_2]$, which would be of great interest because of its double bond.

A further unsaturated silicon fluoride, $Si_{10}F_{16}$, was obtained in 1954 by M. Schmeisser [13] by fluorinating $Si_{10}Br_{16}$ with zinc fluoride in benzene solution. It is a white solid, which retains zinc fluoride in the ratio 1 mol ZnF_2 to 2 mol $Si_{10}F_{16}$.

References:

[1] H. Sainte-Claire Deville (Ann. Chim. Phys. [3] **49** [1857] 62/78, 76). – [2] L. Troost, P. Hautefeuille (Ann. Chim. Phys. [5] **7** [1876] 452/76, 458/9, 464/5). – [3] A. Jaquerod, M. Tourpaian (J. Chim. Phys. **11** [1913] 3/28, 19). – [4] O. Ruff (Die Chemie des Fluors, Berlin 1920, p. 119). – [5] W. C. Schumb (Chem. Rev. **31** [1942] 586/95, 589).

[6] M. Schmeisser (Z. Angew. Chem. **66** [1954] 713/4). – [7] D. C. Pease (U.S. 2840588 [1958] from C.A. **1958** 19245). – [8] M. Schmeisser, K.-P. Ehlers (Z. Angew. Chem. **76** [1964] 781/2). – [9] P. L. Timms, R. A. Kent, T. C. Ehlert, J. L. Margrave (Nature **207** [1965] 187/8; J. Am. Chem. Soc. **87** [1965] 2824/8). – [10] H. P. Hopkins, J. C. Thompson, J. L. Margrave (J. Am. Chem. Soc. **90** [1968] 901/2).

[11] V. M. Khanna, R. Hauge, R. F. Curl, J. L. Margrave (J. Chem. Phys. **47** [1967] 5031/7). – [12] J. L. Margrave, D. L. Perry (Inorg. Chem. **16** [1977] 1820/2). – [13] M. Schmeisser (Silicon Sulphur Phosphate Colloq., Münster 1954 [1955], pp. 28/31).

2.4.6 Further Fluorosilanes

The fluorination of silanes, which had been unsuccessfully attempted by H. Moissan, was taken up again in 1905 by O. Ruff, C. Albert [1], this time with metallic fluorides and silicochloroform. They obtained the best results with titanium tetrafluoride TiF_4 in a sealed tube at 100 to 120°C. These workers obtained vapor density and molecular weight values which were somewhat too high for the reaction product expected, the gaseous **silicofluoroform $SiHF_3$** (trifluorosilane), which burns in air. They also observed ready thermal decomposition of the compound, which proceeded to a point where elementary silicon appears, and published some physical properties as well as some of its chemical reactions. These data were first improved almost thirty years later when H. S. Booth, W. D. Stillwell [2] took up the fluorination experiments with silicochloroform again and learned how to fluorinate stepwise using antimony trifluoride with antimony pentachloride as catalyst. In this way they prepared pure silicofluoroform (as well as the possible mixed halides) and observed that the supposed decomposition of silicofluoroform down to elementary silicon is in fact a disproportionation reaction and starts already at liquid nitrogen temperature. Ten years later H. J. Emeléus, A. G. Maddock [3] investigated the mechanism of the fluorination and were in a position to determine characteristic data for the gas on a very pure preparation.

Difluorosilane SiH_2F_2 and **Monofluorosilane SiH_3F**. The first preparation of difluorosilane by fluorinating the corresponding chlorosilane with antimony trifluoride was reported in 1939 by A. G. Maddock, C. Reid, H. J. Emeléus [4]. They gave the boiling point of the compound, which is gaseous under normal conditions and burns in air, as −77.5°C and the melting point as −119.1°C, remarking that it is difficult to separate from SiF_4, which is formed simultaneously. Four years later, H. J. Emeléus, A. G. Maddock [3] were successful in preparing monofluorosilane SiH_3F for the first time and published fresh physical data for it and for SiH_2F_2. They found the boiling point to be unexpectedly high due to association in the liquid phase.

References:

[1] O. Ruff, C. Albert (Ber. Deut. Chem. Ges. **38** [1905] 53/64, 56/63). – [2] H. S. Booth, W. D. Stillwell (J. Am. Chem. Soc. **56** [1934] 1531/5). – [3] H. J. Emeléus, A. G. Maddock (J. Chem. Soc. London **1944** 293/6). – [4] A. G. Maddock, C. Reid, H. J. Emeléus (Nature **144** [1939] 328).

2.4.7 Fluorosilicic Acid H_2SiF_6

As was pointed out earlier, in 1768 A. S. Marggraf recognized the acid character of the distillation product which had dissolved in the water of the receiver during the distillation of fluorspar with vitriolic acid, and had separated out a difficultly soluble salt by neutralization, though he had not investigated this more closely [1]. C.W. Scheele took this up again three years later in his first *Chemische Untersuchung des Flußspats und dessen Säure* [2] and was able to show that this difficultly soluble product could not be pure silica, since it "melted on heating" and when the cooled melt was "powdered and put in the cellar – it deliquesced – and the deliquesced material behaved like genuine waterglass" ["gepulvert und in den Keller gesetzt . . . zerfloß es . . . das Zerflossene sich aber als wahre Kieselflüssigkeit [das ist: Wasserglas] verhielt"]. P. C. Abilgaard, the next to repeat the experiment with fluorspar mentioned only the formation of "particular neutral salts" ["sonderliche Mittelsalze"] without going into them more closely [3]. T. Bergman, in his major study on silica, was the first to be more specific [4]: "But what by the two fixed alkalis [potash and soda] is precipitated [after filtering off the silica deposited] is not pure silica but a special triple salt composed of silica, fluorspar acid, and alkali. It dissolves, even if with difficulty, in hot water, especially if it is prepared with vegetable alkali [potash]. It is, however, easily decomposed by lime water and fluorspar thus results again. And for this reason [this precipitate] is to be carefully distinguished and not confused with that which is precipitated by volatile alkali [ammonia]" ["Quod autem utroque alkali fixo deturbatur, non purum exhibet siliceum, sed peculiarem salem triplicem, silicea terra, acido fluoris & alkali compositum, qui aqua fervida solvitur, quamvis aegre, praesertim alkali vegetabili genitus, aqua vero calcis facile decomponitur, fluorem mineralem regeneratum demittens, ideoque probe distinguendus & minime cum illo confundendus, quod alkali volatili dejicitur"]. In his third publication in 1786 *"Neue Beweise der Eigenthümlichkeit der Flußspathsäure"*, C.W. Scheele [5] states explicitly that this salt decomposes on heating: "If, though, the acid [obtained from a glass apparatus] is saturated with fixed alkali, a clear precipitate likewise results, which falls to the bottom more rapidly [than the silica obtained with ammonia] and, after it has become well dried, has a white appearence like chalk. This precipitate weighs twice as much as the siliceous precipitate [with ammonia] and is a neutral salt saturated with excess acid. Consequently it is soluble in water [though a great deal is required] and consists of silica, fixed alkali and fluorspar acid. It easily becomes liquid in the furnace and, if melting is continued, a clear glass results which is alkaline and gives off a true silica humor" ["Wird die [aus einer Glasapparatur erhaltene] Säure aber mit fixem Alkali gesättigt, so entsteht gleichfalls ein klarer Niederschlag, welcher geschwinder zu Boden fällt [als die mit Ammoniak erhaltene Kieselerde] und nachdem er wohl getrocknet worden, ein weißes Aussehen wie Kreide hat. Dieser Niederschlag wiegt doppelt so viel als der kieselichte Niederschlag [mit Ammoniak], und ist ein mit überschüssiger Säure gesättigtes Neutralsalz; folglich im Wasser (obwohl sehr viel dazu erfordert wird) auflöslich, und besteht aus Kiesel, fixem Alkali und Flußspatsäure. Es ist im Feuer leicht flüssig, und wird die Schmelzung fortgesetzt, so entsteht ein klares Glas, so alkalisch ist, und eine wahre Kieselfeuchtigkeit abgiebt"].

No one else involved in the discovery of silicon fluoride concerned himself with this salt, and still less with the acid contained in it. The first to become involved with it was John Davy in 1812 [6]. He did so very thoroughly and remained for many years the only one to take an interest in this compound. As a result, L. Gmelin in the 2nd Edition of this *Handbook* in 1821 [7] was obliged to quote him as the sole source. He wrote: "Fluorosilicic acid. – Forms in the aqueous state on decomposing silicon fluoride gas with water. The water absorbs and decomposes at the most 265 [in the original, 263] times the amount of silicon fluoride gas, when it reaches a consistent composition. If this material comes into contact with hydrochloric acid gas, aqueous hydrochloric acid is formed, all the silica disappearing and gaseous silicon fluoride being evolved. J. Davy. – This salt is prepared by passing gaseous silicon fluoride by means of a wide tube into mercury over which there is water. If the mouth of the tube is in the water it is rapidly blocked by silica. It has a very acid taste similar to aqueous hydrochloric acid. It irritates the skin like hydrochloric acid and reddens litmus strongly. Vaporizes on heating with white fumes. Vaporizes slowly in air, leaving behind a little silica. J. Davy. – Contains ¾ of the silica which the silicon fluoride gas produces in its decomposition by water, combined with all of the hydrofluoric acid that is formed in this way. J. Davy. – On distillation the liquid evolves silicon fluoride gas and aqueous fluorosilicic acid. In the neck of the retort some gelatinous hydrated silica deposits. A little or no silica remains in the retort. Boric acid forms aqueous fluoroboric acid with the liquid, silica being deposited. Oil of vitriol generates silicon fluoride gas and deposits silica. Ammonia and caustic soda form a hydrofluoric acid salt with it, with precipitation of all the silica, J. Davy" ["Saure flußsaure Kieselerde. – Bildet sich in wässriger Gestalt bei der Zersetzung des Fluorsiliciumgases durch Wasser. Das Wasser absorbirt und zersetzt höchstens 265 [im Original: 263] Maaße Fluorsiliciumgas, wobei es zu einer consistenten Masse wird. Kömmt diese Masse mit salzsaurem Gase in Berührung, so bildet sich wässrige Salzsäure unter Verschwinden sämmtlicher Kieselerde und unter Entwicklung des Fluorsiliciumgases. J. Davy. – Man bereitet dies Salz, indem man das Fluorsiliciumgas mittelst einer weiten Röhre in Quecksilber leitet, über welchem sich Wasser befindet. Steht die Mündung der Röhre im Wasser, so wird sie schnell von der Kieselerde verstopft. Schmeckt sehr sauer, der wässrigen Salzsäure ähnlich; reizt die Haut wie die Salzsäure; röthet stark Lakmus. Verdampft beim Erhitzen in weißen Dämpfen; verdampft an der Luft langsam, unter Zurücklassung von wenig Kieselerde. J. Davy. – Enthält ¾ von der Kieselerde, welche das Fluorsiliciumgas bei seiner Zersetzung durch Wasser erzeugt, mit sämmtlicher dabei gebildeter Flußsäure verbunden. J. Davy. – Die Flüssigkeit entwickelt bei der Destillation Fluorsiliciumgas, und wässrige saure flußsaure Kieselerde; in den Hals der Retorte setzt sich etwas gallertartiges Kieseldehydrat an; in der Retorte bleibt wenig oder keine Kieselerde. – Boraxsäure bildet mit der Flüssigkeit, unter Niederschlagung der Kieselerde, wässrige Flußboraxsäure. – Vitriolöl entwickelt Fluorsiliciumgas, und schlägt Kieselerde nieder. – Ammoniak und Natron bilden mit ihr ein flußsaures Salz unter Fällung aller Kieselerde, J. Davy"].

The acid was next investigated three years later by J. J. Berzelius [8], who confirmed J. Davy's results and occasionally rendered them more precise, as well as preparing and characterizing numerous salts of the acid. Some years later A. Maus [9] discussed methods for preparing the acid that were known up to then and improved them somewhat, but it was in 1863 that F. Stolba [10] described for the first time a procedure which yielded a pure acid, so that he was in a position to determine the maximum concentration of the solution that could be obtained under ordinary conditions. Another preparative method which had already been used for industrial purposes, and which avoided the troublesome removal of precipitated silica, was published by Kessler [11] in 1880. In it gaseous silicon fluoride was passed into sufficiently concentrated hydrofluoric acid. Under certain conditions he obtained colorless and fairly hard crystals with the composition $H_2SiF_6 \cdot 2H_2O$, which melted at 19°C and decomposed to HF and SiF_4 on warming. Four years later C. Truchot [15] prepared a monohydrate which melted at −20°C by passing silicon fluoride into hydrofluoric acid with one equivalent of H_2O. If the hydrofluoric acid contained four to six equivalents of H_2O, a tetrahydrate melting at 0°C was isolated. Twenty years later, E. Baur [16] also obtained crystals from a 30% solution of H_2SiF_6 by cooling to −50°C which appear to be a water-rich hydrate. In order to make very pure solutions of fluorosilicic acid E. Baur, A. Glaessner [12] in 1903 decomposed barium fluorosilicate with sulfuric acid. They were able to show, by measuring the vapor density of the product, that the complete vaporization of the acid without residue, which those who had first prepared the acid observed, took place with at least 50% decomposition. A somewhat cumbersome method for preparing a pure product was suggested by F. Fischer, K. Thiele [13]. They dissolved silicon dioxide in dilute hydrofluoric acid, following J. J. Berzelius' procedure [8], precipitated impurities by careful addition of lead salts and prepared a solution of lead silicofluoride by further addition of lead carbonate. They precipitated the lead from this with hydrogen sulfide and removed excess of the precipitating agent by blowing in air. In 1923 C. A. Jacobson [14], in attempting to prepare anhydrous H_2SiF_6, obtained an acid with 60.6 wt% H_2SiF_6 after several days from a solution which was about 25%, by careful distillation at room temperature and reduced pressure. This appeared to be stable. In 1910 E. Ebler, E. Schott [25] thought they had found a possible way of preparing the anhydrous compound. In it the hydrazine nitrogen in anhydrous crystalline monohydrazinium hexafluorosilicate, $(N_2H_5)HSiF_6$, was thought to be driven off under the action of a stream of dry halogen, leaving H_2SiF_6 behind. The publication of the details of these experiments, which had been expected, never took place.

The first preparation of solutions of fluorosilicic acid on an industrial scale can be attributed to C. M. Tessié du Motay [17] in the second half of the 19th century. A briquetted mixture of fluorspar, sand, carbon, and clay was melted down with coke in a shaft furnace and the resulting silicon fluoride absorbed in a wooden condensation vessel on glass plates over which water was flowing. Solutions with about 9% H_2SiF_6 could be obtained in this way. Today the compound is obtained exclusively as a byproduct of the manufacture of superphosphate from the waste gases escaping from the decomposition reactor, which contain mainly silicon fluoride and hydrofluoric acid, either in an absorption plant as an aqueous solution [18] or by converting them to fluorosilicates with dry metallic compounds [19]. The acid may also be isolated as a byproduct in other technical processes. Thus, for example, in 1898 F. A. Hoppen [20] suggested obtaining it from the hydrofluoric acid used in the purification of unrefined graphite or, as Prior reported in 1903 [21] it may be made from hydrofluoric acid used to achieve greater porosity in clay vessels.

In 1850 H. Rose [22] recommended its application in quantitative chemical analysis, a possibility which J. J. Berzelius had already pointed out in the course of his extensive investigation of the metallic salts of the acid. Ten years later F. J. Otto [23] reported that it had replaced tartaric acid, which had risen sharply in price, for tinning pins in France, where the acid was prepared on a large scale for this purpose. According to a patent by Rousseau [24], it could also be employed in preparing mordants for dyeing and calico printing, again in place of tartaric acid. The acid was also used extensively, usually as a 1 to 2% solution, for eliminating the growth of moulds on walls and for preserving stable manure and seeds because of its antiseptic properties. The solutions appeared in commerce as Montanin [21], Keramyl and also under other trivial names. They were much used in brewing for completely sterilizing copper and brass vessels (though not those of iron, tin, or zinc, which are corroded) [18]. In modern times, too, many applications have been suggested, but they are so numerous and varied that it is impossible to discuss them further here; see in this connection "Silicium" B, 1959, p. 643, for example.

References:

[1] A. S. Marggraf (Hist. Acad. Roy. Sci. Belles Lettres Berlin **1768** 3/11). – [2] C. W. Scheele (Kgl. Svenska Vetenskaps Akad. Handl. **33** [1771] 120/38 in: Sämmtliche physische und chemische Werke [German by S. F. Hermbstädt], Vol. 2, Berlin 1793, pp. 3/32, 24). – [3] P. C. Abilgaard (Schriften der Societät der Wissenschaften Kopenhagen Year 1779 from Neueste Entdeckungen Chem. [Crell] **2** [1781] 168/70). – [4] T. Bergman (Diss. de Terra Silicea 1779 in: Opuscula Physica et Chemica, Vol. 2, Uppsala 1780, pp. 26/53, 34). – [5] C. W. Scheele (Chem. Ann. Crell **1786** I 3/17 in: Sämmtliche physische und chemische Werke [German by S. F. Hermbstädt], Vol. 2, Berlin 1793, pp. 402/22, 421/2).

[6] J. Davy (Phil. Trans. Roy. Soc. [London] **102** [1812] 352/69). – [7] L. Gmelin (Handbuch der theoretischen Chemie, 2nd Ed., Vol. 1, Frankfurt a. M. 1821, pp. 504/5). – [8] J. J. Berzelius (Ann. Physik Chem. [2] **1** [1824] 169/230, 176/204). – [9] A. Maus (Ann. Physik Chem. [2] **11** [1827] 83/4). – [10] F. Stolba (J. Prakt. Chem. **90** [1863] 193/200).

[11] Kessler (Compt. Rend. **90** [1880] 1285/6). – [12] E. Baur, A. Glaessner (Ber. Deut. Chem. Ges. **36** [1903] 4215/7). – [13] F. Fischer, K. Thiele (Z. Anorg. Allgem. Chem. **67** [1910] 302/16, 307). – [14] C. A. Jacobson (J. Phys. Chem. **27** [1923] 577/80). – [15] C. Truchot (Compt. Rend. **98** [1884] 821/4).

[16] E. Baur (Z. Physik. Chem. **48** [1904] 483/503, 497). – [17] C. M. Tessié du Motay (from R. Biedermann in: A. W. Hofmann, Die Entstehung der chemischen Industrie während des letzten Jahrzehntes, Pt. 1, Braunschweig 1875, pp. 287/321, 316). – [18] A. Schloss (Ullmanns Enzykl. Tech. Chem. 2nd Ed. **5** [1930] 411). – [19] M. Schmeisser, P. Voss (Ullmanns Enzykl. Tech. Chem. 3rd Ed. **15** [1964] 745). – [20] F. A. Hoppen (Ger. 105734 [1898] from C. **1900** I 79).

[21] Prior (Z. Angew. Chem. **16** [1903] 195). – [22] H. Rose (Ann. Physik Chem. [2] **80** [1850] 403/6). – [23] F. J. Otto (Lehrbuch der Chemie, Vol. 2, Sect. 3, Braunschweig 1860, p. 279 from Jahresber. Chem. Technol. **7** [1861] 114). – [24] Rousseau (Bull. Soc. Encouragement **1861** 565 from Jahresber. Chem. Technol. **7** [1861] 581). – [25] E. Ebler, E. Schott (J. Prakt. Chem. [2] **81** [1910] 552/6).

2.5 Silicon-Chlorine Compounds

2.5.1 Silicon Tetrachloride. Tetrachlorosilane. Chlorsilicon $SiCl_4$

The first observation on the formation of compounds between silicon and chlorine was made in 1824 by J. J. Berzelius [1] while studying the properties of so-called amorphous silicon, which he was the first to prepare. He wrote: "If silicon is heated in a stream of chlorine it ignites and burns away... The combustion product condenses and is a liquid which is yellowish with excess of chlorine but which appears to be colorless when freed of it. This liquid is very volatile and free flowing, vaporizing at once in the open air disseminating white fumes and leaving silica. It has a suffocating smell somewhat like that of cyanogen. When shaken with water it floats on the surface, dissolving for the most part and leaves behind some undissolved silica. If the amount of water is small, e.g., a drop on as much chlorsilicon, the latter spreads over it and silica remains in a swollen semi-transparent state" ["Wird Silicium in einem Strom von Chlor erhitzt, so entzündet es sich und brennt fort... Das Produkt der Verbrennung condensirt sich und stellt eine Flüssigkeit dar, die mit Überschuß an Chlor gelblich ist, aber davon befreit, farblos zu seyn scheint. Diese Flüssigkeit ist sehr flüchtig und leicht fließend, verdampft an freier Luft im Augenblick unter Verbreitung eines weißen Dampfes und mit Hinterlassung von Kieselerde. Sie hat einen erstickenden Geruch, der dem des Cyans einigermaßen gleicht. In Wasser geschüttet, schwimmt sie auf dessen Oberfläche, löst sich größtentheils auf, oder hinterläßt etwas Kieselerde unaufgelöst; ist die Quantität des Wassers geringe, z. B. ein Tropfen auf ebenso viel Chlorsilicium, so breitet letzteres sich auf ihm aus, und die Kieselerde bleibt in einem aufgeschwollenen, halbdurchsichtigen Zustande zurück"].

As early as the beginning of October of the following year H. C. Oersted (1777–1851) in letters [2] to the editors of scientific journals announced a new and simpler method of preparing "Chlorkiesel" which, according to G. Martin [3] was to become the only useful method for preparing the compound to be used exclusively on a technical scale. He had found that direct chlorination of silica, in which H. Davy [29] had been unsuccessful in 1810, was easily possible if chlorine was passed over red hot silica and carbon. It is notable that he stumbled on this method in the course of making "Chlorargillium, i.e., the compound of chlorine with the combustible base of clay" [$AlCl_3$] ["Chlorargillium, d. h. die Verbindung des Chlors mit dem brennbaren Grundstoff der Thonerde"]. In the original publication [4], in which chlorination of silica and alumina is dealt with in only a few lines, he reported that chlorides of alumina and silica are obtained together if china clay and carbon are used, though "Kieselchlorid" is very volatile and must be collected in a receiver which is cooled with a freezing mixture. He also says that Candidate of Pharmacy and Philosophy Koester has determined the density of the colorless liquid as 1.5 and its boiling point as 50°C. In the following year J. Dumas [5] had already made use of this convenient new method for preparing the volatile compound to determine the atomic weight of the element from the liter weight of this and other readily volatilized silicon compounds, though he found too high a liter weight for $SiCl_4$ because of insufficient purification. That Dumas did not mention the discoverer of the preparative method by name excited F. Wöhler's indignation. When the latter reported on his own experiments with the method, in which the temperature at which the reaction begins clearly interested him most, he wrote to J. J. Berzelius on 22nd October, 1826 [6]: "... I have since started some experiments on the preparation of the chlorine compounds by Oersted's method. This method seems to me to be highly ingenious and in many respects of such great importance that it is quite shameless of Dumas to have used it without mentioning Oersted by name. And it cannot be doubted that he knew of Oersted's work as well as he knew of boron chloride, the discovery of which he also appropriated... I first made the experiment with powdered rutile and carbon powder (when reaction set in even on warming with a spirit lamp)... With silica (separated from "Fluorkiesel" [SiF_4]) no chloride was obtained over the spirit lamp" ["... Ich habe nachher einige Versuche angestellt über die Darstellung der Chlorverbindungen nach Oersteds Methode. Diese Methode scheint mir höchst ingeniös und in vieler Beziehung von so großer Wichtigkeit, daß es in der That von Dumas eine große Unverschämtheit ist, sich derselben bedient zu haben, ohne Oersted zu nennen. Und es ist gar keinem Zweifel unterworfen, daß er Oersteds Versuche gekannt hat, so gut wie er das Chlorbor gekannt hat, dessen Entdeckung er sich ebenfalls zueignet... Zuerst machte ich den Versuch mit gepulvertem Rutil und Kohlepulver (wobei schon beim Erwärmen mit einer Spirituslampe die Reaktion einsetzt)... Mit Kieselsäure (aus Fluorkiesel [SiF_4] abgeschieden) entstand über der Spirituslampe kein Chlorkiesel"]. Twenty years later J. J. Ebelmen [7] noted that "to obtain a reaction it is necessary that the silica be prepared by chemical means and be soluble in alkalis. Quartz, however finely powdered, gives no appreciable amount of silicon chloride" ["zur Erzielung der Reaktion nöthig ist, daß die Kieselerde auf chemischem Wege bereitet und in Alkalien löslich ist. Der Quarz, auch noch so fein gepulvert, giebt keine erkleckliche Quantität von Siliciumchlorür"].

In 1930, V. Spitzin [8] attempted to clarify the question of the reaction mechanism after, as early as 1859, R. Weber [9] had established that silica can be chlorinated without addition of carbon, though only at a white heat. These and a series of further investigations on the chlorination of silica which met with varying degrees of success but which will not be enumerated here led E. Gruner, J. Elöd [10] in 1931 to a study of the behavior of silica modifications and varieties in the process in question. Some patents for the technical application of the method have been filed which differ in the addition of various chlorides to the solid reaction mixture [11] or, as in the first patent of this sort (with a priority of 11.3.1921) recommend addition of sulfur chloride to the gaseous chlorine [12]. There are, however, other patents which were filed earlier and are relevant. That of E. Jüngst, R. Mewes [25] in 1902, where there was a protected reaction between a difficultly volatile metal chloride (such as $CaCl_2$) and elementary silicon, was concerned more with the metal silicides obtained than with silicon chlorides. Towards the end of the first world war the patents of O. Hutchins [26] and R. W. Moore [27] related to the treatment of silicon carbide alone or mixed with elementary silicon with dry chlorine at 1000°C. With suitable apparatus the heat of reaction is said to be sufficient for the process to be carried through without the application of further heat. Difficulties which arise and possible remedies were described in 1926 by P. S. Brallier [28]. It may also be mentioned that chlorination of aluminum silicates to obtain both possible chlorides, which had already been examined by Oersted, has apparently been used technically: at least a patent for it was filed in 1927 [13]. Berzelius' original preparative method [1] is also used occasionally, though not for technical purposes and has also been more closely investigated. Thus, L. Troost, P. Hautefeuille [14] measured the heat of formation of the compound from this reaction and, in 1920, G. P. Baxter, P. F. Weatherill, E. O. Holmes [15] prepared the starting material for their determination of the atomic weight of silicon. They thought they could obtain pure silicon tetrachloride in this way. It was known that with "crude silicon" ["rohes Silicium"] (obtained by reduction of sand with magnesium) as the result of a proposal by L. Gattermann [16] in 1889, a significant amount of higher chlorides is also formed. Experiments have naturally been undertaken to introduce the cheaper and more readily prepared ferrosilicon in obtaining silicon chloride. Thus, in 1889, H. N. Warren [17] suggested the use of a ferrosilicon with only 15% Si for this purpose.

The obvious importance of the compound, which is shown by the large number of patents on its preparation that have been filed, lies not only in the fact that it is the starting substance for the preparation of most silicon compounds and the purest form of the element, see p. 56, but also in a property, by which those who first prepared it were impressed. Silicon tetrachloride vapor in moist air, either alone or in combination with gaseous ammonia, forms a thick white fog. As these aerosols are quite stable they were already used in the first world war 1914–1918 for screening ships, trenches and other things by smoke, especially on the part of the Americans [18, 19, 20]. The English admiralty also introduced them to camouflage a fleet which was to attack Zeebrügge [21]. G. A. Richter [20] reported technical details on screening with $SiCl_4$. It is not therefore surprising to hear that in the late summer of 1917 the American Ordonance Department simultaneously commissioned three electrochemical plants at Niagara Falls to work out processes for the preparation on a full technical scale of the compound which until then had been prepared only by a laboratory method [22].

It was already known to early investigators that the fog from silicon chloride consisted of silica or of compounds which readily went over to it. The compound therefore serves on a considerable scale for the preparation of silica sols and gels when it is decomposed hydrolytically or otherwise by the most diverse methods, though this cannot be discussed in detail here. Two quite opposite decomposition processes will, however, be referred to briefly. The proposal was made by H. Stelling in 1926 [23] to use silicon tetrachloride for fireproofing organic materials and fabrics. A moist fabric was "exposed to a silicon chloride atmosphere in such a way as to lead to precipitation of a silica skeleton which penetrated the fine structure of the fabric" ["einer Siliciumchlorid-Atmosphäre derart aussetzte, daß das Niederschlagen eines die Stoffe auch in allen Feinheiten durchdringenden Kieselsäureskeletts erreicht wird"]. If the organic matter which had been treated in this way was then carefully calcined the unburnt material remained which had the same structure as the starting material and which may be used as a filter for hot gases and liquids and as a support for catalysts. A quite different process for decomposing silicon tetrachloride was suggested in 1935 by the Corning Glass Works, to obtain fine silica aerosols. For this, silicon tetrachloride vapor with excess oxygen was blown into a hot flame and so burnt [24].

Purification of the silicon tetrachloride (and the other chlorosilanes) is critical for their use in atomic weight determinations and also for the preparation of high purity silicon on a technical scale. According to R. C. Hughes [30], repeated distillation, e.g., with fractionating columns filled with glass beads, does not suffice. The addition of elementary sulfur followed by boiling under reflux in which, according to B. C. Meyers [31], high boiling compounds with the impurities which can be separated by distillation are formed, is also insufficient. The reason for this lack of success is thought by R. Rost [32] to be that a large proportion of the impurities form azeotropic mixtures with the halide or possess boiling points close to that of $SiCl_4$. G. Rossenberger [33] found that the required high degree of purity could be reached if the $SiCl_4$ was subjected to extraction with methylcyanide (acetonitrile, CH_3CN), which has the property of being a very good solvent for most of the impurities contained in the silicon halides. Investigations have shown that all of the foreign substances in the crude $SiCl_4$ are found again in the extract. Methyl cyanide itself dissolves only to a very small extent in $SiCl_4$ and may be separated completely by distillation.

For further methods of purification applicable to $SiCl_4$, see p. 134.

References:

[1] J. J. Berzelius (Ann. Physik Chem. [2] **1** [1824] 169/230, 218/9; Kgl. Svenska Vetenskaps Akad. Handl. **1824** 46/98, 57/8). – [2] H. C. Oersted (Letter to Hansteen in: Magasin Naturvidenskaberne **3** [1825] 176 from Ann. Physik Chem. [2] **5** [1825] 132; Letter to Schweigger, Schweiggers J. Chem. Physik **45** [1825] 368). – [3] G. Martin (J. Chem. Soc. **105** [1914] 2836/60, 2836). – [4] H. C. Oersted (Kgl. Danske Videnskab. Selskab Oversigt **1824/25** 13/6 in K. Meyer, H. C. Ørsted, Scientific Papers, Vol. 2, Kopenhagen 1920, pp. 464/6). – [5] J. Dumas (Ann. Chim. Phys. [2] **33** [1826] 337/91, 367).

[6] O. Wallach (Briefwechsel zwischen J. Berzelius und F. Wöhler, Vol. 1, Leipzig 1901 [Reprint, Sändig, Wiesbaden 1966], pp. 149/50). – [7] J. J. Ebelmen (Ann. Chim. Phys. [3] **16** [1846] 129/66, 142; Liebigs Ann. Chem. **57** [1846] 319/55, 332). – [8] V. Spitzin (Z. Anorg. Allgem. Chem. **189** [1930] 337/66, 352/5). – [9] R. Weber (Ann. Physik Chem. [2] **107** [1859] 375/98). – [10] E. Gruner, J. Elöd (Z. Anorg. Allgem. Chem. **195** [1931] 269/87).

[11] K. E. Jackson (Chem. Rev. **25** [1939] 67/119, 107/9). – [12] Consortium für Elektrochemische Industrie G.m.b.H. (Brit. 176811 [1922] from C. **1922** IV 132; C.A. **1923** 857). – [13] E. R. Wolcott Texas Co. (U.S. 1633835 [1921/27] from C. **1927** II 1069; C.A. **1927** 2763). – [14]

L. Troost, P. Hautefeuille (Bull. Soc. Chim. France [2] **13** [1870] 213/20, 226). – [15] G. P. Baxter, P. F. Weatherill, E. O. Holmes (J. Am. Chem. Soc. **42** [1920] 1194/7).

[16] L. Gattermann (Ber. Deut. Chem. Ges. **22** [1889] 186/97). – [17] H. N. Warren (Chem. News **60** [1889] 158). – [18] K. Wolf (Ullmanns Enzykl. Tech. Chem. 2nd Ed. **9** [1932] 505). – [19] J. Meyer (Der Gaskampf und die chemischen Kampfstoffe, Hirzel, Leipzig 1925, pp. 1/424, 413). – [20] G. A. Richter (Trans. Am. Electrochem. Soc. **35** [1919] 323/33, 326/30).

[21] R. Witzinger (Chemische Plaudereien über Gaskrieg Atomzertrümmerung Vitamine und viele andere Gegenwartsprobleme, Bonn 1934, p. 188). – [22] O. Hutchins (Trans. Am. Electrochem. Soc. **35** [1919] 309/21). – [23] H. Stelling (Ger. 429918 [1924/26] from C. **1926** II 12201). – [24] F. L. Ebenhöch, W. Ziese (Ullmanns Encykl. Tech. Chem. 3rd Ed. **15** [1964] 744). – [25] E. Jüngst, R. Mewes (Ger. 157615 [1902/04] from C. **1905** I 194).

[26] O. Hutchins (U.S. 1271713 [1917/18] from C.A. **1918** 1914; Can. 184354 [1918] from C.A. **1918** 1500). – [27] R. W. Moore (U.S. 1350932 [1920] from C.A. **1920** 3300). – [28] P. S. Brallier (Trans. Am. Electrochem. Soc. **49** [1926] 257/66, 257/8). – [29] H. Davy (Phil. Trans. Roy. Soc. [London] **101** [1811] 1/35), J. Davy (The Collected Works of Sir Humphry Davy, Vol. 5, London 1840, pp. 312/48, 334). – [30] R. C. Hughes (J. Opt. Soc. Am. **33** [1943] 49/60, 53).

[31] B. C. Meyers (U.S. 1413958 [1947] from C.A. **1947** 2865). – [32] R. Rost (Silicium als Halbleiter, Berliner Union, Stuttgart 1966, pp. 1/367, 20/1). – [33] G. Rossenberger (Ger. 955415 [1957] from C.A. **1959** 4673).

2.5.2 Higher Homologs of Silicon Tetrachloride

There are some gaps in the higher members of the homologous series Si_nCl_{2n+2}, but among those known are the longest inorganic chain molecules with which we are so far acquainted.

Hexachlorodisilane Si_2Cl_6. The hexachloride of silicon was first prepared in 1871 by C. Friedel [1] by heating hexaiododisilane with corrosive sublimate $HgCl_2$ and submitting the product passing over between 144 and 148°C to purification. Five years later L. Troost, P. Hautefeuille [2] published their preparative method in which the vapor of silicon tetrachloride was passed over molten silicon and quickly cooled. They isolated a fraction boiling in the temperature range mentioned, which solidified at −14°C to crystalline plates reminiscent of boric acid. They determined the density $D_o = 1.58$, the vapor density of the compound, which they called silicon sesquichloride, giving Si_2Cl_6 as the formula. In 1894 L. Gattermann, K. Weinlig [3] identified the compound in the mixture which they had obtained in chlorinating "crude silicon" ["Rohsilicium"] by Gattermann's method [4]. They found the melting point to be −1°C and the boiling point 145 to 146°C, the proportion of the compound amounting to 20% of the mixture [5]. When A. Besson, L. Fournier [6] in 1909 submitted the vapor of trichlorosilane $SiHCl_3$ to a silent electric discharge in presence of hydrogen an oily liquid resulted from which they were also able to isolate "silicoperchloroethane" Si_2Cl_6. Five years later G. Martin [7] reverted to direct chlorination, though he started with the cheaper ferrosilicon (with 50% Si). Putting in 50 kg of ferrosilicon and 143 kg of chlorine he obtained (following careful separation) 54 kg $SiCl_4$, 3 kg Si_2Cl_6, 200 g Si_3Cl_8, 3 g Si_4Cl_{10}, 2 g Si_5Cl_{12}, and 0.5 g Si_6Cl_{14}. He determined the density of Si_2Cl_6 as $D_4^{15} = 1.5624$, the boiling range as 144 to 145.5°C and the solidification point as −3°C. For the melting point a value of −1°C was found, the substance having been shown by chemical analysis to be pure. The Besson-Fournier method was modified in 1925 by A. Stock, A. Brandt, H. Fischer [8] by operating an arc between zinc electrodes in silicon tetrachloride vapor. Contrary to expectations, the yield of higher homologs was small and furthermore a quantity of admixed oxychloride $(SiCl_3)_2O$ was present. This could not be separated, even by vacuum distillation because it had almost the same volatility as Si_2Cl_6 at low temperatures. More recently

H. Kautsky, H. Kautsky Jr. [9, 10] prepared a number of homologs of this series by a method similar to the above based on the Svedberg method for colloidal dispersion, using high voltage short circuit sparks between silicon powder in nonconducting silicon tetrachloride. They obtained a number of homologs of this series and, by using an extremely slow separation process which they called "pseudodistillation", were able to separate these. Chemical analysis showed Si_2Cl_6 to be among them.

Octochlorotrisilane Si_3Cl_8. This compound was first isolated in 1894 by L. Gattermann, K. Weinlig [3] in separating the reaction products from chlorination of so-called "crude silicon" ["Rohsilicium"] and was characterized as a liquid boiling between 210 and 215°C and freezing at −12°C. A. Besson, L. Fournier [6] were also able to isolate it from the mixture arising in their process and report some physical data but G. Martin [7] in 1914 was the first to determine the characteristic data on preparations of satisfactory purity.

Decachlorotetrasilane Si_4Cl_{10}. A. Besson, L. Fournier [6] thought at first that they could not detect this compound in the mixture which had resulted from passing a silent discharge through silicochloroform vapor but, later [11], they had to announce that the oily liquid which they had held to be the homolog with n = 5 was really Si_4Cl_{10}. How small the proportion of this and the higher members of the series is when silicon or ferrosilicon is chlorinated was shown by G. Martin [7]. It is for this reason that the use of other silicides, and particularly of calcium silicide, was recommended [12] but the yield is again quite small [13].

Dodecachloropentasilane Si_5Cl_{12}. As early as 1909 A. Besson, L. Fournier [6] thought that they had prepared this compound as an oily liquid but they had soon to acknowledge [11], that they had had in their hands the member of the series with n = 4. At the same time they believed they were able to say that they had now actually succeeded in isolating the member with n = 5 as a very viscous liquid. As R. Schwarz, H. Meckbach [14] rightly objected in 1937, however, an exact determination of the chain length was lacking in this investigation and measurement of the molecular weight had been omitted altogether. In 1914, G. Martin [7] reported the isolation of 2 g of the supposedly liquid compound and recently H. Kautsky, H. Kautsky Jr. [9, 10] announced the discovery of the compound though G. Urry [13] thought that it was very difficult to compare the data given by the latter authors because of unusual separation methods used. A new preparative method for the compound was found in 1961 by A. Kaczmarczyk, M. Millard, G. Urry [21] in the careful pyrolysis of Si_6Cl_{14} in presence of trimethylamine. They obtained Si_5Cl_{12} (in addition to $SiCl_4$) as crystals which could be sublimed in vacuum and which had a molecular weight and chemical analysis in keeping with the formula given. There was a sharp melting point at 347°C, the melt solidifying slowly to a glass which at room temperature recrystallized in the course of months. Two years later A. Kaczmarczyk, M. Millard, J. W. Nuss, G. Urry [15] reported that in the amine-induced disproportionation of Si_2Cl_6 and Si_3Cl_8, Si_5Cl_{12} was mainly produced (in addition to $SiCl_4$ in the first case and Si_2Cl_6 in the second). They obtained the compound as white needle-shaped crystals sublimable in vacuum at 70°C and melting sharply at 345 ± 2°C. Molecular weight determinations and analyses gave accurately the values expected for Si_5Cl_{12}. The compound reacted above 100°C with dry gaseous hydrogen chloride forming silicochloroform and tetrachlorosilane. While the latter investigators assumed that the Si_5Cl_{12} they had obtained had a chain structure like that of n-pentane, iso-pentane and neo-pentane, E. Wiberg, A. Neumaier [20], after repeating the amine-induced disproportionation of Si_2Cl_6 first observed by C. J. Wilkins [18], proposed a completely different structure for this molecule, on the basis of its reactions with HCl, Cl_2 and the results of hydrogenation with LiAl-hydride, which yields SiH_4 and $4SiH_2$. This does not agree with any of the structures which had been assumed previously. The colorless moisture-sensitive crystals melted at 306°C with decomposition. More recent spectroscopic studies on Si_5Cl_{12} and its adducts with $SiCl_4$ and their methyl derivatives, however, favor a neo-pentane-like structure [22]. The readily decomposed addition compound $Si_5Cl_{12} \cdot SiCl_4$ may be obtained by dissolving Si_5Cl_{12} in tetrachlorosilane and cooling to −45°C [13].

Tetradecachlorohexasilane Si_6Cl_{14}. A. Besson, L. Fournier [6] thought they had prepared this compound in 1909, as did G. Martin [7], at least in small amounts, and later H. Kautsky, H. Kautsky Jr. [9], but the criticisms that are mentioned above are also valid in this case.

When, however, A. Kaczmarczyk, G. Urry [19] in 1960 repeated Wilkin's disproportionation of Si_2Cl_6 a white microcrystalline solid with the composition $(Si_{1.00}Cl_{2.36})_n$ resulted after removal of $SiCl_4$ which sublimed at 125°C under very low pressure and then condensed at low temperatures to beautiful clear crystals. On reverting to room temperature these again yielded the white solid. Molecular weight determinations corresponded very well with the formula Si_6Cl_{14}. The high temperature form melts sharply at 318±3°C. If disproportionation is carried out in a sealed apparatus at room temperature, i.e., the tetrachlorosilane is not removed, one obtains transparent cubic crystals of an adduct $Si_6Cl_{14} \cdot SiCl_4$ which is unstable at normal pressure above −45°C. The equation for the reaction is accordingly $5Si_2Cl_6 \rightarrow Si_6Cl_{14} \cdot SiCl_4 + 3SiCl_4$ [13]. Recent investigations by A. Kaczmarczyk, J. W. Nuss, G. Urry [16] have yielded more precise physical data and enable polymorphism involving four different forms to be recognized clearly. Spectroscopic studies on the molecule in the gas phase and in solution lead to the conclusion that it has a neo-hexane structure [22].

Higher Members. In 1937 R. Schwarz, H. Meckbach [14] were able to obtain a still higher member of the series Si_nCl_{2n+2} with n = 10 and a measured molecular weight M = 1032 (theoretical M = 1060) by the action of heat on $SiCl_4$ vapor in a hydrogen atmosphere in a so-called hot-cold tube. Their apparatus consisted of a silicon rod heated electrically to about 1100°C which was surrounded by a water-cooled double-walled copper tube a small distance away. The purified product was a highly viscous oil which dissolved in organic solvents without decomposition and which, unlike the known lower members of the series, ignited in the air on contact with a flame, "decrepitating with lively fire phenomena" ["knisternd mit lebhafter Feuererscheinung"] to produce a white smoke of SiO_2. The chain length given was confirmed by hydrolysis. G. Urry [13] later thought that further characterisation of this substance was desirable.

The highest member of this series with the formula $Si_{25}Cl_{52}$ and a molecular weight M = 2546 (the value found) was made in 1947 by R. Schwarz, C. Danders [17] together with $Si_{10}Cl_{22}$ as a solid deposit on the hot-cold tube described above when $SiCl_4$ vapor in a nitrogen atmosphere was brought into contact with a silicon rod heated to 1250°C. The compound was extracted with ether and remained after evaporation of the latter as a resinous, plastic mass which could be stretched. From the hydrolysis products it was concluded that it had a chain structure. Further results to characterize this substance were again thought by G. Urry [13] to be desirable.

References:

[1] C. Friedel (Compt. Rend. **73** [1871] 1011/3). – [2] L. Troost, P. Hautefeuille (Ann. Chim. Phys. [5] **7** [1876] 452/79, 461/3). – [3] L. Gattermann, K. Weinlig (Ber. Deut. Chem. Ges. **27** [1894] 1943/8). – [4] L. Gattermann (Ber. Deut. Chem. Ges. **22** [1889] 186/97). – [5] L. Gattermann, E. Ellery (Ber. Deut. Chem. Ges. **32** [1899] 1114/6).

[6] A. Besson, L. Fournier (Compt. Rend. **148** [1909] 839/42). – [7] G. Martin (J. Chem. Soc. **105** [1914] 2835/60). – [8] A. Stock, A. Brandt, H. Fischer (Ber. Deut. Chem. Ges. **58** [1925] 643/57, 648/51). – [9] H. Kautsky, H. Kautsky Jr. (Ber. Deut. Chem. Ges. **89** [1956] 571/81, 575/81). – [10] H. Kautsky, H. Kautsky Jr. (Z. Naturforsch. **9b** [1954] 235/6).

[11] A. Besson, L. Fournier (Compt. Rend. **149** [1909] 34/6). – [12] W. Schumb, E. L. Gamble (in: H. S. Booth, Inorganic Syntheses, Vol. 1, McGraw-Hill, New York 1939, pp. 1/197, 42). – [13] G. Urry (J. Inorg. Nucl. Chem. **26** [1964] 408/14). – [14] R. Schwarz, H. Meckbach (Z. Anorg. Allgem. Chem. **232** [1937] 241/8, 241 footnote 1). – [15] A. Kaczmarczyk, M. Millard, J. W. Nuss, G. Urry (J. Inorg. Nucl. Chem. **26** [1964] 421/5).

[16] A. Kaczmarczyk, J. W. Nuss, G. Urry (J. Inorg. Nucl. Chem. **26** [1964] 427/33). – [17] R. Schwarz, C. Danders (Ber. Deut. Chem. Ges. **80** [1947] 444/8). – [18] C. J. Wilkins (J. Chem. Soc. **1953** 3409/12). – [19] A. Kaczmarczyk, G. Urry (J. Am. Chem. Soc. **82** [1960] 751/2). – [20] E. Wiberg, A. Neumaier (Angew. Chem. Intern. Ed. Engl. **1** [1962] 517).

[21] A. Kaczmarczyk, M. Millard, G. Urry (J. Inorg. Nucl. Chem. **17** [1961] 186/8). – [22] J. W. Nuss, G. Urry (J. Inorg. Nucl. Chem. **26** [1964] 435/44).

2.5.3 Unsaturated Silicon Chlorides. Subchlorides

Silicon Dichloride $SiCl_2$. In the course of studies on the preparation of silanes and chlorosilanes from silicon tetrachloride and hydrogen in a glow discharge in 1937 R. Schwarz, G. Pietsch [1] obtained, in addition to other compounds which in part were well defined, very small quantities of a white substance which was solid at room temperature, soluble in absolute methanol and dioxane and which disproportionated at 400°C to $SiCl_4$ and silicon. They spoke of it as silicon dichloride, though "with all reserve" ["mit allem Vorbehalt"]. That unsaturated compounds occur, at least as intermediates, had already been assumed in 1876 by L. Troost, P. Hautefeuille [2] when they had observed "transport of silicon in a silicon tetrachloride atmosphere" ["apparente volatilisation du silicium"] at high temperatures and had tried to interpret this in terms of "subchlorides of silicon" ["Sous-chlorures du Silicium"]. They recognized that, with slow cooling, the transport phenomena resulted from decomposition of the subchloride, while the temperature range for decomposition could be passed over by rapid chilling. Equilibrium in the reaction $Si + SiCl_4 \rightarrow 2SiCl_2$ was subsequently investigated in 1953 by H. Schäfer, J. Nickl [3].

Other Silicon Subchlorides. In 1951 K. A. Hertwig, E. Wiberg [4] were engaged in a detailed study of Schwarz and Pietsch's method mentioned above. They obtained 90 to 100% of the chloride introduced as a solid red-brown to yellow product with a composition varying between $SiCl_{0.5}$ and $SiCl_{2.6}$ according to the ratio of hydrogen to silicon chloride vapor used. The silicon subchloride obtained was thought to be a mixture of highly polymeric homologs which probably resulted from coupling of SiCl-, $SiCl_2$-, and $SiCl_3$-radicals formed in the breakdown of the starting mixture. In addition to compounds which were solid and extraordinarily finely powdered, resinous and oily products also occurred with increasing Cl-content which differed from the material poorer in chlorine by being soluble in benzene and ether. Thus, for example, a resinous subchloride with the composition $SiCl_{1.46}$ gave a yellow solution of $SiCl_{1.45}$ in benzene for which a molecular weight of 1680 was determined cryoscopically. From these and other properties K. A. Hertwig, E. Wiberg [4] felt able to conclude "with some certainty" that molecules were present which represented sections of the diamond lattice of silicon. If a chain structure, such as occurs in the "sub"chlorides from Si_2Cl_6 (considered as a subchloride: "$SiCl_3$") to Si_6Cl_{14} ("$SiCl_{2.3}$") or $Si_{25}Cl_{52}$ ("$SiCl_{2.1}$") which are known so far, is present, it is to be expected that $SiCl_2$ (with infinitely long chains) would be insoluble. E. G. Rochow, R. Didtschenko [5] obtained a highly viscous yellow oil with the composition $SiCl_{2.61}$ on passing $SiCl_4$ vapor over very pure silicon heated at 1000°C. R. Schwarz, R. Thiel [6], on the other hand, got a solid which was orange-red when hot and yellow when cold, when the compound $Si_{10}Cl_{20}H_2$ was decomposed thermally at normal pressure in an inert atmosphere and the resulting liquid silicon chloride was distilled off. The yellow flaky material, which Debye-Scherrer diagrams showed to be amorphous and which ignited in dry oxygen at 98°C was insoluble so that nothing could be said about the molecular size [7]. The color change mentioned was reversible and took place at 180 to 200°C and it seems therefore to be a question of a highly polymeric compound with alternating double bonds. The compound was also thought to be homogeneous since its hydrolysis product was homogeneous [8]. A. Kaczmarczyk, G. Urry [9] in 1960 produced a yellow substance, which they described as a silicon subchloride with the composition $SiCl_{1.80}$, by another route involving the disproportionation of disilicon hexachloride at higher temperatures catalyzed by traces of trimethylamine. They were convinced that the material they had obtained was different from any 'subchloride' described before. H. Kautsky, H. Kautsky Jr. [10] obtained a material which "approximated to the formula $(SiCl_2)_n$ in average composition" as the final product of the rearrangement of Si-Cl compounds when liquid $SiCl_4$ was subjected to a high voltage discharge under certain conditions; they did not describe it more precisely.

Silicon Chlorides with Ring Structures. In the course of attempts to prepare silicon monochloride $(SiCl)_x$ by thermal decomposition of the compound $Si_{10}Cl_{20}H_2$ at temperatures between 260 and 270°C, R. Schwarz, A. Köster [11] were successful in 1952 in getting what, from the chemical analysis and molecular weight (average value found M = 935), they considered to be a single substance. It was described as yellowish, vaseline-like, transparent, and liquifying at 200°C. Their data corresponded with the formula $Si_{10}Cl_{18}$ (theoretical molecular weight M = 919) and it was assumed to have a double ring structure corresponding with that of dekalin.

References:

[1] R. Schwarz, G. Pietsch (Z. Anorg. Allgem. Chem. **232** [1937] 249/56). – [2] L. Troost, P. Hautefeuille (Ann. Chim. Phys. [5] **7** [1876] 452/79, 456/9). – [3] H. Schäfer, J. Nickl (Z. Anorg. Allgem. Chem. **274** [1953] 250/64). – [4] K. A. Hertwig, E. Wiberg (Z. Naturforsch. **6b** [1951] 336/7). – [5] E. G. Rochow, R. Didtschenko (J. Am. Chem. Soc. **74** [1952] 5545/6).

[6] R. Schwarz, R. Thiel (Z. Anorg. Allgem. Chem. **241** [1939] 395/415). – [7] R. Schwarz (Angew. Chem. **51** [1938] 328/31). – [8] R. Schwarz, U. Gregor (Z. Anorg. Allgem. Chem. **241** [1939] 395/415). – [9] A. Kaczmarczyk, G. Urry (J. Am. Chem. Soc. **82** [1960] 751/2). – [10] H. Kautsky, H. Kautsky Jr. (Ber. Deut. Chem. Ges. **89** [1952] 2/15).

[11] R. Schwarz, A. Köster (Z. Anorg. Allgem. Chem. **270** [1952] 2/15).

2.5.4 Further Chloro Compounds of the Silanes

Silicochloroform. Trichlorosilane $SiHCl_3$. In 1857, H. Buff, F. Wöhler [1] obtained a liquid reaction product on leading gaseous hydrogen chloride over crystalline silicon at temperatures below a visible red heat. This was "usually turbid and apparently always a mixture of several compounds" ["gewöhnlich trübe und, wie es scheint, stets ein Gemenge von mehreren Verbindungen"]. From it they isolated on fractional distillation a substance to which they assigned the formula $Si_2Cl_3 \cdot 2HCl$ (Si = 21). Some years later F. Wöhler [2] doubled the formula in order to bring out the analogy between this compound and a substance which had then been prepared from "silicocalcium" ["Kieselcalcium"] – obtained by melting crystalline silicon, calcium chloride, and sodium together – and by decomposing it with hydrochloric acid. This had been called leukon (formulated as $Si_6H_4O_{10}$ and later [3] as "silicon hydroxide" $Si_2O_3H_2$). He wrote $Si_6Cl_{10}H_4$ (Si = 14) as he believed leukon and the substance resulting from decomposition of the chloride with water at 0°C to be identical. His paper concluded: "If, therefore, the existence and characteristics of all these silicon compounds are established, there is still doubt as to their structure which can only be resolved by new investigations. Because of lack of time I must leave this to others who can obtain the compounds in a completely pure state for analysis by discovering new reagents and routes and who will perhaps be more fortunate than I have been in spite of every endeavour" ["Wenn also einerseits das Dasein und die Eigenthümlichkeit aller dieser Siliciumverbindungen wohl erwiesen ist, so bleiben doch noch Zweifel über ihre Zusammensetzungsweise, die nur durch neuere Untersuchungen zu lösen sind. Diese muß ich aus Mangel an Zeit Anderen überlassen, die in der Auffindung von Mittel und Wegen, diese Körper zur Analyse im Zustande vollkommener Reinheit zu erhalten, vielleicht glücklicher sind, als ich es trotz aller Bemühungen gewesen bin"]. C. Friedel, A. Ladenburg [4] applied themselves to this task some years later. They criticized the work saying "that neither a vapor density determination nor any clear evidence for the magnitude of the molecule weight" ["weder eine Dampfdichtebestimmung noch sonst ein sicherer Anhalt über die Moleculgröße"] had been put forward by Wöhler and Buff, while the boiling point of the compound (42°C) was surprisingly low for the composition assumed in relation to that of the tetrachloride (59°C). Furthermore, the "tetratomicity" (i.e., the tetravalency) which had been recognized in the meantime and analogy with carbon led them to expect silicochloroform $SiCl_3H$ (Si = 28). After careful fractionation of the crude product obtained by the Buff-Wöhler method [1] they were able to substantiate their suggestion. C. Pape [5] in 1884 was astonished at the extent to which the yield in this reaction depended on the temperature: at a red heat he obtained virtually only tetrachloride. He also noticed the ready inflammability of the compound's vapor when admixed with air. In place of silicon, L. Gattermann [6] used "crude silicon" ["Rohsilicium"] which he had prepared from sand and magnesium and which was essentially a Mg-silicide. The cheaper ferrosilicon (with 15% Si) was used in the same year, 1889, by H. N. Warren [7] and somewhat later C. Combes [8] introduced a copper silicide (with 20% Cu) while, in 1902, H. Moissan, Holt [9] reacted a vanadium silicide with the formula VSi_2 with hydrogen chloride. In 1905, O. Ruff, K. Albert [10] compared the processes using metal silicides with one another and concluded that the best results could be obtained with Cu-silicide. A. Stock, F. Zeidler [11] then also employed this method of obtaining the compound, which they purified very carefully, determining physical data and undertaking a series of investigations on its chemical behavior. In 1945, D. T. Hurd [14] studied the reduction of tetrachlorosilane with hydrogen and aluminum at 400°C, obtaining chiefly silicochloroform as well as 3 to 5% of dichlorosilane.

Silicochloroform obtained by one or other of these methods must undergo further careful purification if it is to be the starting material for preparing the purest silicon, since the difficulties mentioned for tetrachlorosilane, see p. 128, again arise. In addition to the methods described there, hydrated silicic acid, which precipitates in the hydrolysis of $SiHCl_3$ and is characterized by a high capacity for absorption is often used to remove impurities. G. Rosenberger [22] proposed that water be introduced in such a way that the hydrolysis product is distributed uniformly over the total volume. This can be done by adding either 1% of a water-containing halide, e.g., $CaCl_2 \cdot 6H_2O$, or powdered cellulose saturated with water vapor. After a sufficient interval it is filtered and then fractionated in the usual way. According to R. Rost [23] very pure silicochloroform which can be used for the purpose under discussion can even be obtained by the original Buff-Wöhler method if one starts with aluminum of low silicon content (maximum 2%) at 210 to 330°C, by using of hydrogen chloride diluted with an inert gas (Ar, N_2, H_2) at 6 atm pressure in a fluidized bed. Troublesome "reaction pockets" ["Reaktionsnester"], which always cause the presence of $SiCl_4$, are prevented from forming in this way and aluminum chloride produced simultaneously serves as a trap (by absorption) for other troublesome impurities such as boron and phosphorus, which are particularly undesirable. As a result, very pure silicochloroform is obtained.

Dichlorosilane SiH_2Cl_2. Monochlorosilane SiH_3Cl. According to A. Stock, C. Somiesky [12], both dichlorosilane and monochlorosilane may have been obtained as early as 1857 by H. Buff, F. Wöhler [1] as byproducts in the initial preparation of silicochloroform, though they were not isolated, see above. In 1909 A. Besson, L. Fournier [13] believed they had isolated both dichlorosilane and monochlorosilane as liquids boiling at +12°C or −10°C and formed by passing hydrogen chloride over heated graphitic silicon. They tried without success to obtain both compounds by reducing silicon tetrachloride with calcium hydride. Ten years later A. Stock, C. Somieski [12] criticized the lack of analytical data in the publication by the French investigators after the Germans had been successful in preparing both compounds by chlorinating monosilane with hydrogen chloride using aluminum chloride as the catalyst. They also separated them by their process of vacuum distillation. The dichloro compound, which was particularly sensitive to moisture and grease, had a liquid range of −122 to +8.5°C but the boiling point of the monochloro compound was at −30.5°C. It was not spontaneously inflammable in air under normal conditions and yielded disiloxane $(SiH_3)_2O$ instantly with water. When D. T. Hurd in 1945 reduced silicochloroform in the manner described above he obtained dichlorosilane in addition to small amounts of monochlorosilane [14].

Chlorides of Higher Silanes. Chlorination of **disilane** by hydrogen chloride with aluminum chloride as catalyst occurs readily, as A. Stock, K. Somiesky [20] were able to show in 1920, even at 100°C. Several chloro compounds are always produced together, the equilibrium which is set up favoring products with medium chlorination. It was not, however, possible to prepare the separate components in a state of purity either because they occurred in too small quantities (Si_2H_5Cl) or because of almost identical volatilities ($Si_2H_4Cl_2$ and SiH_3Cl_3).

Mixtures of several **chlorotrisilanes**, which were not spontaneously inflammable but fumed strongly in air, having at the same time a pungent smell of hydrogen chloride and the mustiness of silanes, were obtained by A. Stock, P. Stiebeler [15] in 1923 when they brought together a totally oxygen-free chlorohydrocarbon such as chloroform or tetrachloromethane with liquid silanes in presence of aluminum chloride. The chemical behavior of silanes towards chlorohydrocarbons had already been studied 20 years earlier by H. Moissan, S. Smiles [16] and had resulted only in violent explosions since, as A. Stock, P. Stiebeler [15] showed, the two investigators had not taken into account the high solubility in the chlorohydrocarbons of oxygen, which decomposes the silanes.

A chlorosilane with the formula $Si_5Cl_{10}H_2$ was isolated in 1947 by R. Schwarz, C. Danders [21] as an easily liquifiable fraction from the preparation of higher silicon chlorides from $SiCl_4$ in hydrogen in the hot-cold tube described earlier. It passed over at 60 to 70°C on vacuum fractionation and the formula assigned was based on chemical analysis.

A very long-chain incompletely chlorinated silane was obtained in 1938 by R. Schwarz, R. Thiel [17] on exposing $SiCl_4$ vapor in presence of hydrogen to temperatures above 1000°C in their hot-cold tube. The formula $Si_{10}Cl_{20}H_2$ was attributed to it from chemical analysis and molecular weight determination (value measured in benzene: 924; theoretical: 991). After boiling off low molecular impurities in vacuum what was considered to be a homogeneous product remained as a colorless viscous oil. On attempting to distil it at normal pressure in an indifferent atmosphere it decomposed to liquid silicon chlorides, which boiled off, and a residual yellow substance which was considered to be a subchloride.

Chlorosilicic Acid. In preliminary work for a determination of the atomic weight of silicon by means of tetrasilane in 1905, W. Becker, J. Meyer [18] stated: "Removal of hydrogen chloride from silicon tetrachloride appears to be no easy matter. If we assume, and there is much to be said for this assumption, that hydrogen chloride and the silicon chloride combine in small amounts to form chlorosilicic acid, it must dissociate extraordinarily readily into its components" ["Die Entfernung des Chlorwasserstoffs aus dem Siliciumtetrachlorid scheint keine leichte Sache zu sein. Nehmen wir an, und es spricht manches für diese Annahme, daß sich der Chlorwasserstoff mit dem Siliciumchlorid in geringer Menge zu Siliciumchlorwasserstoffsäure verbindet, so muß dieselbe außerordentlich leicht in ihre Komponenten dissoziieren"].

A. Stock, C. Somieski, R. Wintgen [19], however, were able in 1917 to remove hydrogen chloride smoothly from an $SiCl_4$-HCl mixture by distillation at −80°C, so that this conjecture of the existence of an analog of fluorosilicic acid in the case of the chloro derivatives appears to be incorrect.

References:

[1] H. Buff, F. Wöhler (Liebigs Ann. Chem. **104** [1857] 94/109, 94/8). – [2] F. Wöhler (Liebigs Ann. Chem. **127** [1863] 257/74, 269/70). – [3] H. v. Fehling, C. Hell (Neues Handwörterbuch der Chemie, Vol. 4, Braunschweig 1886, p. 83). – [4] C. Friedel, A. Ladenburg (Liebigs Ann. Chem. **143** [1867] 118/28). – [5] C. Pape (Liebigs Ann. Chem. **222** [1884] 345/74, 358/9).

[6] L. Gattermann (Ber. Deut. Chem. Ges. **22** [1889] 186/97, 190). – [7] H. N. Warren (Chem. News **60** [1889] 158). – [8] C. Combes (Compt. Rend. **122** [1896] 531/3). – [9] H. Moissan, Holt (Compt. Rend. **135** [1902] 78/81). – [10] O. Ruff, K. Albert (Ber. Deut. Chem. Ges. **38** [1905] 2222/43).

[11] A. Stock, F. Zeidler (Ber. Deut. Chem. Ges. **56** [1923] 986/97). – [12] A. Stock, C. Somieski (Ber. Deut. Chem. Ges. **50** [1917] 1739/64, 1739). – [13] A. Besson, L. Fournier (Compt. Rend. **148** [1909] 555/7). – [14] D. T. Hurd (J. Am. Chem. Soc. **67** [1945] 1545/8). – [15] A. Stock, P. Stiebeler (Ber. Deut. Chem. Ges. **56** [1923] 1087/91).

[16] H. Moissan, S. Smiles (Compt. Rend. **134** [1902] 1549/52). – [17] R. Schwarz, R. Thiel (Z. Anorg. Allgem. Chem. **235** [1938] 247/53). – [18] W. Becker, J. Meyer (Z. Anorg. Allgem. Chem. **43** [1905] 251/66, 260). – [19] A. Stock, C. Somieski, R. Wintgen (Ber. Deut. Chem. Ges. **50** [1917] 1754/64, 1761). – [20] A. Stock, K. Somieski (Ber. Deut. Chem. Ges. **53** [1920] 759/69, 760/4).

[21] R. Schwarz, C. Danders (Ber. Deut. Chem. Ges. **80** [1947] 444/8). – [22] G. Rosenberger (Ger. 1028543 [1958] from C.A. **1960** 15873). – [23] R. Rost (Silicium als Halbleiter, Berliner Union, Stuttgart 1966, pp. 1/367, 22).

2.6 Silicon-Bromine Compounds

Silicon Tetrabromide. Tetrabromosilane $SiBr_4$. The first silicon-bromine compound was prepared in 1831 by G. S. Sérullas [1]. He wrote: "The well known occurrence of a compound of chlorine with silicon has without doubt allowed all chemists to suppose that one could prepare a similar compound of bromine with silicon, but no one has to my knowledge gone about obtaining it" ["L'existence connue d'un composé de chlore et de silicium a, sans aucun doute, fait supposer à tous les chimistes qu'on pouvait produire un composé semblable de brome et de silicium, mais personne que je sache, ne s'est occupé de l'obtenir"]. He was then easily able to obtain the analogous bromine compound by the method used by Oersted for tetrachlorosilane, see p. 126. He described it, after purification, as a colorless liquid which gave off thick white fumes in air, crystallizing at −12 to −15°C and boiling at 148 to 150°C. The preparative method and physical data were confirmed 16 years later by J.-I. Pierre [2]. This method, starting from silicic acid and carbon, was improved 40 years later by J. E. Reynolds [3] who introduced very finely powdered starting materials. Two years later L. Gattermann [4] started from "crude silicon" ["Rohsilicium"] obtained by reduction of sand with magnesium. When in 1903 M. Blix [5] substituted "completely pure silicon" ["völlig reines Silicium"] he was surprised that his compound had a constant melting point of 5°C and also that the constant boiling point was higher than his predecessors had specified. Conditions for a quantitative yield by this method were published in 1935 by H. Rheinboldt, W. Wisfeld [6]. Following the work of E. Lay [7] in 1910, E. Pohland [8] replaced the more expensive pure silicon by ferrosilicon and determined physical data for a carefully purified specimen. A. Besson, L. Fournier [9] obtained the tetrabromide in addition to higher bromides when they subjected silicobromoform vapor to a silent electric discharge. The latter had been made in just a 10% yield by passing hydrogen bromide over amorphous silicon at a dull red heat, the residue consisting of $SiBr_4$.

Silicon Hexabromide. Hexabromodisilane Si_2Br_6. In 1880 C. Friedel, A. Ladenburg [10] reported on experiments which they had undertaken two years before the 1870/71 war. They had obtained hexabromodisilane by decomposing hexaiododisilane with bromine, when iodine separated out instantly. The compound was produced from solution as rhombohedral plates, whose boiling point they found to be about 240°C. H. Moissan, Holt [11] made the compound in 1902 by decomposing a vanadium silicide with bromine at a red heat though they did not go into its properties more closely. Eight years later A. Besson, L. Fournier [9] were able to provide more exact data, giving 95°C as the melting point of the white crystals and 265°C as the boiling point of the liquid. They had obtained their material by fractionation of the reaction products resulting from the passage of a silent electric discharge through silicobromoform vapor.

Higher Silicon Bromides. A. Besson, L. Fournier [9] synthesized octabromotrisilane Si_3Br_8 and decabromotetrasilane Si_4Br_{10} in the same way as hexabromodisilane. By vacuum sublimation they were able to obtain the first of the two compounds as small highly lustrous crystals with a melting point of 133°C. The second, which was likewise sublimable in vacuum, melted at about 185°C though it started to decompose.

Silicobromoform. Tribromosilane $SiHBr_3$. The first to prepare this compound were H. Buff, F. Wöhler [12] who, in 1857, passed hydrogen bromide over red hot silicon, naming it "Siliciumbromürbromwasserstoff" and characterizing the "colorless liquid fuming very strongly in air" ["farbloses, an der Luft sehr stark rauchendes Liquidum"] only by reporting the density. They were almost certainly right in their view that it was contaminated with tetrabromide. Thirty years later L. Gattermann [4] embarked on the same reaction with "crude silicon" ["Rohsilicium"] prepared from magnesium and sand, containing Mg-silicides. Yields were higher and separation more successful. A similar good result was achieved by C. Combes [13] (1896) when he substituted copper silicide. The reaction between crystalline silicon and hydrogen bromide was fully investigated in 1930 by W. C. Schumb, R. C. Young [14], who determined the most favorable conditions for formation of silicobromoform. A. Stock, C. Somieski [15] in 1917 started from monosilane. Its reaction with elemental bromine at room temperature and normal pressure was explosive and they had to use a large excess of monosilane and work at low temperatures. They were able [20] to carry out the reaction in such a way that all four bromides of monosilane occurred together, though in proportions which varied with the experimental conditions, and also separated them by fractionation. Silicobromoform has a boiling point of 111.8°C and is a liquid which supercools readily, being still liquid at −73°C. A new method for the technical preparation of silicobromoform needed for preparing the element in a pure state was published in 1978 by L. M. Woerner, E. B. Moore [24]; they passed a mixture of hydrogen and tetrabromosilane through a fluidized bed of finely powdered unrefined silicon at 650°C.

Dibromosilane SiH_2Br_2. Monobromosilane SiH_3Br. Both compounds were separated in 1910 by A. Besson, L. Fournier [9] from the reaction products of bromination of amorphous silicon with hydrogen bromide at a dull red heat. The first was described as a colorless liquid boiling at 75°C which was spontaneously inflammable in air while the second compound was only stated to boil between 30 and 40°C. When A. Stock, C. Somieski [16] were successful in 1918 in improving their method for obtaining bromosilanes by direct bromination of monosilane, which has been described above, by replacing elementary bromine by hydrogen bromide and by using aluminum bromide as catalyst they obtained 25% SiH_3Br and 33% SiH_2Br_2 from the monosilane introduced without the occurrence of secondary products in significant quantities. They were able to confirm in the case of SiH_2Br_2 that it was a spontaneously inflammable liquid as those who first prepared it had claimed. In the case of SiH_3Br they showed it to be a gas which ignited in air with a loud crack.

It may be mentioned at this point that A. Stock, C. Somieski [15] extended the use of their catalytic method to the bromination of disilane and were able to isolate *monobromodisilane* Si_2H_5Br and *dibromodisilane* $Si_2H_4Br_2$.

Unsaturated Silicon Bromides. The first observation of what was very probably a silicon subbromide was made as early as 1910 by A. Besson, L. Fournier [9] in the course of their attempts to prepare various silicon-bromine derivatives in a state of purity. They found that a reddish yellow residue remained on distillation. In 1934, H. Rheinboldt, K. Schwenzer [17], in the course of a systematic study of the reduction of inorganic halides in ethereal suspension by magnesium obtained by using silicon tetrabromide a yellow-brown amorphous substance which they recognized as a silicon compound but did not investigate quantitatively. Twenty two years later M. Schmeisser, M. Schwarzmann [18] had this experiment repeated by K. Grahamer, found that the old results were confirmed and established that they had obtained a polymeric *silicon monobromide* $[SiBr]_x$. The same compound, which was described as a "brownish yellow flaky residue" was obtained a year later by A. Pflugmacher, I. Rohrmann [23] when they decomposed disilicon hexabromide by prolonged heating at 300°C according to the equation $3Si_2Br_6 \rightarrow 2SiBr + 4SiBr_4$ with removal of $SiBr_4$ from the equilibrium. Analysis gave an Si : Br ratio = 1:1. Decomposition of the hexabromide at higher temperatures led to an inhomogeneous product richer in silicon. The monobromide was highly water-sensitive and insoluble in benzene. M. Schmeisser [22] described his polymeric $(SiBr)_x$, which he was able to isolate from the reaction products from $SiBr_4$ and red hot silicon as a greenish black benzene-soluble substance. A silicon dibromide $[SiBr_2]_x$ was obtained by M. Schmeisser, M. Schwarzmann [18] when $SiBr_4$ vapor was passed over very strongly heated silicon. This gave in addition to Si_2Br_6 a brown material which was completely transparent in thin films and brittle, resembling colophony. It was very sensitive to water and readily soluble in most nonpolar solvents. It was also stable towards oxygen at room temperature when moisture was excluded. It first began to glow above 100°C and burnt to SiO_2. Cleavage to Si_2Br_6 and subbromides with the general composition $[SiBr_{<2}]_x$ could be observed in vacuum at 200°C while, at 350°C $[SiBr]_x$ was formed which was identical with that obtained by the method described above. Cracking was complete at 500 to 600°C and elementary silicon was found. It was not possible to determine the exact structure but the good solubility in organic solvents suggested the presence of rings of limited size. Cryoscopic experiments in benzene gave M values of the order of 3000. Smaller molecular weights for $[SiBr_2]_x$ of about 2000 had been measured in 1959 by W. Herrmann [19], who had obtained the compound as the main product together with $[SiBr]_x$ and Si_2Br_6 on reducing of tetrabromosilane with magnesium in a circulatory system. Another method for preparing highly polymeric $(SiBr_2)_x$ was described by P. Voss [21] in 1962. This involved catalytic disproportionation of disilicon hexabromide with trimethylamine according to the equation $xSi_2Br_6 \rightarrow xSiBr_4 + (SiBr_2)_x$. The polymer obtained in this way is a powder with the color of egg yolk which is insoluble in all solvents and shows no softening point though it decomposes at 180°C to $(SiBr)_x$ and $SiBr_4$. It still retains trimethylamine. M. Schmeisser, M. Schwarzmann [18] were able to bring about a reaction of the ether-soluble $(SiBr_2)_x$ with magnesium, when an insoluble brown powder with the composition $(SiBr_{1.46})_x$ was formed. A further unsaturated compound of bromine and silicon with the formula $Si_{10}Br_{16}$ was isolated by M. Schmeisser [22] from the reaction products of the reduction of $SiBr_4$ with red hot silicon. This yielded a yellow solid, and at about 400°C a dark brown to red substance with variable Si : Br values ($SiBr_{1.2}$ to $SiBr_{0.5}$) was obtained from it. The $Si_{10}Br_{16}$ was decomposed by water and ignited in gaseous bromine to form $SiBr_4$.

References:

[1] G. S. Sérullas (Ann. Chim. Phys. [2] **48** [1831] 87/99, 87/90). – [2] J.-I. Pierre (Ann. Chim. Phys. [3] **20** [1847] 5/53, 28/32). – [3] J. E. Reynolds (J. Chem. Soc. **51** [1887] 590/2). – [4] L. Gattermann (Ber. Deut. Chem. Ges. **22** [1889] 186/97, 189/90). – [5] M. Blix (Ber. Deut. Chem. Ges. **36** [1903] 4218/20).

[6] H. Rheinboldt, W. Wisfeld (Liebigs Ann. Chem. **517** [1935] 197/211, 206). – [7] E. Lay (Diss. München T.U. 1910 from [8]). – [8] E. Pohland (Z. Anorg. Allgem. Chem. **201** [1931] 265/81). – [9] A. Besson, L. Fournier (Compt. Rend. **151** [1910] 1055/7). – [10] C. Friedel, A. Ladenburg (Liebigs Ann. Chem. **203** [1880] 241/55, 253).

[11] H. Moissan, Holt (Compt. Rend. **135** [1902] 78/81). – [12] H. Buff, F. Wöhler (Liebigs Ann. Chem. **104** [1857] 94/109, 99). – [13] C. Combes (Compt. Rend. **122** [1896] 531/3). – [14] W. C. Schumb, R. C. Young (J. Am. Chem. Soc. **52** [1930] 1464/9). – [15] A. Stock, C. Somieski (Ber. Deut. Chem. Ges. **50** [1917] 1739/54).

[16] A. Stock, C. Somieski (Ber. Deut. Chem. Ges. **51** [1918] 989/96). – [17] H. Rheinboldt, K. Schwenzer (J. Prakt. Chem. [2] **140** [1934] 273/90, 286). – [18] M. Schmeisser, M. Schwarzmann (Z. Naturforsch. **11b** [1956] 278/82). – [19] W. Herrmann (Wiss. Z. Univ. Rostock Math. Naturw. Reihe **9** [1959/60] 401). – [20] M. Schmeisser, P. Voss (Fortschr. Chem. Forsch. **9** [1967/68] 165/205, 170/1).

[21] P. Voss (Diss. Aachen T.H. from [20]). – [22] M. Schmeisser (Silicon Sulphur Phosphates Colloq., Münster 1954 [1955], pp. 28/31). – [23] A. Pflugmacher, I. Rohrmann (Z. Anorg. Allgem. Chem. **290** [1957] 101/2). – [24] L. M. Woerner, E. B. Moore (Brit. Appl. 2028289 [1978] C.A. **92** [1980] No. 218730).

2.7 Silicon-Iodine Compounds

Silicon Iodide. Tetraiodosilane SiI_4. The question of whether J. J. Berzelius was also able to prepare silicon-iodine compounds in the same way as he prepared silicon chloride, see p. 125, is answered in different ways in the chemical literature. In the major publication reporting the discovery of the element [1] one reads at one point [2]: "Silicon burns in iodine" ["Le silicium brûle dans l'iode"] which signifies compound formation, whereas at another point [3] Berzelius' original text [1]: "When silicon is heated in a stream of iodine it cannot be induced to combine with it" ["Silicium in einem Strom von Jodgas erhitzt, konnte nicht dazu gebracht werden, sich mit demselben zu verbinden"] is correctly reproduced. Berzelius [4] confirmed this lack of success in a letter to H. Davy on 21st April, 1824: "I have not succeeded in combining it [silicon] with iodine" ["Je n'ai point réussi à le [le silicium] combiner à l'iode"]. One must, however, note that in the German translation of the Swedish text in *Poggendorffs Annalen der Physik und Chemie* [5] quoted above, the word "kalium" is found instead of "silicium" which is found in the Swedish text, which is surprising and completely out of place in this context, though it is corrected in the version quoted. The text which had become so incomprehensible is probably the reason why L. Gmelin, whose reporting of chemical research in his *Handbuch* was highly esteemed by J. J. Berzelius in a letter [6] to J. Liebig, gave no information [7] on "silicon and iodine" in the third edition, which appeared in 1827. The reason for Berzelius' failure could lie in insufficient exclusion of atmospheric oxygen, especially as he says nothing about the carrier gas for the stream of iodine vapor. The earliest successful preparation of a silicon-iodine compound was therefore in 1857, when H. Buff, F. Wöhler [8] established the necessary reducing conditions for reaction between the two elements when they passed hydrogen iodide over red hot silicon. The crude product was described as a dark red friable material which was soluble in carbon disulfide. From this solution red crystals were obtained which they considered to be a single compound which they called "Siliciumjodür-Jodwasserstoff", that is silicoiodoform. In fact, as C. Friedel [9] was able to show twelve years later, they had tetraiodosilane contaminated with less than 8% of silicoiodoform in their hands. C. Friedel had obtained the pure compound as a white material melting at 120.5°C and boiling at about 290°C by passing iodine vapor carried on carbon dioxide over red hot silicon and had established its composition by chemical analysis and measurement of the vapor density. Twenty years later the compound was prepared by L. Gattermann [10] by the same route and oxygen-free nitrogen was introduced as the carrier gas by R. Schwarz, A. Pflugmacher [11] in 1942. In 1959 L. V. McCarty [25] operated in the same method in his major investigation on the effect of the purity of SiI_4 on the effectiveness of silicon prepared from it for semiconductors. Two years previously G. Szekely [26], in work which was similar to that of B. Rubin [15] on the possibilities of purifying SiI_4, had introduced argon as the carrier gas.

Formation of silicon tetraiodide from monosilane and iodine vapor at higher temperatures was shown to occur by R. Mahn [12] in 1870. That the compound can also be obtained by thermal decomposition of silicoiodoform was observed by O. Ruff in collaboration with E. Geisel [13] in 1908. Silicoiodoform $SiHI_3$ was boiled for a long time, when the temperature of the liquid and vapor rose gradually to 300°C. On vacuum sublimation of the decomposition product, silicon tetraiodide was obtained as regular octahedra or cubes. The first modern crystallographic study was undertaken in 1931 by O. Hassel, H. Kringstad [14], who established a cubic structure from Laue-, rotation-, and Debye diagrams. As silicon tetraiodide is favored for use as a starting material in the preparation of high purity silicon, see p. 55, and can be satisfactorily purified for this purpose by recrystallization from toluene, by fractional distillation, and by zone melting [15], several processes for producing the compound on a technical scale have been worked out and patented [16].

Hexaiodosilicon, Silicon Hexaiodide, Hexaiododisilane Si_2I_6. This was first prepared in 1869 by C. Friedel, A. Ladenburg [17] by heating silicon tetraiodide in a sealed tube with finely divided metallic silver. The new compound was obtained from the reaction mixture after removal of unchanged tetraiodide as colorless crystals which melted at about 250°C with partial decomposition. A fuller report first appeared in 1880 [18] (on account of the outbreak of the Franco-German war). This method of preparation was studied more fully in 1942 by R. Schwarz, A. Pflugmacher [11] and the experimental conditions for high yields were published.

The existence of *higher saturated silicon iodides* was thought by R. Schwarz, A. Pflugmacher [11] to be unlikely because of the thermal instability of Si_2I_6.

Silicoiodoform. Triiodosilane $SiHI_3$. H. Buff, F. Wöhler [8] thought they had found the first method for preparing this compound, which they called "Siliciumjodür-Jodwasserstoff" when they passed hydrogen iodide gas over red hot silicon. They described the compound as follows: "The iodide forms a dark red friable mass which fumes strongly in air, becoming initially bright vermillion and finally snow-white. It melts readily and solidifies on cooling in a crystalline form. When more strongly heated it starts boiling and distils over. Whether, as appears, its vapor is colored we were not able to observe with certainty. It is only slowly decomposed by water, in which it instantly becomes vermillion in color. It is soluble in large amounts in carbon disulfide to give a blood red color. When this solution is concentrated by distillation dark red crystals separate out" ["Das Jodür bildet eine dunkelrothe, spröde Masse, die an der Luft stark raucht und dabei anfangs lebhaft zinnoberroth, zuletzt schneeweiß wird. Es ist leicht schmelzbar und erstarrt beim Erkalten krystallinisch. Stärker erhitzt, geräth es ins Sieden und destillirt über. Ob, wie es schien, sein Gas gefärbt ist, konnten wir nicht mit Sicherheit sehen. Von Wasser, worin es sich augenblicklich zinnoberroth färbt, wird es nur langsam zersetzt. In Schwefelkohlenstoff ist es in großer Menge in blutrother Farbe löslich. Wird diese Lösung durch Destillation concentrirt, so scheiden sich beim Erkalten dunkelrothe Krystallen aus"]. When twelve years later C. Friedel [9] followed this method of preparation closely he was led to the conclusion that the product was actually silicon tetraiodide with less than 8% of $SiHI_3$. He tried to raise the yield of the desired compound by passing a mixture of hydrogen and hydrogen iodide over red hot silicon. He was now actually successful in isolating a highly refractive liquid from the reaction product which boiled at about 220°C and had a density of D° = 3.362. C. Combes [19] was able to show in 1896 that the method for obtaining $SiHHal_3$ compounds by passing the gaseous hydrogen halide over metal silicides, which is quite successful for the other halogens, cannot be used for the preparation of silicoiodoform. A new preparative method was worked out in 1908 by O. Ruff in collaboration with E. Geisel [13]. They brought silicohydrocyanic acid SiHN or silicohydrotrianilide $SiH(NH \cdot C_6H_5)_3$ into contact with hydrogen iodide in benzene or carbon disulfide and obtained a colorless liquid solidifying at 8°C which could not be distilled at normal pressure without decomposition.

Diiodosilane SiH_2I_2. Monoiodosilane SiH_3I. A. Stock's prediction [20] that the silanes and hydrogen iodide would react to form compounds, which was made in 1933, was confirmed six years later by A. G. Maddock, C. Reid, H. J. Emeléus [21, 22]. They obtained the two compounds (in addition to SiI_4 and $SiHI_3$) by decomposition of monosilane SiH_4 with hydrogen iodide at 80°C with aluminum triiodide as catalyst and described them as colorless liquids, SiH_3I freezing at −57.0°C and SiH_2I_2 at −1°C. Their vapors burned readily in air.

Disilanemonoiodide Si_2H_5I was obtained in 76% yield in 1959 by L. G. L. Ward, A. G. MacDiarmid [27] by treating disilane with hydrogen iodide in presence of aluminum triiodide. After careful purification it was possible to determine the physical properties of the gas, which disproportionates even at room temperature.

Unsaturated Silicon Iodides. As early as 1869 C. Friedel, A. Ladenburg [17] had noticed that a yellow substance was produced in the thermal decomposition of silicon hexaiodide which they called silicon subiodide SiI_2 and attempted to characterize. Their analysis did not, however, correspond with the formula which had been assumed. Experiments by R. Schwarz, A. Pflugmacher [11] in 1942 showed that the yellow material was a highly polymeric silicon-iodine compound which they investigated more closely, coming to the conclusion that, in both its mode of formation and its behavior, it corresponded with silicon monochloride $(SiCl)_x$ produced in the decomposition of the silicon chlorides, see p. 132. In 1964 M. Schmeisser, F. Friederich [23] prepared a *silicon subiodide* with the composition $[SiI_{2.2}]_x$, in 60 or 70% yield by the action of a glow discharge on silicon tetraiodide vapor in high vacuum. This was a yellowish red solid, which X-ray diagrams showed to be amorphous and which was insoluble in all the usual solvents, softened at 120°C, and had an oily consistency at 180°C, where it started to decompose with evolution of SiI_4 and Si_2I_6. On pyrolysing at a temperature of 230 to 240°C a residue remained which was a brittle solid, dark red in color and which showed a metallic glance at a surface of fracture. It started to soften at 150°C and was soluble only in a naphthaline-benzene mixture. The formula $[SiI_2]_x$ was attributed to it. If it was pyrolyzed, the resulting red liquid, which became more yellow with rising temperature, decomposed again with formation of SiI_4 and Si_2I_6 and the liquid decomposition residue solidified in several hours at 350°C to a solid yellow colored porous substance with the formula $[SiI]_x$ [23, 24].

References:

[1] J. J. Berzelius (Kgl. Svenska Vetenskaps Akad. Handl. **1824** 46/98, 58). – [2] R. Callas (in: P. Pascal, Nouveau Traité de Chimie Minérale, Vol. 8, Silicium, Paris 1965, p. 413). – [3] J. W. Mellor (A Comprehensive Treatise on Inorganic and Theoretical Chemistry, Vol. 6, Longmans Green, London 1925 [Reprint 1957], pp. 1/1024, 982). – [4] J. J. Berzelius (in: H. G. Söderbaum, Jac. Berzelius Lettres, Vol. 2: Correspondance entre Berzelius et Sir Humphry Davy, Uppsala 1912, p. 71). – [5] J. J. Berzelius (Ann. Physik [2] **1** [1824] 169/230, 219).

[6] J. Carrière (Berzelius und Liebig, ihre Briefe von 1831–1845, 2nd Ed., München 1898 [Reprint Wiesbaden 1967], Letter of 13.12.1831, p. 21). – [7] L. Gmelin (Handbuch der theoretischen Chemie, 3rd Ed., Vol. 1, Varrentrapp, Frankfurt a.M. 1827, p. 738). – [8] H. Buff, F. Wöhler (Liebigs Ann. Chem. **104** [1857] 94/109, 99/100). – [9] C. Friedel (Liebigs Ann. Chem. **149** [1869] 96/101, 97). – [10] L. Gattermann (Ber. Deut. Chem. Ges. **22** [1889] 186/91, 190).

[11] R. Schwarz, A. Pflugmacher (Ber. Deut. Chem. Ges. **75** [1942] 1062/71). – [12] R. Mahn (Jenaische Z. Med. Naturw. **5** [1870] 158/66, 164). – [13] O. Ruff (Ber. Deut. Chem. Ges. **41** [1908] 3738/44). – [14] O. Hassel, H. Kringstad (Z. Physik. Chem. B **13** [1931] 1/12). – [15] B. Rubin, G. H. Moates, J. R. Weiner (J. Electrochem. Soc. **104** [1957] 656/60).

[16] M. Schmeisser, P. Voss (Ullmanns Enzykl. Tech. Chem. 3rd Ed. **15** [1964] 744). – [17] C. Friedel, A. Ladenburg (Compt. Rend. **68** [1869] 920/4). – [18] C. Friedel, A. Ladenburg (Liebigs Ann. Chem. **203** [1880] 241/55). – [19] C. Combes (Compt. Rend. **122** [1896] 531/3). – [20] A. Stock (Hydrides of Boron and Silicon, Cornell Univ. Press, Ithaca, N.Y., 1933, pp. 1/250, 27).

[21] A. G. Maddock, C. Reid, H. J. Emeléus (Nature **144** [1939] 328). – [22] H. J. Emeléus, A. G. Maddock, C. Reid (J. Chem. Soc. **1941** 353/8). – [23] M. Schmeisser, F. Friederich (Z. Angew. Chem. **76** [1964] 782). – [24] M. Schmeisser, P. Voss (Fortschr. Chem. Forsch. **9** [1967/68] 181/2). – [25] L. V. McCarty (J. Am. Electrochem. Soc. **106** [1959] 1036/42).

[26] G. Szekely (J. Am. Electrochem. Soc. **104** [1957] 663/7). – [27] L. G. L. Ward, A. G. MacDiarmid (J. Am. Chem. Soc. **82** [1960] 2151/3).

2.8 Silicon-Sulfur Compounds

2.8.1 Silicon Sulfides SiS_2 and SiS

In 1826, at the conclusion of a long series of investigations *"Über die Schwefelsalze"* [1], J. J. Berzelius in summarizing his work, but without saying where his experiments with the separate elements had been described, wrote: "It is already known from old experiments that *boron, kiesel (silicium P.* [the latin name made up by the editor Poggendorff]) and *titan* do give no sulfur salts in the wet way. It is likely that they can be produced by melting (the components) together. For the present, however, they have no interest that could compensate for the difficulties of their investigation" ["daß *Bor, Kiesel (Silicium P.* [lateinische Benennung vom Herausgeber Poggendorff ergänzt]) und *Titan* auf nassem Wege keine Schwefelsalze geben, ist aus älteren Versuchen schon bekannt. Es ist glaublich, daß sie auf trockenem Wege durchs Zusammenschmelzen erzeugt werden können. Sie haben indessen für jetzt kein Interesse, das die Schwierigkeiten ihrer Untersuchung aufwiegen könnte"]. In a letter to P. L. Dulong [2] in Paris dated 24th April, 1824, he wrote that he had no success with this method for silicon: "Silicon does not combine with sulfur, that is to say, I have not succeeded in making the compound, though I have no doubt that it exists. However, silicon [prepared] from potassium [and which still contains potassium] burns vigorously when heated with sulfur, but when [the product] is dissolved it leaves pure silicon" ["Le silicium ne se combine pas avec le soufre, c'est-à-dire, je n'ai point réussi à faire cette combination, quoique je ne doute point qu'elle existe. Cependant, le silicium de potassium chauffé avec du soufre, brûle vivement, mais laisse, lorsqu'on le dissout le silicium pure"]. However, only a little later, in announcing the first preparation of the element, he was able to describe the preparation and properties of a silicon sulfide and of one of its double salts [3]: "When silicon is heated to a full red heat in *sulfur vapor* or when the vapor of *sulfur* is passed over white hot silicon, it ignites and burns, although less vigorously than in oxygen gas. This is also the case with noninflammable silicon. Sulfiding usually proceeds just as incompletely as oxidation, and one obtains a slag-like dark gray mass.

It may sometimes happen, especially if the experiment is carried out in a vessel which has been made free of air before the sulfur is vaporized, that the silicon is fully sulfided, at least a part of it. It then forms a white earthy material, which is instantly dissolved in contact with water with evolution of "Hydrothiongas" [hydrogen sulfide]. At the same time the silicon oxidizes to silica, which dissolves in the water. If the amount of water is small it is possible to obtain so concentrated a solution that, after some evaporation it gelatinizes and, on drying, leaves behind silica as a transparent elastic material. Incompletely sulfurized silicon also decomposes water vigorously, evolving hydrogen sulfide and yielding a solution of silica in water. The silicon which is not combined with sulfur separates unchanged. In free air the silicon sulfide emits a strong smell of hydrogen sulfide, losing its total sulfur content after a short time. It may be stored in artificially dried air. It may be calcined at a red heat and gives sulfurous acid and silica. This occurs slowly, however, so that, after ignition for some moments, it still retains the property of decomposing water. *Silicon potassium* combines readily with sulfur with luminescence, but when the material is dissolved in water much silicon remains undissolved if the product has not again been subjected to a white heat, in which case the silicon combines with sulfur at the expense of potassium previously sulfurized to a higher degree. This compound is now a true *double sulphuret* [that is: double sulfide] and has a dark brown almost black color. It forms a molten mass which will dissolve in water" ["Wird Silicium bis zum vollen Glühen in *Schwefelgas* erhitzt, oder werden Dämpfe von *Schwefel* über weißglühendes Silicium geleitet, so entzündet es sich und brennt, wiewohl weniger lebhaft als in Sauerstoffgas; dieß ist auch mit nicht entzündlichem Silicium der Fall. Die Schwefelung geht dabei gewöhnlich eben so unvollständig vor sich, wie die Oxydation, und man erhält eine schlackige dunkelgraue Masse. Bisweilen geschieht es indeß, besonders wenn man den Versuch in einem Gefäße vornimmt, welches vor der Verwandlung des Schwefels zu Gas, luftleer gemacht wurde, daß sich das Silicium vollkommen schwefelt, wenigstens einem Theil seiner Masse nach. Es stellt alsdann einen weißen erdigen Körper dar, der in Berührung mit Wasser augenblicklich mit Entwickelung von Hydrothiongas [Schwefelwasserstoff] aufgelöst wird. Dabei oxydirt sich das Silicium zu Kieselerde, welche sich im Wasser auflöst, und ist die Menge des Wassers geringe, so kann man eine so concentrirte Auflösung erhalten, daß sie, nach einiger Verdunstung, gelatinirt, und nach der Eintrocknung Kieselerde als eine durchsichtige gesprungene Masse hinterläßt. Auch das unvollkommen geschwefelte Silicium zersetzt das Wasser mit Heftigkeit, entwickelt Hydrothiongas und liefert eine Auflösung von Kieselerde in Wasser. Das nicht mit Schwefel verbundene Silicium wird dabei unverändert abgeschieden. An freier Luft stößt das Schwefelsilicium einen starken Geruch nach Hydrothiongas aus, und verliert nach kurzer Zeit seinen ganzen Schwefelgehalt. In künstlich getrockneter Luft kann es aufbewahrt werden. Durch Glühen läßt es sich rösten und giebt schweflige Säure und Kieselerde; dieß geschieht aber langsam, so daß es, einige Augenblicke lang geglüht, noch die Eigenschaft behält Wasser zu zersetzen. *Silicium-Kalium* verbindet sich leicht unter Erglühen mit Schwefel, löst man aber die Masse in Wasser auf, so bleibt viel Silicium unaufgelöst zurück, wenn nicht die Masse von neuem der Weißglühhitze ausgesetzt wird, bei welcher sich alsdann das Silicium auf Kosten des vorher zu einem höhern Grade geschwefelten Kaliums mit Schwefel verbindet. Diese Verbindung ist nun ein wahres *Doppel-Sulphuret* [das ist: Doppelsulfid], und hat eine dunkelbraune, fast schwarze Farbe. Es bildet eine geschmolzene Masse, welche sich in Wasser auflöst"].

It must have surprised J. J. Berzelius when, five years later, he learned from Niels Gabriel Sefström (1787–1845), the director of the Fahlun Mining Academy, who had done much to develop iron and steel production and had also discovered the element vanadium [4] "that silica in a carbon crucible can be reduced by hydrogen sulfide gas in his blast furnace [a new construction which allowed higher than otherwise normal temperatures to be obtained] to silicon sulfide. The latter volatilizes and, where it burns, produces the special sublimate of silica which is obtained from the blast furnace, among other substances. This fact will contribute a great deal towards explaining a number of hitherto incomprehensible phenomena [in metallurgical processes]" ["daß die Kieselerde sich in seinem Gebläseofen [einer Neukonstruktion, die höhere als die sonst üblichen Temperaturen erreichen ließ] im Kohletiegel durch Schwefelwasserstoffgas zu Schwefelsilicium reduciren läßt. Letzteres verfliegt, und wo es verbrennt, erzeugt es das besondere Sublimat von Kieselerde, welches man unter anderem aus den Hochöfen erhält. Diese Thatsache wird Vieles zur Erklärung einer Menge vorher unbegreiflicher Erscheinungen [bei Hüttenprozessen] beitragen"]. It was in this way that Berzelius informed J. C. Poggendorff, the editor of the *Annalen der Physik und Chemie* of this observation in a letter and the latter reprinted the extract from the letter under the heading *"Schwefelsilicium"*, but the writer of the letter did not mention the occurrence of the compound under these conditions in any of the editions of his *Lehrbuch* which appeared subsequently, in spite of the significance that he attached to the observation [5]. The short-note [4] had, as will be apparent, been completely overlooked, for there is no reference to it in the Catalogue of the Royal Society [6] or in a modern bibliography of the Swedish chemist [7], and it was not until 1964 that J. R. Partington [8] discovered it. It may be added that, in 1927, M. Thimann [9], without being aware at the time of Sefström's observation, prepared by this method a very pure silicon disulfide which could be excited to fluorescence by carbon.

What the nature of the "incomprehensible phenomena" in metallurgical processes in the Sefström-Berzelius note probably was, which was later explained by the observation of a silicon sulfide, is indicated by an experience which K. E. Schafhäutl (1803–1890), the Munich professor for geography, mining, and metallurgy, had during a stay of several years in England [10]. He observed the formation of feathery silicon dioxide in a rolling mill in the explosive combustion of a substance present in the iron, which had at first been taken for elementary silicon, see p. 44 for original discussion, but had then been recognized as silicon sulfide. He noted in conclusion that a similar observation had been described in 1819 [11], that is, long before the existence of a silicon-sulfur compound was known. It was not until thirty years later that iron metallurgists again became aware of the significance of silicon sulfide in blast furnace operation, see below.

The next attempt to prepare silicon sulfide was made in 1848 by I. Pierre [12], who passed a mixture of dry hydrogen sulfide and tetrachlorosilane through a red hot porcelain tube. From the condensed reaction products he first isolated a water-clear liquid boiling at just over 100°C, which he believed to be "Siliciumchlorosulfür $SiSCl_2$" [twenty years later, C. Friedel, A. Ladenburg [13] showed that this compound has the formula $SiCl_3SH$ and named it "dreifach gechlortes Siliciummercaptan"]. I. Pierre's distillation residue was a mixture of sulfur and silicon sulfide, which he freed from sulfur by heating in a stream of nitrogen, thus obtaining the compound in a pure form. Another method for preparing SiS_2 was put forward in 1852 by E. Fremy [14]: "I prepared this substance with the greatest ease when I passed carbon disulfide vapor over small lumps composed of carbon and gelatinous silicic acid contained in a porcelain tube at a bright red heat" ["Ich habe diesen Körper mit der größten Leichtigkeit dargestellt, indem ich Schwefelkohlenstoffdampf über Kügelchen aus Kohle und gallertartiger Kieselsäure, welche in einer Porzellanröhre zu lebhafter Rotglut gebracht waren, hinwegleitete"]. On closer study of the compound he noticed that, in its decomposition in water, the resulting silicic acid remained "completely in solution" in the hydrogen sulfide containing liquid. The mineralogist A. Descloizeaux (1817–1897) had, shortly before, found both substances together in solution in water from the eruption of Icelandic geysers [15], and this led E. Fremy [16] to wide ranging geological speculation. In keeping with ideas then current as to the nature of the oxide, he assumed the formula of the sulfide to be SiS_3. The preparative method of J. J. Berzelius was first taken up again in 1880 by P. Sabatier [17] and, because of the small yield, an attempt was made to improve this by replacing hydrogen sulfide by carbon disulfide. In this experiment he observed a yellow coloration in the boundary zone of the sublimate, which he attributed to the formation of another substance, a "subsulfide".

All these preparative methods for silicon sulfide were found by W. Hempel, v. Haasy [18] in 1900 to give unsatisfactory yields and also to be much too difficult. As a result, they sought a simpler route and found this in the melting together of the two elements which J. J. Berzelius had tried without success in 1824 [2], though they carried it out in a special way. Amorphous, reactive silicon was mixed intimately with three times the amount of sulfur, the mixture was melted in a Hessian crucible at 150°C and the melt was introduced in portions into another crucible at a red heat: "A minute after the introduction, combination of the components occurs with a clearly audible reaction. If reaction takes place at once, the temperature is too high and it must then be reduced. When the additions have been completed, the temperature is raised somewhat to separate excess sulfur from the silicon sulfide formed. After cooling the crucible contents, there is a molten reaction mass of SiS_2 on which excess sulfur rests in a sharply separated layer, so that it can easily be separated mechanically. Silicon sulfide prepared in this way is 92 to 95% (pure) with a yield of 80 to 90% of the silicon used. The impurities consist of unchanged silicon and silica, which probably stems for the most part from the crucible material". They purified the raw product by sublimation at a bright red heat and reduced pressure and obtained beautiful long white needles, which were sharply separated from an orange red substance, which also sublimed, to which the formula SiS was attributed. In 1957, M. Schmeisser, H. Müller, W. Burgemeister [19] published a simplified form of this method, in which the reaction of a silicon-sulfur mixture wetted with sulfur chloride may be initiated by priming with a mixture of Mg_2Si and S_2Cl_2 which has already started to react after warming with a Bunsen flame to form SiS_2. In 1964, M. S. Silverman, J. R. Soulen [20] examined the direct synthesis of silicon disulfide from the very pure elements at very high temperatures (700 to 2300°C) and pressures (25 to 80 kbar) in an suitable apparatus, which is described by H. T. Hall [21]. They obtained in this way a 95% yield of colorless, transparent and birefringent crystals with a density 9% higher than that calculated from the X-ray data of normal disulfide, and this was treated as a new high-pressure modification of SiS_2. Precise crystallographic work by C. T. Prewitt, H. S. Young [22] confirmed this.

The preparative method of Pierre which has already been mentioned, was modified in 1903 by M. Blix, W. Wirbelauer [23] in that they first prepared by thermal treatment of a mixture of the gases $SiCl_4$ and H_2S the "Siliciumchlorosulfhydrat" $SiCl_3SH$ and from this, by a further thermal treatment, they isolated silicon sulfochloride $SiCl_2S$ as large colorless prisms. The latter decomposes on dry distillation at normal pressure and at the temperature of a water bath to tetrachlorosilane and silicon disulfide. The authors refer especially to the fact that "silicon disulfide has, for the first time, been obtained at such low temperatures and, so to say, by a wet route". Another synthesis of SiS_2 was published in 1926 by E. Tiede, M. Thimann [24]. They allowed crude aluminum sulfide prepared by Biltz's method to react with quartz sand in a porcelain tube at the temperature of a silite furnace and sublimed the silicon sulfides formed out of the reaction zone in a stream of nitrogen, observing the monosulfide as well as the disulfide. The latter, after doping with small quantities of carbon, showed a green phosphorescence. For doping use was made of organic substances such as terephthalic acid, which are added during sublimation to the nitrogen stream. A silicon sulfide which gave a yellow luminescence also occurred, the sulfur content of which was considerably below that of the disulfide without it being possible to assign a definite formula to it [9, 25]. The temperature dependence of the yields of the two compounds was investigated in 1931 by X. Siebers, E. J. Kohlmeyer [26], who found that at a temperature of 1350°C the reaction product consisted of about 33% SiS and 61% SiS_2 and that at 1450°C 93.6% SiS_2 was obtained. They worked in a carbon crucible with a stream of nitrogen.

On the question of the most suitable method for preparing silicon disulfide, M. Schmeisser, H. Müller, W. Burgemeister [19] saw several answers as correct. For continuous production they recommend the reaction of hydrogen sulfide with pure silicon at 1500°C as, in this way, pure SiS_2 results without any extra expense. For the occasional preparation of small quantities it is convenient to pass hydrogen sulfide or sulfur vapor over silicides such as Ca_2Si and Mg_2Si at 1000°C. These silicides also react very vigorously with sulfur chlorides such as S_2Cl_2, even on warming with a Bunsen flame, forming SiS_2. They found a completely new way for preparing SiS_2 in the thermal decomposition of silicic acid thio esters such as $Si(SC_2H_5)_4$, which takes place quantitatively even at 200°C on addition of elementary sulfur. The ethyl sulfane produced can be removed readily by solution in benzene.

All of the authors who have been concerned with the preparation of silicon disulfide described it as consisting of white, extraordinarily beautiful needles, which are long though thin, have a silky glance and are also flexible. They possess a not inconsiderable tensile strength. It was very difficult to make further observations because of the great sensitivity of the compound to water vapor. In spite of this, some physical data were published as early as 1911, together with a characterization of the crystal habit [27]. This, however, did not come to light since, at the time, the data were attributed in error to aluminum sulfide as was realized fourteen years later [28]. They reemerged in the literature in 1950 [29]. There were diverse views on the unusual crystal structure of the compound, which remains completely unchanged to the eye if it is decomposed to silica by atmospheric moisture, rather as in the little understood phenomena in iron smelting described earlier. W. Bruhns (in: W. Biltz, F. Caspari [27]) attributed a hexagonal character to the crystals while Johsen and Seifert (in: E. Tiede, M. Thimann [24]) supposed them to be isomorphous with α- or β quartz until W. Bussem, H. Fischer, E. Gruner [30] were able to show in a detailed X-ray analysis that the compound crystallizes in the rhombohedral system. This also shows that the lattice of silicon disulfide can be considered as a packed assembly of one dimensional macro molecules of unlimited length, the atoms of which are held together by forces which can be described in terms of chemical valencies, while van der Waals forces must be assumed to be holding the chain molecules together. E. Zintl, K. Loosen [31] therefore considered silicon disulfide as a structural analog of organic fibrous materials, such as cellulose.

The significance that the formation of silicon-sulfur compounds at the temperatures of the blast furnace can have for many processes in the production of iron and steel is indicated by the short note by the iron metallurgist N. G. Sefström [4] which was quoted earlier and was completely overlooked. About fifty years passed before it was recognized by iron metallurgists and studied. "A. Ledebur, the old master of German iron metallurgy reported in 1877 on a piece of pig iron from the collection of the Mining Academy at Freiberg [32], which has particularly excited his attention". This was how P. Dolch [33] began his description of the way in which this investigation developed. "On shattering the piece, there was a cavity in its interior filled with beautifully developed octahedral crystals which were covered with silica, which was pure white in color, like velvet. The silica coating could not be removed by rubbing with the finger but dissolved very readily in caustic potash".

In interpreting this phenomenon, A. Ledebur [34] suggested the possibility that gaseous silicon compounds might play a part in the formation of such blow holes, particularly silicon sulfide. He therefore repeated Fremy's experiment [14], which is mentioned above, and became convinced that the production of the high silica content in the smoke at the top and pool of the shaft furnace is probably due to volatilization of silicon sulfide. The heavy fume which is given off on tapping spiegel-ferromanganese from the blast furnace, the high silica content and the strong smell of sulfur dioxide also indicate that here too silicon sulfide is being set free. A. Ledebur [35] found that his view was supported by new observations on the blast furnace and saw them as confirmed when, in 1888, the English iron metallurgist T. Turner reported experiments to the May meeting of the Iron and Steel Institute in London [36] which had as their object the elucidation of the relationship between silicon and sulfur in pig iron and cast iron. T. Turner summed up the results of his research in the words: " . . . silicon has the power of expelling sulphur from cast iron". The lecture attracted a great deal of attention and at once led iron metallurgists to object, as they believed they had a sufficient basis for assuming other mechanisms than the formation of silicon sulfide for explaining the desulfurization of iron in the blast furnace. Subsequently, an extensive controversial literature resulted which continued until quite recent times, though it will not be considered here, see, however, for example, "Gmelin-Durrer: The Metallurgy of Iron", Vol. 4a, 1972, pp. 96/103.

The next experimental investigation to clarify the question of the formation of a silicon sulfide in the blast furnace was published in 1903 by F. Wüst, A. Schüller [37], who heated mixtures of ferrosilicon and iron sulfide in the corresponding proportions and observed formation of silicon monosulfide. As far as the desulfurization of iron in the blast furnace is concerned, they thought that detectable volatilization first sets in at a silicon content above 6% and a sulfur content of 3%. In 1934, P. Bardenheuer, W. Geller [38] believed they could conclude from these experiments that the assumption that a silicon sulfide could play a part in desulfurization in the blast furnace could practically by excluded. Five years later, W. Oelsen, H. Maetz [39] disputed this idea on the basis of similar experiments of their own under changed conditions. It was also the opinion of F. Osann [40] that participation of a gaseous compound in desulfurizing iron is indeed possible but is proved by no means. This was true until the most recent times. Even at the start of the century, however, the question of the desulfurizing of iron by silicon at very high temperatures began to acquire practical significance with the introduction of the electric furnace [41]. The extensive desulfurization associated with its use was first attributed to the basic furnace lining and then to the very high temperatures attained. It was even suggested that "the molecular vibration due to the alternating current drives off sulfur dioxide resulting from oxidation of the sulfide by the iron oxide" [42], a view which was at once sharply rejected [41]. Soon, however, observations were made which could not be explained in the manner indicated and, in 1924, R. Durrer [43] observed desulfurization in the gas phase in the form of a sulfur-silicon compound during the operation of an electric furnace. Twenty years later, however, in summarizing, R. Durrer [44] could only say: "In melting extensive desulfurization may occur through silicon sulfide. It is not yet possible to say conclusively if the high temperatures in nonelectric melting suffice or if the very high peak temperatures which occur in these are necessary. It seems, however, that temperatures which can be attained in nonelectric melting are sufficient. It is also not yet altogether clear if sulfur escapes as SiS_2 or SiS. Desulfurization of molten iron by means of silicon sulfide first occurs with higher silicon contents (>3%). Perhaps the silicon, once it is dissolved in the iron is too stable for this purpose and perhaps the situation is similar in the case of sulfur".

Volatile silicon-sulfur compounds also have a certain significance in a process for purifying crude bauxite discovered by T. R. Haglund [45]. In this, bauxite is reacted with carbon and iron pyrites in an electric furnace, when part of the silicon in the silicates present as impurities escapes in the form of Si-S compounds. This was investigated more closely by X. Siebers, E. J. Kohlmeyer [26]. On the basis of this, E. J. Kohlmeyer, X. Siebers [46] protected by a patent a process for volatilizing silicon and silicic acid containing raw materials, zinc sulfide being especially recommended as the metal sulfide. P. Dolch [33] saw the occurrence of the Si-S compounds in the burning of electrodes for electric furnaces and in the production of water gas as a direct confirmation of the following prediction by E. Fremy [14] in 1852: "As silicon sulfide is probably produced in all cases where silica is submitted to the double action of a binary compound which gives up sulfur to it and at the same time takes possession of its oxygen, this sulfide is perhaps not so rare as was thought up to now" ["Comme le sulfure de silicium se produit probablement dans tous les cas où la silice se trouve soumis à la double action d'un composé binaire qui lui cède du soufre et s'empare en même temps de son oxigène, ce sulfure n'est peut-être pas aussi rare qu'on le pensait jusqu'à présent"].

Silicon Monosulfide SiS, was observed by P. Sabatier [17] as early as 1880 as an intensely colored secondary product in the preparation of silicon disulfide and was later also observed by W. Hempel, v. Haasy [18], E. Tiede, M. Thimann [24], and E. Zintl, K. Loosen [31], but hardly studied in any detail. It also played a part in discussions of the desulfurization of pig iron, as indicated above. It was not, however, until 1911 that L. Cambi [47, 48] tried to prepare it in a pure state for the first time. He obtained the compound by heating ferrosilicon (with 90% Si) with sulfur in the ratio 1:2 and believed there were two modifications, one a compact black solid and the other a yellow powder. Later he realized that the black color of one of the substances was caused by finely divided elementary silicon. When R. F. Barrow, W. Jevons [49] later prepared the compound by the action of carbon disulfide on silicon at 1000°C, they observed that the yellow powder is obtained at a sublimation temperature of 850°C but that raising this to 1000°C leads to condensation of a red glass. H.-H. Emons, P. Hellmold, H. Knoll [50] arrived at the same result, investigating the glassy material roentgenographically and with the electron microscope. They established the amorphous nature of the substance with only weak indications of structure in the compound, which was considered as a polymer. That the glassy substance is a high polymer of indefinite composition is also apparent from the results of M. S. Silverman, J. R. Soulen [20], who carried out syntheses at high pressures (25 to 80 kbar) and high temperatures (700 to 2300°C) with very pure silicon and very pure sulfur, and obtained homogeneous glasses, the color of which varied between canary yellow, orange, and green according to the proportions of the components in the charge (1:1 to 1:2.2). All of these colored substances were roentgenographically amorphous.

References:

[1] J. J. Berzelius (Kgl. Svenska Vetenskaps Acad. Handl. **1826** 53/78; Ann. Physik Chem. [2] **8** [1826] 411/26, 424). – [2] H. G. Söderbaum (Jac. Berzelius Lettres, Vol. 2, Pt. 4, Correspondance entre Berzelius et P. L. Dulong, Uppsala 1915, p. 54). – [3] J. J. Berzelius (Ann. Physik Chem. [2] **1** [1824] 169/230, 216/8). – [4] J. J. Berzelius (Ann. Physik Chem. [2] **17** [1829] 379). – [5] J. J. Berzelius (Lehrbuch der Chemie, German translation by F. Wöhler, 3rd Ed., Dresden – Leipzig 1833, Vol. 1, pp. 331/2; 4th corrected Ed., Dresden – Leipzig 1835, Vol. 1, pp. 331/2; 5th revised Ed., Dresden – Leipzig 1843, Vol. 1, p. 325).

[6] Catalogue of Scientific Papers (1800–1863) Compiled and Published by the Royal Society of London, Vol. 1, London 1867, pp. 330/41. – [7] A. Holmberg (Bibliographie de Berzelius, Pt. 1, Stockholm – Uppsala 1933, pp. 64/7). – [8] J. R. Partington (A History of Chemistry, Vol. 4, London – New York 1964, p. 153). – [9] M. Thimann (Diss. Berlin Univ. 1927, pp. 1/39, 17, 27). – [10] K. E. Schafhäutl (Neues Jahrbuch Mineral. Geol. Paläontol. Ref. **1846** 641/95, footnote pp. 689/91).

[11] J. J. Prechtl (Jahrb. K. K. Polytechn. Inst. Wien **1** [1819] 193 from [10]). – [12] I. Pierre (Ann. Chim. Phys. [3] **24** [1848] 286/302; J. Prakt. Chem. **46** [1849] 65/78). – [13] C. Friedel, A. Ladenburg (Liebigs Ann. Chem. **145** [1868] 179/90). – [14] E. Fremy (Compt. Rend. **35** [1852] 27/9; J. Prakt. Chem. **57** [1852] 106/9). – [15] A. Descloizeaux (Ann. Chim. Phys. [3] **19** [1847] 444/70, 464).

[16] E. Fremy (Ann. Chim. Phys. [3] **38** [1853] 312/44, 314/9; Compt. Rend. **36** [1853] 178/81; J. Prakt. Chem. **59** [1853] 11/4). – [17] P. Sabatier (Compt. Rend. **90** [1880] 819/21). – [18] W. Hempel, v. Haasy (Z. Anorg. Allgem. Chem. **23** [1900] 32/42, 39/40). – [19] M. Schmeisser, H. Müller, W. Burgemeister (Angew. Chem. **69** [1957] 781). – [20] M. S. Silverman, J. R. Soulen (Inorg. Chem. **4** [1965] 129/30).

[21] H. T. Hall (Science **128** [1958] 445/9). – [22] C. T. Prewitt, H. S. Young (Science **149** [1965] 535/7). – [23] M. Blix, W. Wirbelauer (Ber. Deut. Chem. Ges. **36** [1903] 4220/8, 4222/4). – [24] E. Tiede, M. Thimann (Ber. Deut. Chem. Ges. **59** [1926] 1703/6, 1704). – [25] E.Tiede, M. Thimann (Ber. Deut. Chem. Ges. **59** [1926] 1706/12).

[26] X. Siebers, E. J. Kohlmeyer (Arch. Erzbergbau Erzaufbereitung Metallhüttenw. **1** [1931] 91/129, 112). – [27] W. Biltz, F. Caspari (Z. Anorg. Allgem. Chem. **71** [1911] 182/97, 189/93). – [28] W. Biltz (Z. Anorg. Allgem. Chem. **146** [1925] 289/90). – [29] E. J. Kohlmeyer, H. W. Retzlaff (Z. Anorg. Allgem. Chem. **261** [1950] 248/60). – [30] W. Bussem, H. Fischer, E. Gruner (Naturwissenschaften **23** [1935] 740).

[31] E. Zintl. K. Loosen (Z. Physik. Chem. A **174** [1935] 301/11). – [32] A. Ledebur (Berg-Hüttenmänn. Ztg. **36** [1877] 279 from [33]). – [33] P. Dolch (Chem. Fabrik **8** [1935] 512/4). – [34] A. Ledebur (Berg-Hüttenmänn. Ztg. **37** [1878] 321/6). – [35] A. Ledebur (Stahl Eisen **4** [1884] 633/41, 638).

[36] T. Turner (J. Iron Steel Inst. [London] **1888** I 28/43, discussion 44/57; Stahl Eisen **8** [1888] 580/2). – [37] F. Wüst, A. Schüller (Stahl Eisen **23** [1903] 1128/33). – [38] P. Bardenheuer, W. Geller (Mitt. Kaiser Wilhelm Inst. Eisenforsch. Düsseldorf **16** [1934] 77/91). – [39] W. Oelsen, H. Maetz (Mitt. Kaiser Wilhelm Inst. Eisenforsch. Düsseldorf **21** [1939] 335/51). – [40] F. Osann (Stahl Eisen **28** [1908] 1017/22).

[41] M. Haff (Elektrochem. Met. Ind. **6** [1908] 96). – [42] A. Schmid (Stahl Eisen **27** [1907] 1613/5). – [43] R. Durrer (Untersuchungen zur Klärung der Frage der elektrischen Verhüttung schweizerischer Eisenerze, 1924, p. 46 from R. Durrer, H. Hellbrügge, B. Marinček, Arch. Eisenhüttenw. **14** [1941] 527/32, 527). – [44] R. Durrer (Die Metallurgie des Eisens, 2nd Ed., Berlin 1942, p. 246). – [45] T. R. Haglund (Fr. 555219 [1922/23] from C. **1923** IV 793).

[46] E. J. Kohlmeyer, X. Siebers (Ger. 534984 [1930/31] from C. **1931** I 3524). – [47] L. Cambi (Atti Reale Accad. Lincei [5] **19** II [1910] 294/300). – [48] L. Cambi (Atti Reale Accad. Lincei [5] **20** II [1911] 433/40). – [49] R. F. Barrow, W. Jevons (Proc. Roy. Soc. [London] A **169** [1939] 45/65, 48). – [50] H.-H. Emons, P. Hellmold, H. Knoll (Z. Anorg. Allgem. Chem. **341** [1965] 78/87).

2.9 Silicon-Carbon Compounds. Silicon Carbide

2.9.1 Early Observations

It is possible in the light of our present knowledge that J. J. Berzelius already had silicon carbide in his hands in 1810 among the reaction products in his attempts to reduce silica by carbon in presence of iron [1], see p. 18, and accordingly this is sometimes quoted in the literature as an early occurrence of the compound [2]. This was not, however, apparently noticed as significant by the Swedish scientist himself. In fact, Berzelius made the first observation of a carbide of silicon later in 1824, when, in his study of the "*Zersetzung der flußspatsauren Kieselerde durch Kalium*" [3] ["*Decomposition of fluorspar acid silica earth by potassium*"], he found that he obtained a reaction product which was darker in color if the potassium metal contaminated by carbon in the course of its preparation had not been purified previously by distillation. The darker product, on burning, yielded not only silicon dioxide but also carbon dioxide. Berzelius wrote, as he also made a prediction as to the composition of the carbide: "That it [the dark powdery substance] burnt in oxygen gas and has given off carbon(dioxide) without increasing in weight, was no longer so difficult to understand, as it is an ordinary phenomenon for oxides which contain 3 atoms of oxygen that their "quadricarburetum" burns without change of weight" ["Daß er [der dunkle, pulverförmige Körper] in Sauerstoffgas gebrannt und Kohle[nsäure] abgegeben hat, ohne an Gewicht zuzunehmen, war nicht mehr so schwer zu verstehen, als es eine gewöhnliche Erscheinung bei den Oxyden ist, welche 3 Atome Sauerstoff enthalten, daß ihr Quadricarburetum ohne Gewichtsveränderung verbrennt"]. According to J. W. Mellor [4] the formula Si_3C_8 for silicon carbide may be deduced from this.

The next observation of silicon carbide, which, however, was wrongly interpreted, occurred in 1849, when C. Despretz [5] attempted to prepare diamonds by melting more or less pure types of carbon in an electric circuit, and described the round granules which sometimes occurred and were considered like iron silicate or something similar. Very probably there may have been some consisting of silicon carbide, see "Kohlenstoff" B 1, 1967, but in no way can one speak of a "first preparation" by this savant as R. Calas [6] did. This is also true when, on one occasion in the course of his experiments, Despretz heated carbon covered with sand extraordinarily strongly: "I surrounded and impregnated needle-shaped rods of carbon with more fusible materials such as silica, alumina, and magnesia in order to see if the presence of substances which can be fused in the heat of the electric battery will facilitate the fusion of carbon. Silica, alumina, and magnesia have dissipated in the form of vapor and the carbon remained with its properties [unchanged]. I fixed a needle-shaped rod of carbon in an earthenware crucible, I filled this crucible with dry sand, and I passed the current through the carbon. It melted and vaporized. I have obtained a sort of very hard fulgurite tube, the interior of which was covered with a glaze of smoky quartz. The inner diameter of this tube was at least ten times that of the carbon" ["J'ai enveloppé, j'ai imprégné des baguettes aciculaires de charbon de matières plus fusibles, de silice, d'alumine, de magnésie, pour voir si la presénce d'un corps fusible au feu de la pile rendrait plus facile la fusion du charbon. Le silice, l'alumine, le magnésie se sont dissipées sous la forme de vapeur, et le charbon est resté avec ses propriétés. – J'ai fixé une baguette aciculaire de charbon dans un creuset de terre, j'ai rempli ce creuset de sable bien sec, et j'ai dirigé le courant à travers le charbon; il s'est fondu, volatilisé. J'ai obtenu une espèce de tube fulminaire très-dur, dont l'intérieur était verni par du quartz enfumé; le diamètre intérieur de ce tube était au moins dix fois celui du charbon"]. O Hönigschmidt [7], in his publication on *Karbide und Silicide* in 1914, took it as certain that the coating of the fulgurite "par du quartz enfumé" must have consisted of silicon carbide, though he emphasized that it was certainly not recognized as such.

In 1882, when R. S. Marsden [8] carried out investigations on the solubility of silica in molten silver, he was able to observe the formation of faintly yellow colored hexagonal lamellae, which he held to be silicon dioxide crystallized similarly to graphite. J. W. Mellor [4], however, thought that they very probably consisted of silicon carbide. The "diamonds" obtained at the time by the same investigator in Edinburgh [9], see "Kohlenstoff" B 1, 1967, were probably also nothing but silicon carbide. A further early observation of the occurrence of silicon carbide in reduction experiments with silica, which even led to a claim of priority in the discovery of the compound, was made in 1887 by C. F. Mabery [10]; for details, see later.

References:

[1] J. J. Berzelius (Ann. Physik **36** [1810] 89/102). – [2] M. C. Parché (Kirk-Othmer Encycl. Chem. Technol. 2nd Ed. **4** [1949] 114/32). – [3] J. J. Berzelius (Ann. Physik Chem. [2] **1** [1824] 169/230, 208/10). – [4] J. W. Mellor (A Comprehensive Treatise in Inorganic and Theoretical Chemistry, Vol. 5, Pt. 1, Longmans Green, London 1924 [Reprint 1960], p. 875). – [5] C. Despretz (Compt. Rend. **29** [1849] 709/24, 720).

[6] R. Calas (in: P. Pascal, Nouveau Traité de Chimie Minérale, Vol. 8, Pt. 2, Masson, Paris 1965, p. 570). – [7] O. Hönigschmidt (Karbide und Silicide, Knapp, Halle a.d.S. 1914, p. 86). – [8] R. S. Marsden (Proc. Roy. Soc. Edinburgh **11** [1881/82] 37/40). – [9] R. S. Marsden (Proc. Roy. Soc. Edinburgh **11** [1881/82] 20/6). – [10] C. F. Mabery (Am. Chem. J. **9** [1887] 11/5, 15).

2.9.2 The First Preparation

Whereas in the cases so far described a silicon carbide has been formed only by chance, on 8th May, 1882, Albert J. Colson, who a year previously had obtained oxygen- and nitrogen-containing silicon-carbon compounds in the laboratory of P. Schützenberger at the Collège de France [1], and had concluded from their composition that a tetratomic radical Si_2C_2 existed, announced to the Paris Academy [2] that he had been successful in preparing a pure silicon carbide: "When some silicon on a thick layer of lamp black is heated to a bright red heat it is easy, after cooling, to separate a small loosely knit lump of silicon from the carbon. If this is pulverized and treated successively with boiling caustic potash solution and then with warm hydrofluoric acid, a small residue is obtained composed of carbon and silicon in variable proportions. In order to obtain a substance of definite composition, I introduced into a porcelain tube two [porcelain] boats charged with silicon and heated to a bright red heat, while hydrogen saturated with benzene [benzine] vapor at 50 to 60°C was passed over them. At the end of three hours, the first boat contained a light black powder and the second a material which was whitish gray (sometimes the two substances were present at the ends of the same boat)" ["En chauffant au rouge vif du silicium sur du noir de fumée fortement tassé, il est facile, après refroidissement, de séparer du charbon un culot peu consistant de silicium. Si l'on pulvérise ce bloc, qu'on le traite par une dissolution bouillante de potasse caustique, puis par l'acide fluorhydrique chaud, on obtient un faible résidu, composé de charbone et de silicium en proportions variables. Pour obtenir un composé défini, je place dans un tube en porcelaine deux nacelles contenant du silicium, et je chauffe au rouge vif en faisant passer un courant d'hydrogène saturé de benzine à 50° ou 60°C. Au bout de trois heures, la première nacelle contient une poudre noire, légère; la seconde, un corps gris blanchâtre (parfois les deux corps se trouvent aux extrémités de la même nacelle)"]. After purification of the black powder, A. Colson obtained results on analysis by combustion with a mixture of lead chromate and lead oxide which appeared to correspond with the formula SiC_2. The whitish substance contained considerable oxygen [2].

Linking up with this preliminary work, ten years later P. Schützenberger [3] again tried to prepare a definitive carbide of silicon. He was able to communicate his results to the Academy in Paris on 16th May, 1892. In a crucible of about 20 to 30 cm^3 capacity made of retort carbon, which was closed by a cover of the same material, a mixture of equal parts of finely powdered crystalline silicon and powdered silica was introduced (the latter only as a diluent). The carbon crucible was put in a second somewhat larger crucible, made of refractory clay, and this in turn in a third fairly large crucible, the gap being filled with lamp black. After the whole had been heated to a bright red heat for many hours, the reaction material in the inner crucible, which was somewhat sintered and greenish in color, and which evolved no hydrogen with boiling caustic potash and therefore contained no more silicon, was boiled with hydrofluoric acid, when all the silica and some silicon nitride was dissolved. There remained a bright green residue, the composition of which corresponded with the formula SiC. P. Schützenberger emphasizes explicitly that the compound was not produced at the expense of the carbon crucible, which remained unattacked, but by reduction of the gaseous carbon monoxide by silicon at a bright red heat. He was thus the first to publish the preparation and analysis of silicon carbide, even if the latter was in an amorphous form [4].

Five years later it became known that the crystalline form of the compound, which has attained great technical significance under the name carborundum had already been observed a year before Schützenberger by H. Moissan. However, he first reported it in 1897 in his book *Der elektrische Ofen* [5]: "In my attempts to prepare crystalline carbon, as early as 1891 I had occasionally been able to find small crystals of silicon carbide in the batches of silicon which had become fused in the blast furnace in an envelope of carbon. At the time, however, I published nothing about this and priority for the discovering crystalline silicon carbide belongs to [E. G.] Acheson" ["Bei meinen Versuchen zur Darstellung von krystallisiertem Kohlenstoff hatte ich schon im Jahre 1891 Gelegenheit, in den Siliciummassen, welche im Gebläseofen in einer Kohlenhülle geschmolzen worden waren, kleine Krystalle von Siliciumcarbid zu finden. Ich habe aber damals nichts darüber veröffentlicht und so gebührt die Priorität der Entdeckung des krystallisierten Siliciumcarbids [E. G.] Acheson"].

Acheson, however, raises a much broader claim, namely who was the first ever to prepare the compound, and in collaboration with the physicist and electrotechnician, Nikola Tesla (1856–1943), who was the first to introduce a technical application [6]. He also refutes the assertion of Schützenberger's priority: "The disturbed state of mind – not to specify it more definitely – which I first experienced upon learning of Mr. Schützenberger's work, was completely relieved when I found the date of his communication [16th May, 1892] was three months later than the date on which Mr. Nikola Tesla exhibited a lamp containing carborundum, before the assembled scientific societies of France. The composition of what I had called carborundum was not, however, known at that date. Mr. Tesla could not state what it was, nor did any one else, to my knowledge, know that it was carbide of silicon responding to the formula SiC". As is apparent from his collected lectures [7], N. Tesla could only surmise on the composition of the material used by him – the editor of the collection considered it to be "a sort of carbon". The carbon filaments of the electric lamps, which were made luminous by high frequency currents which had been discovered by him [and were then named after him later] were, because of the small use, coated with a slurry of tar and carborundum powder, which Acheson had given to him, the former being vaporized off on using the lamp, so that a layer of carborundum was left on the carbon filament. He also molded electrodes from tar and carborundum powder under high pressure which, after destroying the tar, proved to be very durable. These, as well as the lamps, were demonstrated in lectures on his high frequency current which were held at many places in Europe and America. When Acheson had made his discovery he first mentioned in 1911: "Dr. Acheson said that this first carborundum experiment was made in March, 1891" [8].

Acheson's priority was contested in 1900 by C. F. Mabery [9], who claimed it for himself. As early as 1887 in experiments on the reduction of silica [10] in an electric furnace constructed by E. H. Cowles and A. H. Cowles [11], which was almost the same as Acheson's furnace, he claimed to have found "an amorphous substance, greenish yellow, occasionally inclining to deep green in color, with a vitreous luster", which was without doubt silicon carbide but which at the time was believed to be a silicon monoxide, probably because it was located between the reduced silicon and its dioxide [9, 10].

References:

[1] P. Schützenberger, A. Colson (Compt. Rend. **92** [1881] 1508/11). – [2] A. Colson (Compt. Rend. **94** [1882] 1316/8). – [3] P. Schützenberger (Compt. Rend. **114** [1892] 1089/93). – [4] E. Donath (Oesterreich. Z. Berg-Hüttenwesen **51** [1903] 421/3). – [5] H. Moissan (Der elektrische Ofen [German tanslation by T. Zettel], Berlin 1897, pp. 84, 326/7).

[6] E. G. Acheson (J. Franklin Inst. **136** [1893] 194/203, 279/89, 286/7). – [7] T. C. Martin (Nikola Tesla's Untersuchungen über Mehrphasenströme und über Wechselströme höherer Spannung und Frequenz [German translation by H. Maser], Halle a.d.S. 1895, pp. 136, 255/8). – [8] E. G. Acheson (Trans. Faraday Soc. **7** [1911] 217/20, 217). – [9] C. F. Mabery (J. Am. Chem. Soc. **22** [1900] 706/7). – [10] C. F. Mabery (Am. Chem. J. **9** [1887] 11/5, 15).

[11] E. H. Cowles, A. H. Cowles (U.S. 319945 [1884] from J. W. Mellor, A Comprehensive Treatise in Inorganic and Theoretical Chemistry, Vol. 5, Pt. 1, Longmans Green, London 1924 [Reprint 1960], p. 875).

2.9.2.1 Carborundum

Silicon carbide, which its discoverer first thought to be a compound of alumina, has acquired great technical importance under the misleading name carborundum, made up from the latin carbo = carbon and the Indian word for natural crystalline aluminum oxide, *kurund* in Hindi, *kurundum* in Tamil [1]. The announcement of the method for preparing this product on 21st June, 1893, at the meeting of the Franklin Institute by the electrical engineer Edward G. Acheson (1856–1931) is quite often equated with the discovery of the compound, especially as, in his report, the discoverer only expressed himself vaguely [2]. Thus phrases such as: "... original synthesis by Acheson in 1886 ..." [3] are as possible in the literature as the assertion [which is probably more correct] that the discovery was made in 1891 [4]. As the product had been displayed at the World Fair in Chicago in 1893 and duly caused a sensation, some of the visitors became interested in the point in time of the discovery. R. Volkmann [5], for example, reported that H. H. Williams, the representative at the Fair of the Carborundum Company plant at Monongahela, Pa., founded by Acheson in September 1891, gave the discovery date as September 1892. This is inconsistent not only with the date of establishing the factory [6] but also with the use of the material by N. Tesla, which was mentioned earlier, see p. 152. Another visitor to the Fair [6] also reported that, in June 1892, plant production of carborundum was approaching an output of 25 lb a day (sold at 2 to 4 dollars a pound) and that in September 1892 the extensive preliminary experimental work for the production of grinding and cutting tools using carborundum had been taken up. The date for the patent application for the production process, for which 10th May, 1892, is given, and that of the U.S. patent 492767 (28th February, 1893) are also not in keeping with the late date given for the discovery [7]. In his lecture [2], not only does E. G. Acheson give no statements of time but he is also silent on what the representative of his company at the World Fair [5] mentioned namely that Acheson's stimulus for the experiments with carbon went back to his collaboration over several years with Thomas A. Edison [8] who, among other things, had also attempted to prepare diamonds on a factory scale. E. G. Acheson published the following [2]: "For a number of years prior to 1890, I had been keeping a constant watch for anything, that might suggest a method by which carbon could be crystallized. A realization of the importance of abrasive materials, in the industrial arts, together with the known superiority of crystalline carbons over all other substances, acted as a constant stimulus to continue exertion for the solution of the problem. In the year mentioned, having become associated with an electric light company at Monongahela, Pa., I found myself in a position to conduct experiments on a line which I had some years earlier formulated. The scheme was to cause carbon to be dissolved in melted silicate of alumina, or in the metals reduced therefrom, and by cooling the same to the point of solidification cause the contained carbon to crystallize ... The first experiment was with a furnace constructed of an iron bowl lined with carbon, in the central cavity of which was placed a mixture of carbon and clay; through the mixture was passed an electric current of sufficient amount to fuse the mass – a violent reaction following the fusion – the iron bowl, and a rod of carbon suspended in the centre of the mixture, forming the electrodes. After the mass had cooled, it was removed, broken and carefully examined, when a few bright crystals, blue in color and apparently very hard, were found to be in that part which immediately surrounded the carbon electrodes. They were exceedingly small and only served to convince me that more and better arranged experiments would produce the desired results". His next experiments, which were now carried out with alternating current in a small furnace constructed of refractory brick, gave a sufficient yield for the closer examination of the crystals. He found: "that the crystals were not what I had expected to obtain – they were not pure carbon". He continued: "At this point in the work the want of a name for the new product was felt. It had not been analyzed, and I was led to believe, from the materials used, together with the color, sapphire blue and ruby-red, hardness and general form, that the material was composed of carbon and alumina, and in this believe it was decided to construct the name of the new material out of carbon and corundum, and it was named carborundum. The fitness of the name, in the eyes of the chemist, is, in view of the now known composition of the substance, doubtful, while in commerce, although phonetic and of pleasing effect in print, it is, perhaps, a trifle lengthy". O. Mühlhäuser, a co-worker of the discoverer who, in his own words [10], "had shared in the discovery by elucidating the nature of the process and explaining the nature of the main- and secondary-products" gave a description of the development of the discovery which was in agreement with this [9].

The classical process for obtaining silicon carbide on a technical scale which was worked out by Acheson still predominates today. A characteristic feature is an electrically heated rod made of conducting carbon, which transfers its heat to the surrounding reaction mixture and so allows a technically useful reaction rate to be reached. The process has been frequently and very fully described and here only a description of the furnace put together from contemporary reports and of procedures in the discoverer's own work will be quoted [11]: "The production furnaces have an inside diameter of 457 mm with a depth of 305 mm and a length of 1529 mm, and possess four carbon electrodes 51 mm thick and 305 mm long, which can be shifted along the length. The carbon core consists of coarse grained carbon which is charged between the electrodes in a layer 254 mm wide, 25.4 mm thick and 1675 [!] mm long. The furnace is preheated with carbon and oil and is broken up again after each run, as otherwise the yield drops. After partial cooling of the reaction mass, the tightly compacted carborundum cylinder consisting of crystal aggregates with a green glance, which constitutes the bulk of the reaction product, is separated mechanically from graphite, amorphous silicon carbide and unattacked starting material, and the crystalline lumps after crushing are first washed with acid and finally with water. The material purified in this way is stamped to a fine powder and sorted into various grades of powder by elutriation. After drying, these are partly put on the market and partly used for preparing grinding wheels, grindstones, etc. For this purpose, sufficiently fine powder is thoroughly mixed with a hard binding agent and exposed in molds to a pressure of 1 to 100 t. The molded objects are first dried in air on clay supports and are then fired in porous clay vessels in a flame furnace with increasing flame up to the incipient melting, which is maintained for some hours, thirty hours being necessary, after which they are slowly cooled over 30 to 40 hours".

Forty years later, furnaces 6 m in length were described [12] and after further 30 years the furnaces could be up to 20 m long, taking a loading up to 4000 kW [13]. The reaction mixture introduced consisted, and still consists today, of an intimate mixture of carbon, sand, common salt, and sawdust, the sand and the carbon being held exactly to the stoichiometric formula while a factor based on experience must be taken into account for the content of volatile components in the coke and the carbonization residue in the sawdust [12, 13]. For a 12 m long, 1500 kW furnace the Carborundum Co. gave the yield as 6500 to 6800 kg SiC in all. The current yield is then 8.5 kWh/kg SiC [12].

In the year following the discovery, O. Mühlhäuser [10], who had been Acheson's collaborator in working out and investigating the process, gave the high value of 3000°C and above for the temperature necessary in the carborundum furnace for the formation of silicon carbide. O. Hönigschmidt [14] also considered that the compound was first obtained crystalline at temperatures around 3500°C. However, measurements with furnaces in technical use as carried out by S. A. Tucker, A. Lampen [15] in 1906, and four years later by H. W. Gillet [16] and then repeated in 1912 by L. E. Saunders [17] gave appreciably lower values. Formation of "amorphous" carborundum is said to occur at temperatures between 1500 and 1600°C. Transformation to the crystalline form takes place at about 1830°C and at 2240°C the compound is already decomposed into its components. More recent measurements in 1931 by F. J. Tone [18] and a year later by R. R. Ridgway [19] gave similar values.

References:

[1] F. J. A. Estner (Versuch einer Mineralogie für Anfänger und Liebhaber, Vol. 2, Vienna 1797, p. 47 from H. Lüschen, Die Namen der Steine, 1st Ed., Thun-München 1968, p. 257). – [2] E. G. Acheson (J. Franklin Inst. **136** [1893] 194/202, 279/89, 286/7). – [3] J. T. Kendall (J. Chem. Phys. **21** [1953] 821/7). – [4] F. J. Tone (Ind. Eng. Chem. **30** [1938] 232/42, 233). – [5] R. Volkmann (Oesterreich. Z. Berg-Hüttenw. **42** [1894] 115/8).

[6] W. P. Blake (Eng. Mining J. **56** [1893] 270). – [7] M. C. Parché (Kirk-Othmer Encycl. Chem. Technol. 2nd Ed. **4** [1949] 114/32). – [8] R. Szymanowitz (J. Chem. Educ. **33** [1956] 113/5). – [9] O. Mühlhäuser (Z. Angew. Chem. **6** [1893] 637/46, 637/8). – [10] O. Mühlhäuser (Z. Anorg. Allgem. Chem. **5** [1894] 105/25, 105).

[11] C. Friedheim (in: F. Stohmann, B. Kerl, Encyklopädisches Handbuch der Technischen Chemie, 4th Ed., Vol. 7, Vieweg, Braunschweig 1900, Column 1735/60, 1756/7). – [12] H. Dannee (Ullmanns Encykl. Tech. Chem. 2nd Ed. **9** [1932] 485/95, 487). – [13] R. Rieder, G. Wiebke (Ullmanns Encykl. Tech. Chem. 3rd Ed. [1964] 692/7, 695). – [14] O. Hönigschmidt (Karbide und Silicide, Knapp, Halle a.d.S. 1914, p. 90). – [15] S. A. Tucker, A. Lampen (J. Am. Chem. Soc. **28** [1906] 853/8).

[16] H. W. Gillet (J. Phys. Chem. **15** [1910/11] 213/305). – [17] L. E. Saunders (Trans. Am. Electrochem. Soc. **31** [1912] 425/49, 438). – [18] F. J. Tone (Ind. Eng. Chem. **23** [1931] 1312/6). – [19] R. R. Ridgway (Trans. Am. Electrochem. Soc. **61** [1932] 217/32).

2.9.2.2 Silundum or Silicized Carbon. Globar. Silit

From 1900, F. Bölling, who had collaborated for a time with E. G. Acheson [1], had been attempting numerous experiments aimed at using carborundum for electrical resistances. They had been fruitless until 1904, when he was succesful in preparing resistances based on silicon carbide capable of being shaped at will, which were suitable for high temperatures (up to 1700°C). These were not conductors of the second class and did not oxidize in air. Reporting on this, he said [2]: "Up till now silicon carbide was known only in the amorphous and crystalline state and it was assumed that it is formed by silicon and carbon at high temperatures. I now found in my experiments that silicon vapor penetrates into carbon if both substances are at a very high temperature. At the temperature (about 1800 to 1900°C) at which the process takes place silicon is present only in the form of vapor, while carbon provided air cannot enter, is stil unchanged. *Silundum* or *silicized carbon,* as this new material is called, is thus a product which is obtained when carbon is stronly heated in silicon vapor at a high temperature. For the manufacture . . . electric furnaces are used, as they are for the production of carborundum. The pieces of carbon, which are to be transformed into silundum or provided with a coating of silundum, are placed in amorphous carborundum, or in a mixture of carbon and sand, and then exposed to heating by the electric current . . . The siliciding of the piece of carbon depends on the time, the temperature used and the volume of the piece . . . The shape of the piece of carbon does not change and it emerges from the furnace in the shape as when it was inserted, brand-marks with which the pieces were provided remaining completely visible . . . Coke and wood charcoal may be very easily silicized. . . . The same is true of pieces pressed from carbon powder with some carbon-containing bonding material . . . It is necessary to have cheap current for the production of the new material, especially when it is a question of large pieces. For smaller pieces up to a diameter of 4 to 5 mm normal current prices can probably be considered. The material is now [1908] produced by the Chemisch-Elektrische Fabrik "Prometheus" in Frankfurt am Main."

F. Bölling protected the preparation of silundum by several patents [3 to 5]. In describing the properties of silundum, R. Amberg [6] pointed out that the new material also conducts the electric current continuously with rising temperature, its resistance being six times as great as that of carbon but the negative temperature coefficient being small. It withstood a temperature of 1600°C over long periods without crystallizing. It may be raised to a white heat and then quenched in water directly afterwards without breaking or becoming cracked. The chemical properties are those of carborundum. Silundum is used primarily for the construction of heating devices of all sorts and also serves for electrodes in the electrolytic production of metals. In 1931 F. J. Tone [7] stated that the siliciding of carbon and wood had been known to the Carborundum Co. in America as early as 1897 in some faulty batches in the production of graphite by the Acheson method arising from carelessness in charging the furnace. Investigations following the latter led to the preparation of heating elements which were sold unter the trade name "*Globar*".

Elsewhere there was also interest in moldable electrical conductors to withstand high temperatures. Thus, on the basis of investigations by a Dr. Egly in the laboratory of the firm Gebrüder Siemens u. Co. in Berlin-Lichtenberg in the autumn of 1904, a first patent was granted [8] in which the problem posed was solved by making an intimate mixture of finely powdered silicon carbide and elementary silicon, getting this in the right-shape by making it plastic with a suitable carbon-containing bonding agent and then heating at 1400 to 1500°C in a carbon monoxide atmosphere either in an electric furnace or with the aid of generator gas. The object obtained was said to consist of "a lower order silicon carbide which can be distinguished from the starting material by a completely different crystal structure". The product was called "*Silit*", and this was later replaced by a *Silit II* which, according to the specification of patent DRP 257468 of 21st July, 1911, was obtained by preparing a molded body from a mixture of elementary silicon and carbon in paraffin or rosin and heating in an atmosphere of carbon monoxide or carbon dioxide to the temperature specified. This should consist essentially of silicon carbide SiC [9]. Molded bodies which contain "Carbazotsilicium" C_2Si_2N in addition to silicon carbide are referred to as *Silit III,* and are prepared according to DRP 176001 of 4th January, 1905. Molded objects consisting of mixtures of silicon with silicon carbide and carbon-containing bonding agents are fired in a nitrogen atmosphere. They are described as being very dense and hard and also capable of withstanding temperature differences [9].

References:

[1] F. Bölling (Umschau **26** [1922] 81/6, 83). – [2] F. Bölling (Chemiker-Ztg. **32** [1908] 1104/5). – [3] F. Bölling (Ger. 173066 [1904/06] from C. **1906** II 1028). – [4] F. Bölling (Ger. 183133 [1905/07] from C. **1907** II 1136/7). – [5] F. Bölling (Ger. 183134 [1905/07] from C. **1907** II 1137).

[6] R. Amberg (Z. Elektrochem. **15** [1909] 725/8). – [7] F. J. Tone (Ind. Eng. Chem. **23** [1931] 1312/6, 1314). – [8] Gebr. Siemens u. Co. (Ger. 177252 [1904/06] from C. **1907** I 1002/3). – [9] H. Großmann (Chem. Ind. [Berlin] **36** [1913] 304/8).

2.9.3 Other Processes for Preparing Silicon Carbide

Shortly after the discovery of carborundum was announced in 1893, H. Moissan [1, 2] published several methods for preparing crystalline silicon carbide. In his comprehensive book *Der elektrische Ofen,* which appeared some years later, he wrote [3]: "The study of the action of the electric arc on silicon led us to the preparation of silicon carbide in beautiful crystals by four different processes: 1. *Direct combination of silicon with carbon.* When we tried to dissolve carbon in molten silicon in the blast furnace we obtained the compound in beautiful crystals which could reach a length of several millimeters. It is isolated in a free state when the lumps of silicon are dissolved in a boiling mixture of nitric acid monohydrate and hydrofluoric acid. This process shows that silicon carbide is formed readily in a solvent between 1200 and 1400°C." [On this, E. Donath [4] remarks that the statement contradicts one by H. Moissan several pages before [3] describing the formation of graphite: "On heating in the blast furnace, molten silicon dissolves carbon, which it then deposited as graphite in the form of black lustrous and clearly crystalline platelets, which yield a yellow graphite oxide. At the temperature of the electric furnace, the silicon gives no more graphite and crystalline silicon carbide results instead"]. "2. *Preparation in the Electric Furnace.* The same compound may be obtained in a much simpler way if a mixture of 12 parts of carbon and 28 parts of silicon is heated. A mass of crystals is obtained in this way which can be purified readily by boiling with nitric acid monohydrate and hydrofluoric acid and by oxidation with nitric acid and potassium chlorate. The crystals are largely yellow in color, but they may also be quite transparent and may show a sapphire blue color. Transparent crystals are obtained if one operates as quickly as possible in a closed crucible with silicon as free of iron as possible. *Crystals in Molten Iron.* Ferrosilicon is heated in the electric furnace together with excess of silicon. The resulting mass is treated with aqua regia to remove iron completely. The crystalline residue is subjected for several hours to the action of a mixture of nitric acid monohydrate and hydrofluoric acid and then treated 8 to 10 times with an oxidizing mixture of nitric acid and potassium chlorate. A metallic melt which contains crystals of silicon carbide may also be prepared by heating a mixture of iron, silicon, and carbon, or, more simply, a mixture of iron, silica, and carbon, in the electric furnace. – 3. *Reduction of Silica by Carbon.* The same compound may be obtained by reduction of silica with carbon in the crucible in the electric furnace. Crystals of silicon carbide obtained in this way are less colored than those from iron, provided that quite pure silica and carbon have been used. – 4. *Action of Silica Vapor on Carbon Vapor.* A more original process for preparing crystalline silicon carbide consists in allowing carbon and silicon to act on one another in the vapor form. The experiment is carried out in a small carbon crucible, which is elongated in shape and contains silicon. The bottom of the crucible is heated to the highest temperature that can be reached in the electric furnace. Afterwards, prismatic, little colored, very hard and friable needles of silicon carbide are found."

H. Moissan [5] later published a further preparation method. He was successful in getting colorless, or only faintly blue, transparent silicon carbide crystals by reduction of quartz powder with calcium carbide.

Formation of silicon carbide in the thermal decomposition of gaseous silicon compounds was observed for the first time in 1909 by J. N. Pring, W. Fielding [6] when they tried to extend the known method of winning high melting metals by thermal decomposition of their chlorides on a glowing carbon filament to the preparation of silicon. Using a silicon chloride vapor–hydrogen mixture and a filament temperature of 1700°C, they obtained a thin coating of silicon carbide instead of the metal. On raising the temperature, elementary silicon first appeared besides but then, at 1925°C, only the carbide was formed. If they added benzene vapor to their gas mixture they obtained only carbide and carbon was also deposited, which could readily be separated from the silicon carbide by burning it off. This preparative method was first taken up again in 1931 by K. Moers [7], who replaced benzene by toluene and stopped the liberation of carbon in this way. His product consisted of small crystals which were citron yellow in color, or emerald green in transmitted light, and were mostly resting on the prism apex. They were never obtained colorless. In the course of a comprehensive investigation of the deposition of high-melting materials from the gas phase of their compounds in 1949, I. E. Campbell, C. F. Powell, D. H. Nowicki, W. B. Gonser [8] also prepared silicon carbide by the method just described and determined some thermal data for the compound. Four years later, J. T. Kendall [9] used the method for growing both completely pure silicon carbide crystals and also those doped in various ways, the conditions of formation and properties being closely studied. All were yellow in color. Good results were also obtained when trichloromethyl silane Cl_3SiCH_3 vapor alone was introduced into the hydrogen atmosphere instead of the silicon chloride-hydrocarbon mixture [10]. This hot filament method which is usually referred to as the van Arkel process, has been repeated many times and also slightly modified [11]. By replacing the hot filament by a hot graphite surface and using methyldichloro silane $SiHCl_2CH_3$, thick compact blocks of high-purity SiC were successfully prepared [15]. In 1947, R. Iley, H. L. Riley [19] obtained a very pure colorless silicon carbide, which was evidently cubic, in the form of a mold-like coating on vitreosil tubes heated to 1200 to 1300°C, on which ethylene saturated with water vapor was allowed to decompose.

The van Arkel process in its original form yields only very small crystals, although they are very pure. These small crystals are deposited very close together with their apices on the hot filament and, since they cannot conduct the current, they prevent their own growth by the drop in temperature which they cause. The further study of physical, and especially of electrical, properties, however, requires larger crystals and J. A. Lely [12] discovered a way of growing such crystals by sublimation. He used the idea of preparing a suitable sublimation cavity by stacking pieces of commercial pure bright green silicon carbide in a graphite crucible round a smooth rod, so that a more or less closed SiC wall and base about 20 mm thick resulted. The cavity obtained in this way was then covered with a large piece of SiC. When this crucible was heated for 6 to 8 hours in a carbon tube resistance furnace newly developed by W. J. Kroll, A. W. Schlechten, L. A. Yerkes [13] while filled with a protective gas (argon, hydrogen, or carbon monoxide), under conditions such that there is sufficient drop in temperature inside the crucible between the heated exterior and the interior, where heat is conducted away upwards and downwards, sublimation of SiC takes place. In this way tabular, hexagonal crystals 5 mm long and 2 mm thick are successfully obtained [12]. The results which D. R. Hamilton [16] obtained by this method were similarly satisfactory, when he investigated the degree of purification obtained and was able to prepare a silicon carbide with less than 10 ppm each of Al and Fe and traces of Mg and Mn.

In the same year, 1958, R. N. Hall [14] had attempted to make use of H. Moissan's observation [3] of the formation of silicon carbide in a silicon melt saturated with carbon, which is mentioned above, in order to grow larger crystals. However, he only achieved the build up of layers about 0.1 mm thick on the seed crystal introduced. Success in growing larger single crystals of SiC from the melt first came with the introduction of the Travelling Solvent Method by L. B. Griffiths, A. I. Mlavsky [17] in 1966, which was based on a somewhat older patent by K. M. Hergenrother [18]. They allowed thin layers of a molten metal, the most suitable of which for silicon carbide proving to be chromium, to travel through polycrystalline silicon carbide, obtaining in this way single crystals of α-SiC which could be used for semiconductor purposes. Wetting of the compound by liquid chromium was particularly difficult to bring about. It was successfully done by heating it in vacuum to 1300°C in an electron beam and then treating the surface with chromium vapor. Rectifiers prepared from the single crystals functioned up to 540°C. This method was investigated in detail by W. F. Knippenberg, G. Verspui [15] until, ultimately, they understood how to prepare colorless single crystals up to 12 mm long.

It seems appropriate at this point to include some observations on the range of colors shown by silicon carbide. Very pure silicon carbide, as obtained by sublimation by J. A. Lely [12], as single crystals by W. F. Knippenberg, G. Verspui [15] and also occasionally even earlier, forms colorless transparent crystals which are often somewhat gray. The impurity in these amounts to about 10^{-5} at.% [12] and may also increase to 10^{19} foreign atoms per unit volume [15]. The green color of commerical "pure" silicon carbide stems from a nitrogen content of about 10^{-4} to 10^{-3} at.%. Sublimation experiments in pure nitrogen at 1 atm led to almost black crystals with 0.08 at.% N. Doping experiments with phosphorus gave green crystals, while boron and aluminum produced a blue coloration. Two colored crystals with a sharp separation of the layers may be obtained by changing the doping substance in the course of the experiment. A technical "black" product, which is preferably referred to as dark blue, contains up to 0.5% of such impurities as Al, Fe, Mg, and Ca, the aluminum, as D. Lundqvist [20] has shown, being built into the lattice. Iron, on the other hand, is probably present as a foreign substance [12]. W. F. Knippenberg, G. Verspui [15] obtained colorless polycrystalline silicon carbide with layers which were brownish yellow in color, when they allowed it to grow in a helium atmosphere.

References:

[1] H. Moissan (Compt. Rend. **117** [1893] 423/5). – [2] H. Moissan (Compt. Rend. **117** [1893] 425/8). – [3] H. Moissan (Der elektrische Ofen [German translation by T. Zettel, Berlin 1897], pp. 84, 326/7). – [4] E. Donath (Oesterreich. Z. Berg-Hüttenw. **51** [1903] 421/3). – [5] H. Moissan (Compt. Rend. **125** [1897] 839/44, 842).

[6] J. N. Pring, W. Fielding (J. Chem. Soc. **95** [1909] 1497/506). – [7] K. Moers (Z. Anorg. Allgem. Chem. **198** [1931] 233/75, 253/5). – [8] I. E. Campbell, C. F. Powell, D. H. Nowicki, W. B. Gonser (Trans. Am. Electrochem. Soc. **96** [1949] 318/33, 325). – [9] J. T. Kendall (J. Chem. Phys. **21** [1953] 821/7). – [10] J. T. Kendall, D. Yeo (Proc. 11th Intern. Congr. Pure Appl. Chem., London 1947, Vol. 1, pp. 171/5).

[11] W. F. Knippenberg (Philips Res. Rept. **18** [1963] 161/274, 178). – [12] J. A. Lely (Ber. Deut. Keram. Ges. **32** [1955] 229/31). – [13] W. J. Kroll, A. W. Schlechten, L. A. Yerkes (Trans. Am. Electrochem. Soc. **89** [1946] 317/29). – [14] R. N. Hall (J. Appl. Phys. **29** [1958] 914/7). – [15] W. F. Knippenberg, G. Verspui (Philips Res. Rept. **21** [1966] 113/21).

[16] D. R. Hamilton (J. Electrochem. Soc. **105** [1958] 735/9). – [17] L. B. Griffiths, A. I. Mlavsky (J. Electrochem. Soc. **111** [1964] 805/10). – [18] K. M. Hergenrother (U.S. 2996456 [1961] from C.A. **1961** 26714). – [19] R. Iley, H. L. Riley (Nature **160** [1947] 468). – [20] D. Lundqvist (Acta Chim. Scand. **2** [1948] 177/91).

2.9.4 Occurrence of Silicon Carbide in Nature and in Technical Processes

In 1904, H. Moissan [1] found a very small quantity of small hexagonal green crystals, which after thorough investigation, he was able to identify as silicon carbide, when he dissolved a large amount of iron from the meteorites from the Cañon Diablo, which have become famous, in hydrochloric acid. In the following year G. F. Kunz [2] proposed in the course of a lecture to the Academy of Sciences in New York that the silicon carbide found as a mineral be named *Moissanite* in honor of the discoverer, and mineralogists followed this suggestion [3]. Twenty years later, P. Aloisi [4] believed he had also been able to establish the terrestrial occurrence of moissanite on the basis of small rock fragments in the Wadi Tiscé in the north of the (then Italian) Somaliland which, from their appearance and their "chemical and physical character", could be described as silicon carbide. The first find of the mineral to be established with certainty came from investigations of the sand deposits of Siberian Rivers by B. I. Ozernikova which, from their composition, must have originated in Kimberlite deposits. These were also found in 1957 and the mineral samples were analyzed [5]. A year later, A. J. Regis, L. B. Sand [6] found cubic β-silicon carbide in a Green-River formation. There is probably no doubt that in the last two discoveries the silicon carbide owed its formation to volcanic processes, but the position is different in the case of the next occurrence, discovered by N. Gnoevaya, L. Grozdanov [7] in 1965 in N.W. Bulgaria in triassic limestone and in dolomite. In this case it was assumed that, for the formation of moissanite, there had been a reaction of hydrothermal solutions with the organic substances of the limestone. Finally, in 1974, R. G. Gevorkyan, A. G.

Gurkina, F. V. Kaminskii [8] found three granules of SiC in the alluvial sands of the ophyolytic zone of Armenia, the crystal structure of which they determined accurately.

Small quantities of silicon carbide, which in themselves are without significance and usually do not interfere, can form readily from impurities in many technical processes taking place at high temperatures, e.g., in blast furnaces, in the production of calcium carbide or in the electrolysis of melts. These attract attention only if the hardness of the mostly tiny crystals is noticed and they are confused with diamonds, thus leading to a mistaken claim of diamond synthesis. Only one such case, which was checked many times and was held for a long time to be a true diamond synthesis, will be dealt with briefly here (for a full treatment and further cases see "Kohlenstoff" B 1, 1967, p. 237). Stimulated by a diamond find in 1891 by A. E. Foote [10] in the Cañon Diablo iron meteorites mentioned above [this was the so-called "carbonado" form; the find was checked in 1939 by C. J. Ksanda, E. P. Henderson [9] and confirmed], H. Moissan published a first [11], and then, a little later, a further report on a successful diamond synthesis [12]. He had saturated iron with sugar carbon in the electric furnace at temperatures in excess of 3000°C and rapidly cooled the entire melt in the most varied ways. After suitable working up of the metal regulus some very small octahedral crystals were isolated, especially after cooling in liquid lead, which weighed some milligrams and in part were waterclear and optically isotropic. They sank in methylene iodide CH_2I_2 (D ~ 3.3) and scratched steel and corundum. On combustion there was no residue worth mentioning and the analysis was said to correspond with almost 100% C. As the properties stated were most consistent with those of diamond, H. Moissan claimed these crystals as diamonds and the technical world of his time agreed with him [13, 14], although numerous objections to Moissan's discovery and views have been raised, and many checks on his work have proved negative, see "Kohlenstoff" B 1, 1967, p. 241. After making critical comments on the exactness of the combustion analysis at about the turn of the century and establishing nonuniformity of the synthetic product A. Neuhaus [15], for example, stated with surprise that Moissan and his co-workers, as well as those who followed him (who have been passed over here) as well as the "discoverer" of the diamonds in iron and steel had been content with such a rough figure for the specific gravity (heavier than CH_2I_2) and the hardness (harder than corundum, but no test with carborundum). It is no longer possible to make an unequivocal judgement in Moissan's case as the crystal samples needed for later investigations by modern methods (very decisive: the refractive index, X-ray diagrams) have not been preserved. It seems, however, probable that Moissan had in his hands the as yet unknown cubic SiC modification. The same is true of some who discovered "iron diamonds". In many cases, though, it was also a question of corundum, see "Kohlenstoff" B 1, 1967, pp. 241/3.

References:

[1] H. Moissan (Compt. Rend. **139** [1904] 773/80, 778, **140** [1905] 405/6). – [2] G. F. Kunz (Am. J. Sci. [4] **19** [1905] 396/7). – [3] P. Ramdohr (Klockmann's Lehrbuch der Mineralogie, 12th Ed., Enke, Stuttgart 1942, p. 315). – [4] P. Aloisi (Period. Mineral. [Rome] **5** [1934] 191/2). – [5] A. P. Bobievich, V. A. Kalyazhnyi, G. I. Smirnov (Dokl. Akad. Nauk USSR **115** [1957] 1189/92; Proc. Acad. Sci. USSR Sect. Sci. Geol. **112/115** [1957] 757/60).

[6] A. J. Regis, L. B. Sand (Bull. Geol. Soc. Am. **69** [1958] 1633). – [7] N. Gnoevaya, L. Grozdanov (Spis. Bulg. Geol. Druzh. **26** [1965] 89/95 from C.A. **64** [1966] 19205). – [8] R. G. Gevorkyan, A. G. Gurkina, F. V. Kaminskii (Zap. Arm. Otd. Vses. Mineral. Obshchestva **7** [1974] 106/10 from C.A. **84** [1976] No. 7536). – [9] C. J. Ksanda, E. P. Henderson (Am. Mineralogist **24** [1939] 677/80). – [10] A. E. Foote (Am. J. Sci. [3] **42** [1891] 413/7).

[11] H. Moissan (Compt. Rend. **116** [1893] 218/24). – [12] H. Moissan (Compt. Rend. **118** [1894] 320/6). – [13] M. Bauer (Edelsteinkunde, 2nd Ed., Leipzig 1909, pp. 300/1). – [14] C. Doelter (Handbuch der Mineralchemie, Vol. 1, Dresden – Leipzig 1912, p. 43). – [15] A. Neuhaus (Angew. Chem. **66** [1954] 525/36).

2.9.5 Formation Temperature. Dissociation. Other Carbides

It was probably on the basis of the views and reports of those who had first prepared carborundum, and from his experience with other carbides and silicides, that O. Hönigschmidt [1] assumed in 1914 that the crystalline compound was first formed at temperatures around 3500°C, though even then there were reports which made this high value improbable. Thus J. N. Pring [7] observed direct combination of the elements in vacuum even below 1300°C and in 1897 H. Moissan [8] had observed formation of the crystals in liquid silicon at temperatures which were scarcely higher than this. Above all, however, there were measurements in furnaces used technically which were carried out by S. A. Tucker, A. Lampen [2], then, four years later, by H. W. Gillet [3] and finally repeated by L. E. Saunders [4] in 1912. According to these, the formation of "amorphous silicon carbide" should occur at temperatures between 1500 and 1600°C, transformation to the crystalline form taking place at about 1830°C while at 2240°C the compound should be completely decomposed into its components, gaseous silicon and solid graphite. More recent measurements in 1931/32 gave quite similar values [5]. A practical man with decade-long experience of the preparation of carborundum, F. J. Tone [6], described the processes taking place in the furnace thus: "When the sand-carbon mixture is held for about two hours at 1800°C under reducing conditions, the mixture loses about 60% in weight, indicating that the formation of carbon monoxide is practically at an end. The product is not crystalline (on simple visual inspection) and is the same as so-called "amorphous silicon carbide", being a greenish material insoluble in hydrofluoric- and nitric acid. When this low-temperature product is heated at 2200 to 2300°C, it recrystallizes and forms the familiar crystalline product."

In 1912, E. Pollitzer [10] attempted to calculate the formation temperature by means of the Nernst heat theorem, assuming all of the oxygen of the air in the electric furnace to be replaced by carbon monoxide, i.e., that the CO partial pressure amounts to 0.2 atm and arrived at an equilibrium temperature of 1325°C, on exceeding which formation of carbide must set in. O. Ruff, M. Konschak [11] carried the calculation further and obtained 1462°C for a pressure of 1 atm CO, though experimentally they found 1635°C. They therefore concluded that the equation $SiO_{2\,solid} + 3C_{solid} \rightarrow 2CO + SiC_{solid}$ does not determine the formation of SiC but rather the equation $SiO_{2\,solid} + 2C \rightarrow Si_{solid} + 2CO$, for which they cited further evidence from calculations and experiments.

In the case of the formation of silicon carbide crystals on hot wires in the thermal decomposition of suitable gaseous silicon compounds, J. T. Kendall [9] observed in 1953 that black SiC crystals are formed at filament temperatures between 1700 and 1900°C, that between 1900 and 2100°C in the middle of the filament yellow crystals result and at the ends black. The best temperature for building up pure yellow crystals was between 2100 and 2200°C.

The dissociation of silicon carbide into the elements [12] was investigated somewhat more exactly in 1926 by O. Ruff, M. Konschak [11] with vapor pressure measurements and it was shown that the compound cannot be melted under normal conditions. From their data they calculated that there was a dissociation pressure of 1 atm at 2700 ± 40°C. Measurements of the solubility of carbon in liquid silicon by R. I. Scace, G. A. Slack [13] in 1959 led to the conclusion that there was a peritectic point at 2830 ± 40°C and 35 atm. Under these conditions the compound decomposes to a silicon-rich liquid in equilibrium with graphite.

As may be inferred from the above, the composition of carbide at first appeared to be uncertain and this opened up the possibility of the occurrence of several carbides. O. Ruff, M. Konschak [11] paid special attention to the question of whether this was correct. In connexion with their observations in the course of their vapor pressure measurements they wrote: "A carbide with a higher carbon content would have had to be more stable to heat and therefore present in the residues which remain on incomplete vaporization of the carbide. We have not been able to find anything of this sort. In the search for a carbide with a lower carbon content we have vaporized silicon in the silicon carbide crucible through the complete pressure and temperature range and subsequently analyzed the crucible wall. It had the composition of ordinary silicon carbide both before and after, with an excess of 2.24%. This was probably caused by the fact that the SiC of the crucible in use at above 2200°C had already undergone partial dissociation."

References:

[1] O. Hönigschmidt (Karbide und Silicide, Knapp, Halle a.d.S. 1914, p. 90). – [2] S. A. Tucker, A. Lampen (J. Am. Chem. Soc. **28** [1906] 853/8). – [3] H. W. Gillet (J. Phys. Chem. **15** [1910/11] 213/305). – [4] L. E. Saunders (Trans. Am. Electrochem. Soc. **31** [1912] 425/44, 438). – [5] R. R. Ridgway (Trans. Am. Electrochem. Soc. **61** [1932] 217/32).

[6] F. J. Tone (Ind. Eng. Chem. **23** [1931] 1313/6, 1315). – [7] J. N. Pring (J. Chem. Soc. **93** [1908] 2101/8, 2104/5). – [8] H. Moissan (Der elektrische Ofen [German translation by T. Zettel], Berlin 1897, p. 326 [French 1st Ed. 1897]). – [9] J. T. Kendall (J. Chem. Phys. **21** [1953] 819/27, 823). – [10] E. Pollitzer (Die Berechnung der chemischen Affinitäten nach dem Nernstschen Wärmetheorem, Stuttgart 1912, p. 125).

[11] O. Ruff, M. Konschak (Z. Elektrochem. **32** [1926] 515/25). – [12] E. Tiede, E. Birnbräuer (Z. Anorg. Allgem. Chem. **87** [1914] 129/68, 167). – [13] R. I. Scace, G. A. Slack (J. Chem. Phys. **30** [1959] 1551/6).

2.9.6 Crystallography

The first crystallographic study of silicon carbide was due to E. G. Acheson, who sent crystals of carborundum which he had discovered to B. W. Frazier, the mineralogist at Lehigh University in South Bethlehem, Pa. He found the yellowish green component of the small crystals sent to be clearly "rhombohedral" in form, while the blue individuals were clearly "hexagonal-holohedral" [1]. F. Becke [2] in Prag came to a similar result two years later in 1895 with a sample of a different origin, observing hexagonal plates in particular. In 1902 G. B. Negri [3] who, according to N. W. Thibault [4] investigated more than 100 individual crystals, but did not publish his results in full, considered his crystals as "ditrigonal-skalenohedral", distinguishing six types. In 1912 and the years that followed H. Baumhauer [5, 6] made this finding more definite in studying numerous separate crystals of silicon carbide and established that many of these hexagonal or rhombohedral crystals are in fact structurally identical in two dimensions, showing the same distances and order in the units, but are different in the third dimension. This lattice phenomenon was called "Polytypie", i.e., polymorphism in one dimension. Later it was apparent that the polymorphism of silicon carbide does not differ basically from that of other substances [7, p. 162]. Several suggestions were made to classify this wealth of forms, the earliest going back to H. Baumhauer [5, 6] and H. Ott [8]. It was settled about twenty years later by the modifications of L. S. Ramsdell [9] and G. S. Zhdanov [10] on the one hand and R. W. G. Wyckhoff [11] and H. Jagodzinski [12] on the other. A comparison of the range of symbols is given by W. F. Knippenberg [7, p. 189]. During this development many publications on the crystallography of the compound appeared, which brought to light numerous details, among which were X-ray data, which were probably first obtained by F. Rinne [13] in 1915. A summarizing discussion of this literature and his own work was given in 1944 by N. W. Thibault [4, 14], who grouped together the rhombohedral and hexagonal forms under the name β-SiC which, because of the way in which it is produced is also called the high-temperature modification.

The form of silicon carbide which was termed amorphous by those who prepared it was recognized as a special form by L. Baraduc-Muller [15] in 1918, and, because of inadequate optical methods of investigation, was also considered to be amorphous. In the same year, however, X-ray investigations by C. L. Burdick, E. A. Owen [16] appeared which established that this modification has a diamond-like structure. N. W. Thibault [4, 14] introduced the designation α-SiC for this cubic form, which is also frequently called the low temperature modification in the literature.

References:

[1] B. W. Frazier (Addendum of E. G. Acheson, J. Franklin Inst. **136** [1893] 279/89, 287/9). – [2] F. Becke (Z. Krist. **24** [1895] 537/42). – [3] G. B. Negri (Riv. Mineral. Crist. Ital. **29** [1902] 33/89 from Z. Krist. **41** [1905] 269/71). – [4] N. W. Thibault (Am. Mineralogist **24** [1944] 249/78, 251). – [5] H. Baumhauer (Z. Krist. **50** [1912] 33/9).

[6] H. Baumhauer (Z. Krist. **55** [1915] 249/59). – [7] W. F. Knippenberg (Philips Res. Rept. **18** [1963] 161/274). – [8] H. Ott (Z. Krist. **61** [1925] 515/31, **62** [1925] 201/17, **63** [1926] 1/18). – [9] L. S. Ramsdell (Am. Mineralogist **32** [1947] 64/82). – [10] G. S. Zhdanov (Dokl. Akad. Nauk SSSR **48** [1945] 39 from [7, p. 197]).

[11] R. W. G. Wyckhoff (Crystal Structures, Vol. 1, New York – London 1948, p. 25). – [12] H. Jagodzinski (Acta Cryst. **2** [1949] 201/14; C.A. **1950** 1794). – [13] F. Rinne (Ber. Verhandl. Kgl. Sächs. Ges. Wiss. Math. Phys. Kl. **67** [1915] 303/40). – [14] N. W. Thibault (Am. Mineralogist **24** [1944] 327/62). – [15] L. Baraduc-Muller (Rev. Met. [Paris] **7** [1918] 657/834, 672).

[16] C. L. Burdick, E. A. Owen (J. Am. Chem. Soc. **40** [1918] 1749/59, 1758).

2.9.7 Uses of Silicon Carbide

As an Abrasive. The discoverer of carborundum, E. G. Acheson, who had wanted to prepare diamonds in his experiments and, as the given name shows, believed he had obtained a product similar to corundum (= emery), recognized the hardness of the new compound and tested it as an abrasive: "The first test for diamond cutting was made by myself." Subsequently he allowed his product to be tested in some diamond grinding shops in New York and reported that in the three shops, against all expectations, his carborundum could be used as a polishing agent with better results than with the diamond dust used until then. The polish was also reached in a shorter time. On the other hand, the new material failed for rough grinding and for forming the facets of brilliants. His material was also found to be suitable for a range of grinding operations in the metallurgical industry since, to obtain the same effect, either appreciably less time or smaller quantities of carborundum than of emery powder were required [1]. In fact, silicon carbide is often assigned the value 9½ on Mohs scale [2] but elsewhere it has the value 9 on this scale, between diamond (10) and topaz (8) [3]. This and other uncertainties in the Mohs scale of hardness induced R. R. Ridgway, A. H. Ballard, B. L. Bailey [4] in 1933 to suggest an extension of the scale, by which the new technical hard materials would be better characterized. They assigned the number 8 to quartz and the number 15 to diamond, obtaining for carborundum the value 13. A classification system on the basis of impact hardnesses was developed six years later by F. Knoop, C. G. Peters, W. B. Emerson [5] on which the scale runs from 32 for gypsum to 8000 for diamond. Silicon carbide gets the value 2000.

It was not only in the form of powder that Acheson used his carborundum for grinding. As he himself reported, he also tried to bring it into widespread use by the preparation of grinding wheels, such as polishing discs for dental purposes, but he met with little success in this. When

Acheson accompanied by his collaborator at this time, F. Boelling, visited "the largest share holder in the Carborundum Company by the name of Huyler" in New York in 1893, the latter made Acheson aware "that there had been complaints of a not too pleasant nature about his small discs for dental purposes". It is possible to learn how the discs were prepared from the experiments in 1894/95 at the Deutsche Carborundum Gesellschaft established in Iserlohn to which Acheson sold his patent (the transaction was rendered null and void when the grinding discs again did not prove to be effective here). Feldspar and kaolin were mixed with 25 to 30% of sieved carborundum in drums, pressed into suitable molds and then fired in a porcelain furnace. In spite of this some grinding wheel factories which were clearly more successful were established in the following years [6]. According to an anonymous report [7], they appear to have been more successful in Austria, even in 1894. E. Frey "the producer of the elastic Naxosschmirgelräder, patent 'Wiktorin & Co' in Wien" reported that "this new chemical grinding agent has shown itself to be far superior to the best Naxoskorundschmirgel in hardness and cutting power ... The carborundum wheel possesses a quite unsurpassed property in that it shows very low wear as the result of its special cutting ability ..." At all events, interest in the new grinding agent was great right at the start and many firms attempted the preparation of grinding tools [6].

As a Refractory Construction Material. E. G. Acheson [1] did not venture to determine if carborundum, which right at the start he had observed to have special "infusibility and incombustibility", could also be used as a refractory construction material, for it is clear that the preparation of building components suitable for this purpose had not been successful apart from some small buttons for Tesla's experiment quoted on p. 152. Acheson's hesitant caution appeared at first to be justified, for over the period from 1893, when the discovery was announced, until 1898, C. F. Geiger [8] in perusing the literature found no sort of reference to such uses of carborundum, or even to attempted uses, although its thermal conductivity and resistance to heat and temperature change were very promising. In spite of this, it was E. G. Acheson [9] who obtained a patent in December 1898, according to which carborundum powder could be briquetted in a suitable way with iron, iron oxide, and iron salts for the preparation of refractory structural units. In June of the following year, B. Talbot [10] patented a "neutral furnace coating" which consisted of powdered carborundum and a refractory binder while, in 1900, F. A. J. Fitzgerald obtained two patents for refractory construction parts made of carborundum. According to these, bonding of granular material was affected by recrystallization of the molded bodies in the electric furnace [8].

In 1901, E. W. Engels [11] suggested the use of carborundum for the preparation of retorts for the distillation of zinc. These were said to have a life of far more than 100 days, instead of the brief life of ordinary retorts of about fifteen days. He blended finely ground carborundum with clay or chamotte and water glass with the addition of water, molded under high pressure, dried the parts (which needed only 14 days instead of 3 months for pure chamotte) and finally fired them. Carborundum muffles also had the advantage of a considerably increased zinc yield with a smaller consumption of coal because of excellent density and good thermal conductivity of the new material. The greatly extended useful life resulted from the small amount of breakage due to the high mechanical strength, the great resistance to chemical attack by zinc oxide and the secondary components of the ore, and the ability to withstand periodical large changes in temperature without the formation of cracks. These advantages must have been genuine for, from this time on, carborundum retorts were installed almost exclusively in German and Belgian zinc smelters [8].

A comparative investigation of the technological thermal properties of the refractory construction materials in current use by S. Wologdine [12] showed that the advantages claimed in the patent could really be expected. The suitability of silicon carbide for this purpose may be recognized clearly from his numerical data. Numerical values which demonstrate the high thermal shock resistance of carborundum are provided by M. L. Hartmann, O. H. Hougen [17]. The further development of this sort of application of silicon carbide is also apparent from the flood of patents in all the industrial countries [2, 8, 13, 14] into which, however, it is not possible to go more fully here. It was interrupted, in the United States of America at least, by the first world war as, for its duration, practically the total carborundum production was used for grinding materials. The variable proportion of SiC in artificial products induced H. N. Baumann, J. P. Swentzel [15] in a related study to suggest classification according to the SiC content. Thus materials with less than 50% Si were referred to as half stuff, that with 70 to 80% as "Grade B" and with 85% SiC or more as "Grade A". In the last of these groups there is also a product with 96.9% SiC. Very recently it has become possible to prepare successfully structural elements with 99% SiC and a very small porosity [16]. Mixtures of finely powdered α- and β-SiC with the addition of 1% aluminum powder are submitted to pressure sintering in a special apparatus at pressures of 10000 lb/sq.inch and temperatures around 2500°C. The densities of these elements reach the value 3.0 compared with $D = 3.22$ [1] for the starting material.

Difficulties in the supply of carborundum during the first world war probably led S. C. Linberger [18] to dilute the material which had become scarce with a lot of graphite and to bind the two with clay so as to get a usable melting pot. This success induced J. L. Ohman [19] after the war to undertake further experiments, in which he replaced the clay by a binding agent containing carbon. The result was so satisfactory that, in 1930, H. E. White [20] commented that a "super-graphite crucible" had been developed in the period since the first world war. H. N. Baumann, J. P. Swentzel [15] also considered that, from then on, all graphite crucibles contained carborundum as the chief additive, irrespective of which bonding agent was used in their preparation.

A. B. Searle [21] enumerated the materials recommended to be added to carborundum as bonding agents for preparing compact objects, in patents at least up to 1924. Apart from those mentioned above, these were: water glass, borax or boric acid, lime, Portland cement, aluminum sulfate, magnesite, dolomite, zirconium silicate, and aluminum silicate. Later additions were: the hydroxides of the alkaline earth metals, mullite ($3Al_2O_3 \cdot 2SiO_2$), the oxides of metals such as iron, chromium, nickel, manganese, vanadium, and copper which are capable of forming silicides [14]. In subsequent years the number of these "bonding agents" increased still further. Additions of silicon nitride Si_3N_4 proved to be especially beneficial and resulted in marked improvement in all the desirable properties mentioned above. Such mixtures allow working temperatures in an inert atmosphere of up to 2300°C, and, in one that is oxidizing, up to 1650°C so that new types of use, e.g., the lining of combustion chambers and the jets of rockets, arise. As silicon carbide also has a low neutron capture cross section, it is a valuable construction material for reactors. When used in conjunction with silicon as "silicon carbide silicide (SiC-Si)" as a sheath for fuel elements it is able to prevent the diffusion of gaseous fission products to a large extent [22].

As an Electrotechnical Material. It will be apparent from what has been said that, even in the initial period of its technical production, silicon carbide was used for preparing electrical contacts, resistances, and heating elements, and these applications also resulted in a large number of patents, see "Silicium" B, 1959, pp. 774/5. Its use for rectifiers was suggested as early as 1906 [23], and the application of silicon carbide as a component of thermoelements was patented in 1919 [24]. A patent was applied for in 1940 for its use as a light source in the passage of a current.

In these proposed uses polycrystalline material was always used, but a prerequisite for the most recent area of application, its use as a semiconductor, is the possibility of preparing larger silicon carbide crystals. Following the first exploratory investigations of the electrical properties of large single crystals of silicon carbide, such as B. Reuter, H. Knoll [25] carried out in the second world war but were first able to publish only later, SiC was soon recognized as a semiconductor and, in 1955 A. Lely [26], for example, was able to prepare both the n- and the p-types. The great advantage of SiC semiconductors is that they can be used at higher temperatures.

References:

[1] E. G. Acheson (J. Franklin Inst. **136** [1893] 279/89, 283/4). – [2] M. C. Parché (Kirk-Othmer Encycl. Chem. Technol. **2** [1948] 855). – [3] W. F. Knippenberg (Philips Res. Rept. **18** [1963] 161/274, 164). – [4] R. R. Ridgway, A. H. Ballard, B. L. Bailey (Trans. Am. Electrochem. Soc. **63** [1933] 369/93, 392). – [5] F. Knoop, C. G. Peters, W. B. Emerson (J. Res. Natl. Bur. Stand. **23** No. 1200 [1939] 34/61, 60).

[6] F. Boelling (Umschau **24** [1922] 81/6). – [7] F. J. Tone (Oesterreich. Z. Berg-Hüttenw. **42** [1894] 117/8). – [8] C. F. Geiger (J. Am. Ceram. Soc. **6** [1923] 301/6). – [9] E. G. Acheson (U.S. 615648 [1898] from [8]). – [10] B. Talbot (U.S. 628288 [1899] from [8]).

[11] E. W. Engels (Ger. 154536 [1901] from F. Peters, in: F. Stohmann, B. Kerl, Encyklopädisches Handbuch der Technischen Chemie, 4th Ed., Vol. 9, Vieweg, Braunschweig 1921, Column 1120/22). – [12] S. Wologdine (Sprechsaal **42** [1909] 611/4). – [13] H. Daneel (Ullmanns Encykl. Tech. Chem. 2nd Ed. **9** [1932] 494). – [14] A. C. Lea (Trans. Brit. Ceram. Soc. **40** [1941] 93/118). – [15] H. N. Baumann, J. P. Swentzel (Bull. Am. Ceram. Soc. **16** [1937] 419/30).

[16] R. A. Alliegro, L. B. Coffin, J. R. Tinklepaugh (J. Am. Ceram. Soc. **39** [1956] 386/9). – [17] M. L. Hartmann, O. A. Hougen (Trans. Am. Electrochem. Soc. **37** [1921] 707/15, 711). – [18] S. C. Linbarger (U.S. 1277227 [1918] from [15, p. 424]). – [19] J. L. Ohman (U.S. 1356939 [1920] from [15, p. 424]). – [20] H. E. White (Fuels Furnaces **8** No. 1 [1930] 107/10 from [15, p. 424]).

[21] A. B. Searle (Refractory Materials, London 1924 from [14, p. 100]). – [22] W. Schreiter (Seltene Metalle, Vol. 2, Leipzig 1961, pp. 411/2). – [23] H. H. C. Dunwoody (U.S. 837616 [1906] from C.A. **1907** 650). – [24] A. Schwartz (Ger. 336936 [1919/21] from C. **1921** IV 21). – [25] B. Reuter, H. Knoll (Naturwissenschaften **34** [1947] 372).

[26] A. Lely (Ber. Deut. Keram. Ges. **32** [1955] 229/31).

Table of Conversion Factors

Following the notation in Landolt-Börnstein [7], values which have been fixed by convention are indicated by a bold-face last digit. The conversion factor between calorie and Joule that is given here is based on the thermochemical calorie, $cal_{th\,ch}$, and is defined as 4.184**0** J/cal. However, for the conversion of the "Internationale Tafelkalorie", cal_{IT}, into Joule, the factor 4.1868 J/cal is to be used [1, p. 147]. For the conversion factor for the British thermal unit, the Steam Table Btu, BTU_{ST}, is used [1, p. 95].

Force	N	dyn	kp
1 N (Newton)	1	10^{5}	0.1019716
1 dyn	10^{-5}	1	1.019716×10^{-6}
1 kp	$9.8066\mathbf{5}$	$9.8066\mathbf{5} \times 10^{5}$	1

Pressure	Pa	bar	kp/m²	at	atm	Torr	lb/in²
1 Pa (Pascal) = 1 N/m²	1	10^{-5}	1.019716×10^{-1}	1.019716×10^{-5}	0.986923×10^{-5}	0.750062×10^{-2}	145.0378×10^{-6}
1 bar = 10^{6} dyn/cm²	10^{5}	1	10.19716×10^{3}	1.019716	0.986923	750.062	14.50378
1 kp/m² = 1 mm H_2O	$9.8066\mathbf{5}$	$0.98066\mathbf{5} \times 10^{-4}$	1	10^{-4}	0.967841×10^{-4}	0.735559×10^{-1}	1.422335×10^{-3}
1 at = 1 kp/cm²	$0.98066\mathbf{5} \times 10^{5}$	$0.98066\mathbf{5}$	10^{4}	1	0.967841	735.559	14.22335
1 atm = 76**0** Torr	$1.0132\mathbf{5} \times 10^{5}$	$1.0132\mathbf{5}$	1.033227×10^{4}	1.033227	1	760	14.69595
1 Torr = 1 mm Hg	133.3224	1.333224×10^{-3}	13.59510	1.359510×10^{-3}	1.315789×10^{-3}	1	19.33678×10^{-3}
1 lb/in² = 1 psi	6.89476×10^{3}	68.9476×10^{-3}	703.069	70.3069×10^{-3}	68.0460×10^{-3}	51.7149	1

Work, Energy, Heat	J	kWh	kcal	Btu	MeV
1 J (Joule) = 1 Ws = 1 Nm = 10^7 erg	1	2.778×10^{-7}	2.39006×10^{-4}	9.4781×10^{-4}	6.242×10^{12}
1 kWh	$\mathbf{3.6} \times 10^6$	1	860.4	3412.14	2.247×10^{19}
1 kcal	4184.**0**	1.1622×10^{-3}	1	3.96566	2.6117×10^{16}
1 Btu (British thermal unit)	1055.06	2.93071×10^{-4}	0.25164	1	6.5858×10^{15}
1 MeV	1.602×10^{-13}	4.450×10^{-20}	3.8289×10^{-17}	1.51840×10^{-16}	1

1 eV/mol = 23.0578 kcal/mol = 96.473 kJ/mol

Power	kW	PS	kp m/s	kcal/s
1 kW = 10^{10} erg/s	1	1.35962	101.972	0.239006
1 PS	0.73550	1	**75**	0.17579
1 kp m/s	9.8066**5** $\times 10^{-3}$	0.01333	1	2.34384×10^{-3}
1 kcal/s	4.184**0**	5.6886	426.650	1

References:

[1] A. Sacklowski, Die neuen SI-Einheiten, Goldmann, München 1979. (Conversion tables in an appendix.)
[2] International Union of Pure and Applied Chemistry, Manual of Symbols and Terminology for Physicochemical Quantities and Units, Pergamon, London 1979; Pure Appl. Chem. **51** [1979] 1/41.
[3] The International System of Units (SI), National Bureau of Standards Specl. Publ. 330 [1972].
[4] H. Ebert, Physikalisches Taschenbuch, 5th Ed., Vieweg, Wiesbaden 1976.
[5] Kraftwerk Union Information, Technical and Economic Data on Power Engineering, Mülheim/Ruhr 1978.
[6] E. Padelt, H. Laporte, Einheiten und Größenarten der Naturwissenschaften, 3rd Ed., VEB Fachbuchverlag, Leipzig 1976.
[7] Landolt-Börnstein, 6th Ed., Vol. II, Pt. 1, 1971, pp. 1/14.